BIO
EVOLUTION

HOW BIOTECHNOLOGY IS CHANGING OUR WORLD

MICHAEL FUMENTO

ENCOUNTER BOOKS
SAN FRANCISCO

First edition published in 2003 by Encounter Books, an activity of Encounter for Culture and Education, Inc., a nonprofit tax exempt corporation.

Encounter Books website address: www.encounterbooks.com
Manufactured in the United States and printed on acid-free paper.

The paper used in this publication meets the minimum requirements of ANSI/NISO Z39.48-1992 (R 1997)(Permanence of Paper).

Library of Congress Cataloging-in-Publication Data

Fumento, Michael.
BioEvolution : how biotechnology is changing our world / Michael Fumento.
p. cm.
Includes bibliographical references and index.
ISBN 1-59403-057-x (alk. paper)
1. Biotechnology—Popular works. I. Title.
TP248.215.F86 2003
303.48'3—dc22

2003049372

10 9 8 7 6 5 4 3 2 1

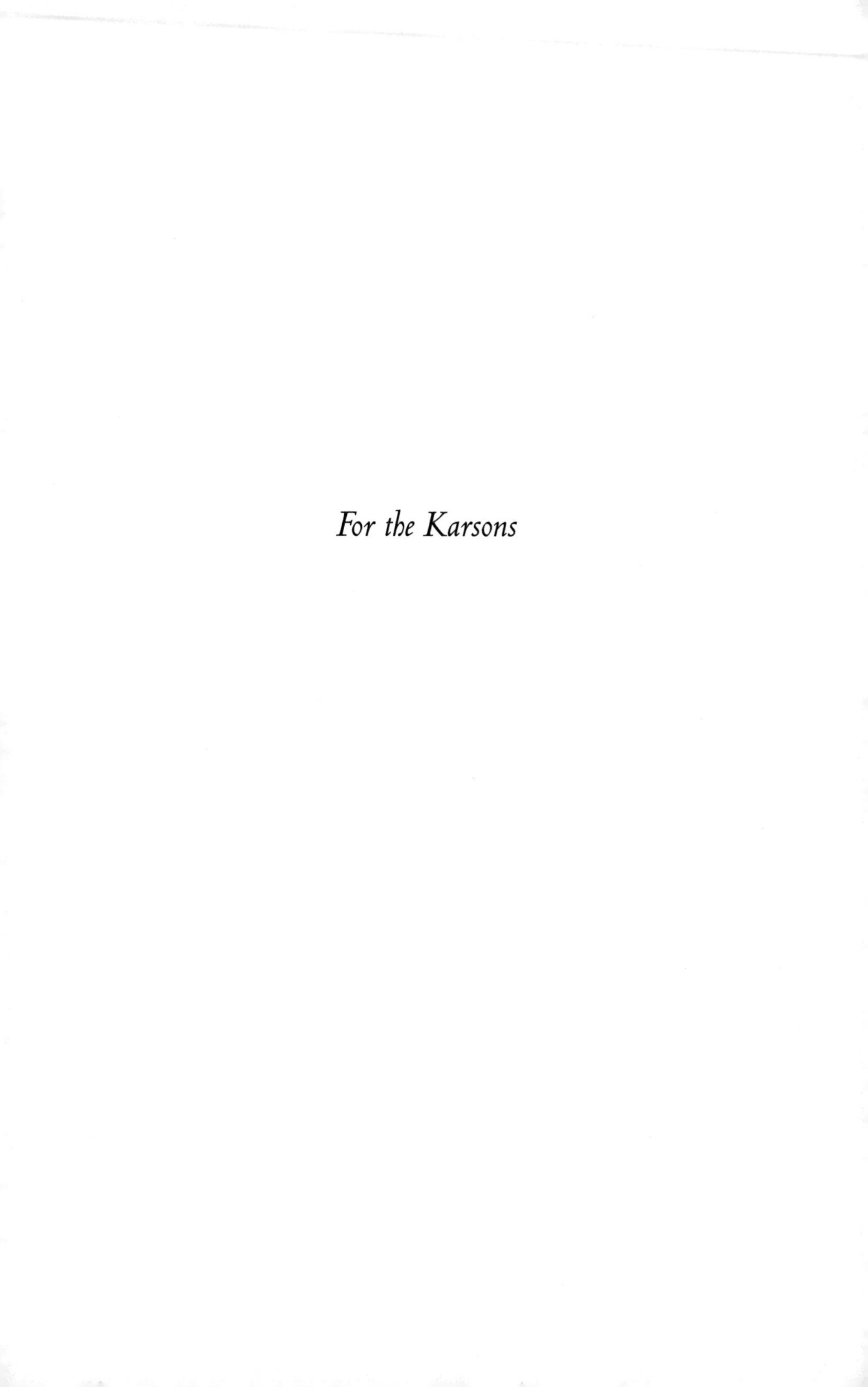

For the Karsons

"In the years to come, the contributions of [the biotechnology] industry will help us to win the war on terror, will help us fight hunger around the world and will help us to save countless lives with new medicines."

—President George W. Bush

CONTENTS

INTRODUCTION

It is 2025. The world's population is growing more slowly than in the past, but growing nonetheless, with almost all of the increase occurring in the underdeveloped countries. Yet rather than cause an apocalyptic tragedy of biblical proportions, this trend goes almost unnoticed, and the famines that have ravaged these nations for centuries are just a bad memory, as is malnutrition and infant mortality. People are being fed on *less* farmland. The rainforests of South America are expanding and—unlike the "Green Revolution" beginning in the 1960s, in which new crop varieties, pesticides and fertilizers staved off famine in much of the world—the crops being grown now require much less pesticide and fertilizer. Toxic waste, including radioactive material, is consumed by voracious bacteria and plants. In virtually every way, the environment is more pristine than at any time since the Industrial Revolution.

Disease still exists, but the great infectious scourges such as malaria, tuberculosis, diarrhea, hepatitis and AIDS have been virtually eliminated. Residents of even the poorest countries now have life expectancies matching those of the industrialized world at the end of the twentieth century, while citizens of those nations are healthier than ever. Alzheimer's is gone, cancer is rare and dying from cancer is almost unheard of. Heart disease rates have also plummeted and life expectancy has soared. The greatest athlete of all time has just retired at age 55, having come off his best professional year. He looks forward to spending the next 50 to 70 years with his wife, grandchildren and great-grandchildren. He and his wife can expect all of those years to be healthy ones, without fear of dementia, arthritis, diminishing eyesight or hearing. His wife will still be attractive, having had her skin rejuvenated every few years so that it's as soft and supple as a newborn's; and while it's too bad she needed to have a heart transplant, at least the heart was grown from her own cells rather than removed from an accident victim.

This scenario may sound like something you would see on the Science Fiction Channel, but it's all science fact. Every development just described and virtually everything else that will be described in this book is on the drawing board or has already moved past it. If it isn't currently being tested on plants, animals and humans, that probably means it's already in use.

Biotechnology won't bring utopia; there will be no Dawning of the Age of Aquarius. Biotechnology won't stop all pain, physical or psychological. It won't be the answer to every medical problem we know today, and it will have to deal with some we don't yet know about. It *can* eventually solve all the world's food problems, but it won't solve all environmental problems. Healthy lifespans will be extended dramatically, but if it's immortality you seek, then seek elsewhere. What biotechnology does offer is a tool that can make a planet unimaginably better than it is now, including:

- An incredible spectrum of cancer therapies that may bring more progress in five years than we've seen since the declaration of the "War on Cancer" in 1971.
- Antibiotics to which no bacterium can evolve defenses and vaccinations that combine the best aspects of live and killed virus vaccines. Many of these vaccines go after the traditional targets of pathogen-borne illness, but dozens in development also target cancer.
- A vast array of medicines and therapies that will not only extend the natural lifespan but—much more importantly—extend the quality of those lives, however long they may be.
- Drugs designed or selected based on our individual genomic structures, rather than for the population as a whole. These will both increase efficacy and decrease side effects.
- Drugs extracted from plants, milk and chicken eggs that will allow proteins to be made in the massive and affordable quantities needed to fight diseases such as cystic fibrosis and multiple sclerosis. Edible vaccines for killer diseases such as cholera and hepatitis B will be incorporated into fruits and vegetables.

- Crops that will provide more nutrition and higher yields using less land, less water, fewer nutrients and fewer chemicals.
- Foods that will stay fresher longer, will no longer cause allergies, and will have offensive tastes (like bitterness) or unwanted ingredients (like caffeine) removed.
- Microbes and plants that can clean up toxic or radioactive waste sites for a fraction of what's now being paid and with no chance of spreading the contamination.

Biotechnology itself is a purely physical science. It is morally neutral, although it has moral overtones, just as nuclear technology does. The ethical questions are highly consequential and must be dealt with now and at every step on the way toward the better world that biotech will bring us. This book presents science in anything but a cold and raw form; it's biotech with a human face. But I should say right at the outset that I don't agree with those who claim that this better world will necessarily also be a brave new world of mind control, mass narcosis and lost humanity. What we have most to fear is fearmongers, the professional futurephobes such as Jeremy Rifkin, author of *The Biotech Century,* who declares that agricultural biotech is "the most radical, uncontrolled experiment we've ever seen."[1] Rifkin insists it's wrong to use genetic engineering even to cure cancer, diabetes or sickle cell anemia.[2] Or take Britain's top biotech critic, Dr. Mac-Wan Ho, who claims that "The large-scale release of transgenic organisms is much worse than nuclear weapons or radioactive nuclear wastes."[3]

We need gadflies, skeptics and pessimists who guard against the abuse of science. But we don't need people who invoke concepts such as the "precautionary principle," demanding that biotech—or any other scientific innovation—be proven absolutely risk-free before we begin to realize the benefits that taking some minimal risks can provide.[4]

On the other side of the ledger, we also need to work to find techniques of biotechnology that *avoid* ethical concerns, rather than ones that provoke the specter of a "posthuman" future. As this book demonstrates, for instance, the inflammatory stem cell controversy is in many ways an unnecessary and unnecessarily polarizing intellectual battle. Adult stem cells, which in most ways already have the advantage over controversial embryonic stem cells, continually show that they may be superior in *all* ways. That is, in comparison with embryonic cells, the adult stem cells are already many years ahead in terms of practical applications, they are safer, they cannot lead to the production of human clones, and they don't raise moral issues concerning the destruction of human embryos. But the aggressive dogma that embryonic cells have far greater potential to become all the types of tissue produced by the body, an arguing point rather than a scientific fact, stands in the way of an approach capable of bearing fruit immediately. Already, three types of adult stem cells have shown the potential to be just as malleable as embryonic cells. Stem cells are the moral argument we don't need to have.

The potential of biotech is so broad and research is advancing so quickly that it can bring us all the miracles that scientists say it will, without necessarily involving processes that many of us find morally or religiously offensive. But to the extent that newer technology cannot accomplish this or has yet to do so, then to use biotech in the best way we must have a fair and proper debate right now. At some points in this book I will take part in that debate; at others, my main contribution will be to ensure that people on both sides possess the scientific facts that permit a reasoned discussion.

The Age of Revolution—or Evolution?

As part of that debate, we need to know where the technology now stands, where it's going, and how fast it's moving. The last is the easiest to answer.

Bill Gates, the president of Microsoft, has written that if he weren't in the computer field, he'd be working in biotechnology. "I expect to see breathtaking advances in medicine over the next two decades, and biotechnology researchers and companies will be at the center of that progress. I'm a big believer in information technology.... But it's hard to argue that the emerging medical revolution, spearheaded by the biotechnology industry, is any less important to the future of humankind. It, too, will empower people and raise the standard of living."[5] Gates' chief rival in the software industry, Oracle's Larry Ellison, agrees. "If I were 21 years old, I probably wouldn't go into computing. The computing industry is about to become boring," he said. Rather, "I'd go into genetic engineering."[6]

Such innovators agree that biotech is where the automobile was a century ago. In retrospect, cars turned out to be a revolutionary development. But really they were just an extension of transportation modes that had been around for millennia, namely horses and horse-drawn vehicles. Similarly, biotech is both a paradigm-jump forward and merely an extension of medicinal, industrial and agricultural practices already in use.

At first, cars often seemed to be more bother than they were worth. They broke down constantly and there weren't many repair shops. Avoiding accidents with cars was different from avoiding accidents with horses, horse-drawn wagons and trolley cars. Horses and wagons could be dangerous, but in ways people were accustomed to. The earliest autos had few advantages over horses, except perhaps as status symbols. Social critics asserted that there was no "need" for cars, and indeed at the time any such need was not apparent. Weird, unscientific claims assailed cars—for example, that if they went much faster than their earliest speeds, their occupants would be crushed. Any possible advantages to non-owners of cars, such as no longer having to wade through piles of disease-carrying equine "exhaust," were ignored. Still, automobiles not only survived as an innovation, but went on to revolutionize the world.

Biotech is presently at the stage of automobiles at the time Henry Ford was building his "quadricycle," and already it offers some advantages over products it's replacing, although these advantages are often incremental and sometimes not yet noticeable. When the doctor gives you a test or a vaccine, for instance, he probably doesn't inform you that you're benefiting from recombinant biotechnology, much less explain what "recombinant" means. But over time, and more particularly as biotech products improve and diversify, the science will sell itself.

Unfortunately, sometimes what comes out of the media pipeline can delay or even derail this process. In just the past few years, more and more media outlets have begun assigning reporters to the "biotech beat." They cover little or nothing else, which is good because covering biotech alone can be terribly demanding since the field is so diverse. One day you're trying to explain a new vaccine, the next day an insect-resistant crop. Paul Raeburn, *Business Week* science writer and head of the National Association of Science Writers, spoke for much of his profession when he admitted, "We don't produce a lot of food in Times Square" and "the closest I ever got to a farm was to lean out the car window and moo at the cows."[7] Thus it's hardly surprising that one biotech vaccine would lead to both of these front-page headlines on the same day:[8]

- "Large Trial Finds AIDS Vaccine Fails to Stop Infection" (*New York Times*)
- "Vaccine for AIDS Appears to Work" (*USA Today*)

We also have to deal with attitudes that vary dramatically from continent to continent. It's not entirely surprising that polls show North Americans to be quite accepting of biotech food while many Europeans would as soon drink bilge water as eat genetically engineered crops.[9] More disturbing is that at least one survey of residents of the fifteen European Union nations showed that their growing fear of biotech food was spilling over into other areas such as medicine.[10]

To find a context for this worry, it's worth recalling that pasteurization had to overcome decades of resistance, and that Edward Jenner's smallpox vaccine (originally made from cowpox virus) caused tremendous controversy and outright fear for over a century. (A famous cartoon from 1802 showed people vaccinated by Jenner sprouting cows from their faces, ears, arms and thighs.)[11] Thomas Edison deliberately sought to terrify Americans out of using his competitor's superior form of electric current (alternating rather than direct) by using AC to fry animals publicly and encouraging the New York State legislature to use AC for the first electric chair. Eventually AC won out anyway.[12]

Ultimately, only two things can defeat such negativism. One is education; the other is the products themselves. This book focuses on biotech advances in human health and longevity, feeding the planet's growing population and cleaning up the environment. It lets biotech speak for itself. It's not about Buck Rogers in the twenty-fifth century, but about what biotech is doing for us right now and will do in the very near future. Rarely do I mention a product or therapy that either isn't already available or has little chance of being so within about five years. There's very little speculation here. There is no science fiction in this book; the world of real biotech is far too interesting for that.

PART ONE

Miracles in Medicine

ONE

The New Biotech Paradigm

In its formative years, the biotech medical industry focused on replacements for drugs that were too expensive to produce, were too hard to obtain, or had become too dangerous because of natural contamination. To countless millions of people worldwide, these drugs were a godsend. Yet today that mission has changed: biotech medicine has been targeting diseases for which there were no known treatments or preventatives. For vaccines that already exist, biotech will provide versions that are more effective, safer and easier to administer. It will deliver vaccines for killer plagues such as malaria and AIDS for which currently there are none. Biotech is also moving us from a paradigm centered on treatment to one that emphasizes cure. Diabetics, for example, are already better off with gene-spliced insulin than with the natural insulin extracted from animals, with its disease-carrying potential.[1] But how much better it would be to prevent or cure the disease completely. There are now people who have actually been cured of diabetes through biotechnology in clinical trials.[2] They offer the hope that all diabetics may some day fare likewise.

Even medical conditions that are arguably not diseases at all, such as aging, will become treatable. Yet as the list of new treatments grows, therapies—both drug and nondrug, biotech and nonbiotech—are in essence becoming victims of their own success in that each year they consume larger portions of income and benefits from employers and government.[3] So it will also fall upon biotech to make therapies cheaper. While consumers complain about the expense of new-generation drugs, biotech products can help slow the rise of medical costs. And the cures that biotech is searching for will ultimately obviate the need for drugs altogether.

Biopharmaceuticals—drugs made from proteins and other organic substances—are the fastest-growing segment of the drug industry, increasing from less than 1 percent of the U.S. market in 1989 to 7 percent in 1998, according to Gillian Woollett, formerly with the Pharmaceutical Research and Manufacturers of America (PhRMA) and now vice president of science and regulatory affairs at the Biotechnology Industry Organization (BIO). About one-third of approximately one thousand drugs now being tested in patient trials are biopharmaceuticals.[4]

By the end of 2002, 234 biotechnology drugs and drug indications (uses for a drug) had been approved in the United States alone, according to BIO, an industry umbrella group. (Since there is no strict definition of what constitutes a biotech drug, other groups use different numbers.) A record number of drugs, 35, were approved in 2002. Among just these 35 there was a remarkable range of applications, from advanced prostate cancer and torn rotator cuffs to a single vaccine that protects against diphtheria, tetanus, pertussis, hepatitis B and polio.[5] In all, these drugs have helped over 325 million patients worldwide, says BIO.[6] Further, there are almost 400 biotechnology medicines from over 140 different companies and the National Cancer Institute in testing for more than 200 diseases. About half of these are for cancer and related disorders, with the next-highest category being infectious diseases other than AIDS.[7] Estimates are that this number could eventually grow to 10,000 drugs.[8]

Of the total number of new medicines approved in the U.S. in the 1990s, 13 percent were biopharmaceuticals, according to *Nature Biotechnology,* and the number in clinical trials continues to climb, from 81 in 1988 to 284 in 1996, and 371 in 2002. Yet all this refers only to the United States. Since the establishment of the European Medical Evaluation Agency in 1995, 34 percent of its approved drugs were biotech-based.[9]

Just what is a biopharmaceutical? For that matter, what is biotech? Definitions vary, but genes are at the bottom of it. These are made up of two strands of protein called DNA, or deoxyribonucleic acid, the "double helix" made famous by Watson and Crick. Genes instruct cells to produce or "code for" proteins; some code for only one protein, while others code for several.[10] These proteins are often, though not always, enzymes. In any case, the proteins always consist of some number and combination of twenty-two different amino acids. Genes are responsible for directing the biological development and subsequent activity of the human body's estimated 100 trillion cells, which come in about 200 varieties. The difference between these cells is not a result of the particular genes they carry, but whether those genes are switched on or off. (The exception to this rule appears to be brain cells, which recent research has shown are often missing large pieces of the genome, for reasons we don't yet understand.)[11]

Genes instruct cells to produce proteins that determine, for instance, what eye color we'll have, whether our hair will be light or dark, and whether we'll have a Bill Clinton–sized nose or a tiny hint of a nose like Nicole Kidman. But we need to be careful to distinguish between traits that are predestined and those that are predisposed. A nose shape is predestined. The age at which you might contract Alzheimer's results from predispositions, combined with outside factors. We'll talk more about this later on. But whether we're concerned with predestiny or predisposition, it's rare when only one gene causes one outcome. Consider that already at least 25 different genes have been found to be involved in the development of tiny little mouse teeth, while the development of asthma may involve 149 genes.[12]

Whether we're talking about medicine or crops, biotech is often equated with "gene splicing," taking a gene or genes from one thing and putting them into another. Because this involves the transferring of genes, it's called "transgenics." Because you're combining a gene from something old with something new, it's also called "recombinant" or "recombinant DNA," abbreviated in turn to "rDNA."[13] "Recombinant" is generally used in reference to medicinal applications, while "transgenic" is generally used more to describe gene-spliced crops and livestock and other nonpharmaceutical applications.

Genomics is a family of technologies that translate information about the genome into knowledge of what genes are present, what each gene does, and how the products of each gene (usually a protein or proteins) contribute to the properties and behavior of cells. Until recently, scientists have been able to study the activities of only an individual gene or at most a few genes at a time. Genomics is an effort to analyze simultaneously what all of the genes in a cell are doing and how their activities are orchestrated, which allows researchers to target individual genes and cells. ("Genetics" is an older term, referring to the study of inherited traits.)[14]

In a sense, biotech is all about observing and manipulating the expression of genes, which insist on doing their own thing no matter what their surroundings. That's why it's possible to remove a gene from one organism and put it into a vastly different one and find that it still expresses that same trait. A gene from a goldfish can be put into a goldfinch. It may do nothing at all there, but if it *does* code for a protein, it will code for the same one that it did in the fish. Researchers have taken a gene from the firefly and put it on a chromosome in a mouse tumor, making the tumor light up so they can readily determine its size as various medicines are tested against it.[15]

Some people, of course, insist that it's unnatural (or "Frankensteinish") to, say, remove a gene from a fish that protects against extreme cold and put it into a plant. But as Matt Ridley writes in his fascinating book *Genome: The Autobiography of a Species in 23 Chapters,* "Wherever you go in the world, whatever animal, plant, bug or blob you look at, if it is alive, it will use the same dictionary and know the same code. All life is one.... We all use exactly the same language."[16] If the gene from the fish actually makes itself at home in the plant and continues to pump out a protein that makes the host organism resistant to cold temperatures, then just how unnatural is it? Not as unnatural as converting petroleum into false teeth, much less mining ore, refining it, shaping it like a joint and using it to replace an arthritic hip.

Genes do what God and nature designed them to do. They don't discriminate on the basis of race, color, creed, religion, species or even kingdom. They can do very nasty things to us, or they can do wonderful things for us.

TWO

Vanquishing Vaccines

One way to see how genes are already saving and improving lives is by focusing on vaccines.

Vaccines are one of the oldest yet most effective technologies. Traditionally they have been used to prevent people or animals from becoming ill, or at least to reduce the severity of their symptoms if they do. Vaccinations date back to the 1700s, when the English doctor Edward Jenner (1749–1823) noticed that persons who contracted cowpox from contact with dairy cattle developed a slight fever that lasted a few days, but thereafter were immune to deadly smallpox. He inoculated some test subjects by working bits of cowpox puss into scratches, and found that when exposed to smallpox, they contracted only mild cases of the otherwise virulent disease if they got it at all. Thus the term "vaccine" comes from the scientific name for cowpox, *vaccinia.*

Vaccines build immunity by using or producing antigens, which are proteins on the surface of the invading pathogen that the immune system uses to identify the germ. The body then produces antibodies to fight the disease. If the person is later exposed to the disease itself, the body is already primed and ready to resist. Some vaccines (like the one for polio) appear to confer lifelong immunity, while others (such as tetanus) require boosters periodically. For still others, such as smallpox (as we discovered in the days after 9/11), we really don't know.

Vaccines vary in the length and degree of protection they confer. Those used against viruses may contain a live but weakened form of the virus, or a killed virus. Live vaccines usually confer longer and stronger immunity but also occasionally produce the disease itself.[1]

The improvements needed for vaccinations are clear. They should be completely safe, unable to induce the disease itself. They should provide a lifetime of protection against diseases that don't mutate much, like the measles, and ultimately against those that mutate rapidly, such as the flu or AIDS. Combination immunizations (called "multivalent") are better, because fewer injections raise the compliance rate and make it easier to immunize people in outlying areas. Moreover, until scientists discover some part of the flu virus or HIV that doesn't mutate, it will be necessary to pack together as many strains of the virus into one vaccine as possible. Current technology doesn't allow fitting more than three strains of flu antigens into a single vaccine, so each year in the United States, the Centers for Disease Control and Prevention will choose the three it believes are most likely to strike North America. Sometimes it guesses wrong.[2] Finally, the higher the protection rate, the better. Efficacy rates of well below 100 percent can be enough to wipe a disease out of an area by reducing its contagiousness below the point at which it can infect new victims. But ideally, we would like to provide full protection to each and every recipient. Only biotechnology offers hope for these kinds of improvements.

Already at the beginning of the twenty-first century, almost a hundred biotech vaccines were under development.[3] Researchers are working on transgenic vaccines to combat influenza, AIDS, cholera, and other diarrheal diseases ranging from those that cause mild discomfort to those that kill.[4] A hepatitis B vaccine made from recombinant yeast, introduced in 1986 by the New Jersey–based pharmaceutical giant Merck & Company, has proven 90 percent effective in preventing this disease, which often leads to debilitating or even fatal liver failure.[5] The very name of that vaccine shows how far we've come since then with recombinant vaccines: it was called Recombivax HB. This is a little like Ford Motors presenting a new vehicle called "The Car."

In 2001 the FDA approved a biotech vaccine that induces immunity to both hepatitis A and hepatitis B.[6] The only approved vaccine for Lyme disease, spread by ticks, is a biotech product.[7] Good progress is being made in developing a vaccine against staph germs, which are the top cause of illness in hospitals.[8] The current anthrax vaccine requires six injections over eighteen months, plus an annual booster; a recombinant vaccine that began safety testing in 2003 would confer immunity with only three injections.[9] Although the Ebola virus has spawned far more scary articles, books and movies than actual victims, its fierce reputation may soon be tamed by a recombinant vaccine that has successfully protected laboratory monkeys.[10] Other targets within biotech's gun sights include hepatitis C, bubonic plague and human papillomavirus (HPV).[11]

HPV not only causes unsightly genital warts but is the main cause of cervical cancer.[12] In the United States, regular pap smears have caused a dramatic thirty-year decline in full-blown cervical cancer cases and resulting deaths.[13] But this prevention is terribly expensive, costing about $3 billion annually for the smears, follow-up biopsies on women with abnormal findings, and other services associated with HPV infection. An effective vaccine would eliminate most of that expense.[14] More importantly, most countries can't afford such preventative care. As a result, cervical cancer kills almost a quarter-million women worldwide each year, making it the world's second-greatest cause of female cancer deaths, after breast cancer.[15]

Merck has developed a vaccine against HPV containing four different antigens—proteins that provoke the immune system into creating protective antibodies.[16] (A study in the *New England Journal of Medicine* that sparked a tremendous media furor concerned an earlier version with only one antigen.)[17] Two of the strains against which the vaccine works cause unsightly genital warts in both men and women, while the other two are thought to cause about 70 percent of all cervical cancers.[18] Judged only by what it may accomplish directly, this vaccine could be a tremendous breakthrough. Yet its potential goes far beyond its specific purpose. Consider, to start, that it's only the second vaccine that can actually prevent cancer. The first was Merck's original vaccine developed against hepatitis B, which often leads to cancer of the liver. Both are made from recombinant yeast. Further, the HPV vaccine would be the first to protect against STDs. "Years ago people did not believe that you could actually prevent an infection of the genital track," Merck's team leader, Dr. Kathrin Jansen, told me. "I always believed it could work, but there was a lot of resistance."[19] Not any more.

As a result of vaccines, root canals may someday be only a topic for essays on forms of medieval torture practiced in the modern era. Scientists are working on using gene-altered microbes as a vaccine to brush or squirt onto people's teeth so the microbe will "elbow out" all other forms of bacteria. That would include *Streptococcus mutans,* which thrives on organic material that coats tooth surfaces and converts food sugars into lactic acid, a corrosive chemical that gradually dissolves enamel and ultimately leads to tooth loss or the dreaded drill.[20]

A recombinant genital herpes vaccine from GlaxoSmithKline is also well along the testing path. About one-fourth of the U.S. population is thought to carry the virus for genital herpes, though many carriers have cases so mild they don't even know it.[21] Strangely and for reasons unknown, the vaccine has absolutely no effect on men. Nor does it help women who have genital herpes and also oral herpes.[22] But by breaking the cycle of transmission it will help persons for whom the vaccine itself is ineffective. It would also prevent the tragic birth of babies with herpes, which can cause complications far more serious than in an adult.

At an earlier stage is a biotech vaccine for allergies. According to the American College of Allergy, Asthma and Immunology, about 38 percent of all Americans are affected by allergies throughout the year.[23] People have allergic reactions when their immune systems overreact to a foreign protein.[24] Allergens induce certain immune cells to fight off an invader with the release of inflammatory chemicals such as histamines, and these make us miserable. Injections have long been used to combat them, but often do not work and sometimes make the patient quite sick.

Researchers at Johns Hopkins University, working with others at the University of California at San Diego and Dynavax Technologies of Berkeley, California, successfully used a synthetic piece of DNA on twenty volunteers.[25] "This study showed that [the DNA vaccine] resulted in the production of an IgG [or so-called "blocking"] antibody response similar to what one sees in conventional allergy shot therapy," said Dr. Peter Creticos of Johns Hopkins. "And none of the individuals experienced any serious adverse reactions."[26]

Malaria is a scourge in some ninety nations, according to the World Health Organization (WHO). Between 700,000 and 2.7 million people die annually from malaria, over 75 percent of them African children.[27] In 1998, WHO listed malaria as the biggest killer in Africa, responsible for one in every four deaths among children under the age of five.[28] For decades, this disease has vexed researchers trying to develop a vaccine.[29] But now, recombinant DNA techniques make the outlook more promising than ever before.

Oxford University scientists have developed the first vaccine that targets the malaria parasite inside the cell. "Instead of using the malaria parasite itself, inactivated in some way, we are actually using DNA, the genetic material," lead researcher Adrian Hill told BBC Radio in September 2000. Rather than futilely attempt to kill the parasite before it gets into the cell, the new vaccine is designed to destroy from the inside. Early trials showed that it was safe and that it produced a strong immune response in recipients. Now it's being tested on far larger numbers of people in the African nation of Gambia.[30]

Antifungal vaccines are also in the works. In late November 2000, scientists at the University of Wisconsin at Madison startled the medical world by announcing that, using recombinant DNA technology, they had created a live vaccine that protects against a fungal infection in mice. The vaccine was reportedly safer than live, nonrecombinant ones and more effective than killed vaccines. "This work achieves what many would have thought highly unlikely, if not impossible," commented Dennis Dixon, chief of the bacteriology and mycology branch at the National Institute for Allergies and Infectious Diseases (NIAID). "It validates the concept of a vaccine approach for disease-causing fungi."[31]

What's so tough about a fungus? It may look simple to us, but compared with viruses and bacteria, a fungus is almost as complicated as the federal tax code. The one used in this study has a genome of 25 million base pairs, the smallest units of DNA. In contrast, the ubiquitous *E. coli* bacterium used in making many therapeutic proteins has only about 4.6 million base pairs. HIV has fewer than ten thousand. Because of its genomic complexity, the best type of vaccine against a fungus appears to be a live one.

The conundrum is that this same complexity makes it all the *harder* to use a live vaccine, because you have to deactivate the part that makes the fungus cause disease. Before biotech, there was no possibility of doing such a thing. But with the relatively new ability to identify and disable single genes, scientists were able to locate the "virulence" gene and knock it out.[32] The fungus in the vaccine was still alive, but much friendlier.

Vaccines for Cancer

Cancer alone is the object of over half of the biotech vaccines being developed.[33] Researchers are developing biotech cancer vaccines targeted at leukemia, cancers of the breast, ovaries, pancreas, lymph glands, brain, colon and virtually every other malignancy.[34] An August 2003 search on the PubMed medical journal database using the terms "cancer" and "vaccine" generated almost eleven thousand articles.[35] These vaccines can be therapeutic or prophylactic, notes Dr. Samir Khleiff, a cancer vaccine researcher at the National Cancer Institute in Bethesda, Maryland. A therapeutic vaccine is one that treats an existing tumor. As such, it's completely different from what we normally envision when we think of vaccines, as something you receive in order to prevent the disease from occurring in the first place. Other biotech cancer vaccines are being developed to prevent cancer from returning in patients who have already been treated and appear to be disease-free. These are called prophylactic or treatment vaccines.[36]

"What the cancer vaccine seeks to do is get the body's immune system to destroy tumor cells, to see those tumor cells, recognize them as dangerous and destroy them in much the same way as it destroys viruses and bacteria when we get an infection," explains Professor Alan Kingsman of Oxford Biomedica, a biopharmaceuticals company based in Oxford, England.[37] "Then the antibodies and cells of the immune system cruise around the body looking for that protein, recognize it as dangerous, and blow those cells away."[38]

Some of these vaccines are "autologous," meaning they are made from a patient's own tumor. The antigenic properties are strengthened outside the body, and then they are reinjected. If the malignancy comes back, the immune system will be prepped for it.

Others, such as Canvaxin from CancerVax in Carlsbad, California, are called "allogeneic," meaning they are not based on an individual's unique antigen expression but are potentially applicable for anybody with certain forms of cancer. Canvaxin contains twenty different antigens from cell lines drawn from three different people. At least one of these antigens will be present in about 95 percent of all those suffering from the deadly skin cancer, malignant melanoma, as well as in many colon cancer patients.[39] In Phase II testing (of the usual three phases of human clinical trials), Canvaxin doubled the five-year survival rate of malignant melanoma patients.[40] To increase the potency further, the company is testing a version that's designed to boost the response of patients with immune systems damaged by chemotherapy.[41]

Looking at cancer as a generalized sort of malady is actually a return to an old theory. At one time, "cancer" was indeed thought to describe essentially one phenomenon that happened to occur in different parts of the body in different ways at the same time, "a group of diseases where abnormal cells divide without control," as one glossary puts it.[42] Then came President Richard Nixon's official declaration of the "War on Cancer" in 1971 (which was supposed to be won within *five years*).[43] Basic research seems to have been sacrificed for quick results, with the consequence that the former suffered even as the latter bore little fruit. In those pessimistic times it became the accepted wisdom that "cancer" was little more than a term of convenience actually representing a spectrum of diseases, often with little in common. Now, because of the advances of biotech, the thinking has swung back: cancer is again being seen as different manifestations of one disease.

Until recently, for example, it was widely believed that anticancer vaccines would have to be specific to each type of cancer because each produces its own specific antigens and it's those antigens that a vaccine uses to recognize and attack an enemy. Yet we now have some experimental vaccines that appear to have at least some impact against numerous cancers, such as one from California-based Cell Genesys that has proven active against every type of cancer tested including those of the prostate, lung and kidney, as well as melanoma. The vaccine consists of irradiated tumor cells that have been genetically modified to secrete a hormone that plays a key role in stimulating the body's immune response to vaccines. What's so special about the "GVAX" vaccine is that it targets multiple antigens on tumor cells. This makes unnecessary the need to pinpoint the specific antigen that may be the most therapeutically important—information that in any case is currently unknown.[44]

One property that virtually all cancers seem to have in common is the need to create blood vessels to sustain themselves, a process called angiogenesis. (This will be discussed in greater detail later.) Another involves the production of an enzyme called telomerase (pronounced te-LOM-er-aze), which rebuilds telomeres (pronounced TEE-lo-mears), the tightly coiled threads of DNA that form a protein cap on the ends of each of our chromosomes.[45] These telomeres shorten each time the cell divides, until the cell cannot divide anymore.[46] Over 85 percent of cancer types, including all the major ones, utilize telomerase to keep dividing until eventually they kill us.

What if we use telomerase as our target? Some researchers have wondered: could we then get a one-size-fits-all cancer vaccine? Possibly. In 2000, researchers at Duke University in Durham, North Carolina, working with others at the Geron Corporation in Menlo Park, California, came up with what the media called a potential "universal" cancer vaccine.[47] Unfortunately, that's a misnomer; if 85 percent of cancers do rely on telomerase, then 15 percent don't. Nevertheless, this breakthrough study found that a telomerase-based vaccine can stimulate an immune response in cancer patients to attack and kill the tumor cells, according to Eli Gilboa, principal investigator of the study and director of the Center for Genetic and Cellular Therapies at Duke.[48] The vaccine slowed tumor growth of human melanoma and breast and bladder cancers implanted into genetically unrelated mice. It did the same against two sets of mouse cells in lab studies. Finally, it was also effective against human kidney cancer cells in the lab.[49] Human trials are still a few years away. Further, said Gilboa, "By itself, telomerase is not a strong antigen, so to make an effective, broadly applicable cancer vaccine we will need to optimize and possibly combine it with other universal antigens."[50]

British and Swiss researchers in lab studies have found a genetic switch that allows cancerous cells to flip the telomerase switch back on after they've been "ordered" by anticancer genes to flip it off. Blocking this process cuts off the enzyme and causes cancerous cells to stop multiplying, the researchers report. They also believe a drug can be found that targets the gene and keeps the telomerase production switched off.[51]

Emerging Diseases

Because of the speed at which they can be developed, biotech vaccines also offer the best solution to so-called "emerging infectious diseases"—diseases newly discovered, or newly discovered in a certain geographic region, or old diseases making a comeback.[52] Some, such as AIDS, have had an incalculable impact on the entire world. AIDS is the classic example of a true emerging disease, regardless of how long it may have existed and what its exact origin.[53] Another is West Nile virus, which was discovered in New York State and then began spreading throughout the country, though it had long made the rounds in Africa, as the name implies, as well as the Middle East and Europe.[54]

In 2000, Acambis, based in both Cambridge, Massachusetts, and Cambridge in England, began to develop a West Nile vaccine using a tried-and-true sixty-year-old yellow fever vaccine.[55] Essentially the company stripped off its molecular protein envelope and replaced it with that of West Nile.[56] "The viruses that are produced have the coat of West Nile but the replication machinery of yellow fever 17D," said Thomas Monath, vice president of research and medical affairs at Acambis. The vaccine is part of a family called ChimeriVax after the Chimera, a creature from Greek mythology with the head of a lion, the body of a goat and the tail of a serpent. The company is using the same technology to develop a vaccine for numerous other terrible mosquito-borne illnesses, such as Japanese encephalitis and dengue fever. "All are members of one virus family in which the proteins are not identical, but they're related," said Monath, and thus "it becomes possible to exchange the proteins from the 17D virus with those of the other viruses."[57] Speed is the key here. Many vaccines have taken well over a decade to develop, but human trials on the West Nile vaccine were expected to begin in 2003.[58]

Other gene-spliced vaccines are also in the works. One was developed by removing key genes from the dengue virus and replacing them with West Nile genes. Because the hybrid vaccine is mostly dengue, it doesn't target the central nervous system as West Nile does. Researchers also used a strain proven not to harm people; the added West Nile genes are just enough to stimulate a response against the disease. When injected into mice, the vaccine protected all of the animals from subsequent exposure to the New York strain of West Nile virus.[59]

Obsolete Needles

Vaccine delivery systems are also going to change, with needles going the way of the slide rule. MedImmune of Gaithersburg, Maryland, has just the thing for needlephobes who need flu shots. Its FluMist is sprayed right into the nostrils and has an excellent protection rate of well over 90 percent. The company received a mid-2003 FDA approval.[60] "Currently, only a small percentage of healthy children and adults are immunized against influenza," said the company's president and CEO, C. Boyd

Clarke. "We think FluMist could eventually provide an attractive new option to these populations."[61] By offering a kinder, gentler delivery mechanism and bringing the vaccine directly to the mucous membrane of the nose, many health experts think FluMist could save billions of dollars in health care costs and lost work time in the United States.[62]

Scientists at Stanford University have found that a simple solution of DNA and water applied to the skin of mice can transmit hepatitis B vaccine, stimulating an immune response just as strong as the traditional injection with a needle. The DNA vaccine was apparently absorbed through the hair follicles.[63] "In principle, the DNA solution could be put on a Band-Aid, or in a spray, or even in shampoo," lead researcher Dr. Paul Khavari told the *New York Times*.[64] "It's an exciting development because it shows that DNA vaccination will get simpler and simpler," said Dr. Stephen Albert Johnston, the director of the Center for Biomedical Inventions at the University of Texas Southwestern Medical Center in Dallas.[65] "We've been injecting common vaccines into the muscle," said Johnston. "This work says that we've been administering vaccines in the wrong place. Soaking it into the skin is a lot simpler and less painful."[66]

Because DNA vaccines have so many practical advantages, other scientists are working to put them into patches and nasal sprays and foods. "DNA is so stable that you can walk around with it in your pocket for months at room temperature," said Khavari. "The lack of need for refrigeration or special medical personnel makes it very attractive, especially in developing countries."[67] Further, several vaccinations could be combined into a single delivery. According to Dr. Dennis Klinman of the Center for Biologic Evaluation and Research at the Food and Drug Administration, "The technology to generate them against multiple different organisms is very promising."[68]

THREE

Miniature Pharmaceutical Factories and Medicinal Milk

How does one go about splicing a gene?

Let's say we've found a gene in a mushroom that makes cells producing a protein that inhibits HIV, the virus that causes AIDS. You can't just feed the mushrooms to people with HIV; the protein would break down in the gut. So you take something called a "restriction enzyme," the so-called "molecular scissors" of gene splicing, which makes it possible to remove a single gene from a strand of DNA.[1] In this case, it's from the mushroom DNA. This gene is then spliced into another organism, mostly into bacteria at present. As the recombined bacteria multiply, they act like a tiny factory to mass-produce the original mushroom gene and its medicinal protein.

Bacteria in general make good hosts for transferred genes because they not only have a chromosome but, unlike animals, they also have a smaller piece of DNA called a plasmid. These have the unique ability to replicate themselves without cell division. When they do so, they make many copies of the desirable genes. Part Three of this book, discussing biotech crops, will explain other ways of transferring genes from one type of organism into another; for now, understanding this type of splicing is enough. Gene-spliced (recombinant) drugs are often derived from a strain of the *E. coli* bacterium that we all have in our intestines, but this type, unlike the nasty one that causes food poisoning outbreaks, doesn't cause disease. Other sources of recombinant drugs currently include yeast, mammalian cells (usually from hamster ovaries) and a virus inserted into insects called a baculovirus.[2] In 2002, corn was added to that list.[3]

Some of biotech's earliest contributions to pharmaceuticals entailed using these little factories to provide proteins originally available only in amounts too small to be useful. Consider leech saliva. Seriously. Somebody got the idea of milking leech jaws for a chemical that would prevent venous thrombosis (blood clots in veins) and treat thrombocytopenia, a platelet deficiency in blood that leads to excessive bleeding.[4] As a therapy, it worked. But the problem is rather obvious. You can feed leeches blood-flavored sourballs all day long but you're still not going to collect much saliva. Splice a clot-busting gene into another organism, however, and you can toss out the leeches and grow just as much of the protein as you need.[5] That's what Genentech's tenecteplase is. By splicing a natural protein (not from leeches) into Chinese hamster ovarian cells, it can reduce the time it takes to treat heart attack victims from ninety minutes to only five, using a single swift injection to dissolve clots that choke off the heart's blood supply. These cells are cultivated outside the body, where they pump out the needed protein. (This is called a "cell line" or "cell colony.")[6]

Similarly, there is a drug for patients with Gaucher's (pronounced Go-SHAYS) disease, an inherited enzyme deficiency that can cause anemia, fatigue, easy bruising and bleeding, and numerous other problems.[7] But it requires processing an astounding ten to twelve *tons* of human placentas a year, making it one of the world's most expensive pharmaceuticals.[8] To make a substitute, researchers found that the best medium for growth was hamster ovarian cells.[9] Unfortunately, for this particular drug, Genzyme's Cerezyme, even this technique is expensive.[10] Therapy costs a whopping $170,000 a year.[11] But scientists are now working to produce the enzyme more economically in tobacco plants.[12]

Some recombinant pharmaceuticals are replacements for natural sources that posed a danger to recipients. For example, children lacking in human growth hormone formerly became dwarves or midgets. With an outside source of the hormone, they can grow to normal size.[13] The same hormone also helps AIDS patients keep their weight up, helping to prevent the awful wasting syndrome that ultimately kills many sufferers.[14] Originally, human growth hormone was extracted from the pituitary glands (located at the base of the brain) of cadavers. In the mid-1980s, though, the horrifying discovery was made that this method was transmitting Creutzfeldt-Jakob, a monstrous, always-fatal disease that eats away the brain.[15] Fortunately, a safe biotech version produced in *E. coli* bacteria was just then becoming available. Now, eight different recombinant human growth hormone products have been approved.[16]

The very first recombinant drug was insulin, developed by Genentech of South San Francisco, and licensed to the Eli Lilly Company of Indianapolis.[17] The FDA approved it in 1982. Prior to this, people with insulin-dependent diabetes mellitus injected themselves with insulin extracted from the pancreases of cattle or pigs. Recombinant insulin gives them a guaranteed safe supply of humanized insulin, and one that almost never provokes allergic reactions.[18]

Yet other recombinants simply provide a supply of a naturally occurring protein that allows individual patients to get enough into their system to combat disease. Rebif, developed by Serono International of Switzerland and sold in the United States since mid-2002 by Pfizer, is a form of interferon beta, a carbohydrate-bearing protein produced by fibroblasts (cells found in skin and connective tissues).[19] Only interferon beta has conclusively been shown to be effective in treating multiple sclerosis, although other types of drugs are being tested.[20] Serono discovered that large amounts greatly relieved symptoms of MS, but the only way to produce these amounts was through gene splicing. They found that patients treated with Rebif had a 90 percent greater chance of remaining relapse-free than those using what had been the state-of-the-art drug.[21] It is injected, but can be self-injected by the patient. Diabetics often inject themselves four times a day; with Rebif it's three times a week.[22]

Blood Work

Some companies are also making great strides in producing transgenic human serum albumin, or HSA. This protein is involved in maintaining fluid balance in the blood, regulating amino acids and fatty acids, and hormone transport. It is given in blood transfusions for shock, serious burns and other emergency surgeries. But HSA is not enough to solve all of our blood supply problems. That supply is safer than ever, in part because there's so much screening of donors with risk factors.[23] The downside is that every time another at-risk group of donors is excluded, the overall pool of possible donors shrinks. For example, in September 2001 when the Red Cross began turning away donors who had spent considerable time in Europe in order to keep so-called "mad cow disease" out of the blood supply, it estimated this would eliminate almost 10 percent of the donor pool.[24] Blood demand in the United States is now increasing at a rate twice that of donations.[25]

Further, because people with some blood types can only receive other specific blood types, there can be shortages of certain types even as the overall blood supply is in good shape. Finally, in the United States alone there are about half a million Jehovah's Witnesses who will not accept blood donations. We have various sorts of artificial blood, but they all do only some of the things human blood does. What's needed is artificial blood so natural in structure that even Dracula couldn't tell the difference.

Efforts to create artificial blood extend back to the seventeenth century when doctors experimented with everything from milk to oil to animal blood.[26] With biotechnology, that centuries-old goal may be within reach. In early 2001, Japanese scientists announced that they had made red blood cells that worked in rats and were ready for testing in humans.[27] Now they're combining these blood cells with transgenic HSA grown in *E. coli.* The artificial blood appears to have the important characteristics of human blood, right down to changing shades of red depending on how much oxygen it's carrying. What's just as important is what it *lacks.* It's type-free, so it can be given to anyone, instantly, without the need for a typing test. And it can never be contaminated, as the U.S. blood supply was as a result of HIV-positive donors.[28]

Meanwhile, scientists in Virginia are showing that perhaps you *can* get blood from a turnip—or some kind of plant, anyway. CropTech Corporation of Blacksburg, Virginia, is working to produce serum albumin for humans in tobacco.[29] CropTech claims that a mere 45,000 acres would be required to supply the entire worldwide demand for the product.[30] (If you're wondering why so much experimentation is being done with tobacco, it's partly because of funding to help tobacco farmers use their crops for something other than smoking and chewing, and also in great part because tobacco contains a high percentage of protein.)

As with tenecteplase, more and more of the new recombinant proteins aren't just superior substitutes to traditional treatments; they do things that previously nothing else could, providing not just palliatives but actual treatments for diseases that heretofore had none, such as psoriasis. As many as seven million Americans suffer from psoriasis, generally believed to be an autoimmune disease.[31] In its most common form, skin cells (called plaque) build up and form elevated, scaly lesions that can be both itchy and painful. About seven million Americans suffer from psoriatic arthritis, a painful joint-swelling form of the disease, according to the National Psoriasis Foundation.[32] Fortunately, about thirty companies have psoriasis drugs in clinical trials; most are coming from the biotech industry.[33] "For us to see so much in the pipeline, it's just amazing," said Jessica Wise, public relations manager at the foundation. "We look at biotech companies as giving a lot of hope in that they're working very close with the genetics and they'll probably be the ones to take what's learned from the genetic side and come up with treatments that are target specific. And we're very encouraged that so many biotech companies have taken an interest in this disease."[34]

The first drug approved for chronic plaque psoriasis was Amevive from Biogen, given the FDA green light in January 2003.[35] Made in Chinese hamster ovarian cell lines, it will probably also be effective against other forms of psoriasis and other autoimmune diseases as well.[36] What's most exciting about Amevive, however, "is that people will have a sustained affect for a very long period of time, perhaps months before they need new therapy," according to Dr. Burt Adelman, executive vice president of research and development.[37] That will put Amevive in the "not quite a cure but the next best thing to it" category, a vital step toward giving patients the benefits of pharmaceuticals without the drawback of regular dosing. Moreover, a twelve-week study reported in November of 2002 used a combination of Amevive and ultraviolet light to achieve "clear" or "almost clear" results on 16 of 21 psoriasis patients, while all were helped significantly.[38]

The New "Aspirin"?

The first psoriasis drug to pop out of the biotech pipeline was Amgen's Enbrel, also made from hamster cell lines.[39] It is the leader of many biotech pharmaceuticals vying to become "miracle drugs" in the same way as aspirin—by treating a spectrum of maladies that seem to have little in common. (Once thought of as little more than a headache remedy, aspirin is now recommended for arthritis symptoms, for reducing the risk of heart attack and stroke, and even as a cancer preventative.)[40]

Enbrel was originally approved for rheumatoid arthritis and then in January of 2002 for psoriatic arthritis. In clinical trials, 70 percent of psoriatic arthritis patients on Enbrel showed improvement after six months and some had almost complete remission of all symptoms.[41] Amgen is seeking FDA approval for Enbrel's use against psoriasis in general and it is being tested for almost two dozen other diseases, from Alzheimer's to cancer.[42]

The drug is what's known as a cytokine inhibitor. Cells have molecules in their outer membranes called receptors that receive messages from various molecules and act as way stations leading to the nucleus. This then tells the relevant genes to switch on or off. A cytokine is one of those receptors. It belongs to a category of white blood cells including macrophages, lymphocytes, interferons, interleukins and tumor necrosis factor (TNF).[43] Cytokines are not evil. They encourage inflammation, which is usually good because it's part of the body's defensive system. The problem comes when they go awry, causing autoimmune system diseases. Referred to sometimes as the "biologic equivalent of friendly fire," autoimmune illness apparently occurs when the body overreacts to a pathogen. Long after the original bacteria is gone or the original virus has been nullified by the immune system, the inflammation remains.[44] "TNF in the joints turns them wild," said Professor Luke O'Neill, director of the Biotechnology Institute at Trinity College in Dublin.[45]

Seattle's Immunex Corporation (now owned by Amgen) made copies of a receptor for TNF. It then used these cloned receptors to test various proteins to see which ones might help prevent the terrible chronic inflammation caused by rheumatoid arthritis and other autoimmune diseases. The best candidate proved to be Enbrel, which is now made in hamster cell lines. It works by "capturing" and inactivating some TNF molecules before they can trigger inflammation—basically like flooding the body with catcher's mitts for TNF baseballs. By interrupting the molecular chain of events that leads to inflammation, Enbrel works with the immune system to help reduce symptoms.[46] It actually "modifies the disease" and is the first truly new response to arthritis in fifty years, said O'Neill.[47]

"My patients are having terrific responses to Enbrel," one rheumatologist told me. "The usual comment I hear is: 'Thank you; you've given me my life back.'" He adds, "Most rheumatologists will tell you they have a renewed purpose in life."[48] "They," by the way, refers not to patients but rather to the *doctors* who for so long had so little to offer them.

Now researchers are discovering excesses of TNF in many other diseases where before they hadn't even thought to look. When injected into the body, Enbrel roams widely, binding to free TNF molecules to prevent them from docking with cell receptors and causing damage. Because high levels of TNF are found in the skin of psoriasis patients, it's a ripe target for Enbrel. Even cancer may be in the drug's gun sights: research shows that TNF may spur the growth of malignancies. The name TNF "is a misnomer," said Ann Hayes, Immunex's senior medical director, referring to the fact that the word represented by the second letter in the acronym, "necrosis," means the death of individual cells or groups of cells, or of localized areas of tissue. "In real life, many tumors use TNF to stimulate their growth."[49] Amgen is now sponsoring early human trials to determine if Enbrel can help in leukemia and lung and ovarian tumors.[50] So far the biggest problem with Enbrel was that Immunex had no idea how wildly successful it would be and supply had fallen far behind demand, despite the company's best efforts to keep ramping up production.[51]

Remicade, which blocks TNF in a slightly different way, could have most of the therapeutic indications of Enbrel.[52] And another recombinant drug, Amgen's Kineret, has also received FDA approval for rheumatoid arthritis.[53] Although clinical studies showed it isn't nearly as effective overall, it works by a different mechanism (blocking a protein named interleukin-I) and thus may offer relief to certain patients not helped by other drugs.[54] All of these recently got competition from Abbott Laboratories' monoclonal antibody (more on monoclonals later), Humira.[55] This is a very new drug, so it's too early to say how well this TNF blocker will alleviate disease. But it does carry the advantage of being easier to take than Enbrel, which requires two shots a week, and Remicade, which requires an infusion in a doctor's office. Humira is self-injected every other week.[56]

Humane Genome Sciences (HGS) of Rockville, Maryland, is trying something similar with Albuferon-alpha, an experimental recombinant hepatitis C virus treatment, which remains active in the bloodstream for up to twenty-eight days, more than twice the duration of the leading treatments now on the market.[57] The company says the drug will require injection once or twice monthly, which would cut the number of annual injections for hepatitis C patients in half.[58]

HGS is also working with a natural human protein called BLyS, or B-lymphocyte stimulator, both to block it *and* to generate more of it. This is based on the concept that an autoimmune disease is the body suffering from too much of a good thing. In this case the suffering appears to be both rheumatoid arthritis and another autoimmune disease, lupus, which causes inflammation, organ damage, infections and sometimes death. About 1.5 to 2 million Americans suffer from lupus, and there have been no real advances in treatment since the 1970s.[59] HGS found that people with rheumatoid arthritis and lupus apparently have ten times the normal level of BLyS in their blood. That could explain why their immune systems become overactive and attack joints and connective tissue, as in the case of lupus. So the company is developing artificial antibodies that would reduce the level at which BLyS is emitted. This could prove a very simple answer to two very cruel and heretofore elusive diseases. The belief is that such artificial antibodies, given as an injection in large doses, could neutralize BLyS in the bloodstream, thereby treating autoimmune illness.[60] The company has entered into a partnership with a British firm to develop these.[61]

Conversely, it appears that many diseases could be alleviated if the body produced *more* BLyS. This includes a form of immune deficiency that causes multiple bouts of pneumonia, bronchitis and sinusitis as well as other infections. HGS has also signed an agreement with Dow Chemical Company of Midland, Michigan, to co-develop a drug to increase the amount of BLyS in the system so as to attack cancers such as leukemia, multiple myeloma and non-Hodgkin's lymphoma, albeit without giving them enough of the protein to cause autoimmune illness.[62] The companies will work together to form a "radio-labeled" form of BLyS by using a Dow technology that allows attaching radioactive material to the protein. In theory, the radioactive metal would bind to the cancer cells and kill them.[63] The very BLyS that cripples millions of Americans may be turned right around to save the lives of millions more.

Fighting Bacteria

In the past, a minor scratch could be life-threatening. President Calvin Coolidge's sixteen-year-old son became very ill after playing tennis and developing a small blister that became infected. The president famously crawled on hands and knees on the White House grounds to catch a rabbit for the boy to hold in his arms as he died. "In his suffering he was asking me to make him well," said the president. "I could not."[64]

Today such painful episodes are distant memories because of antibiotics. One of the most useful antibiotics is erythromycin, which treats a wide range of bacterial illnesses including skin infections, sexually transmitted diseases and respiratory infections—including those caused by the notorious anthrax. Like all antibiotics, it has no effects on viral infections, but it greatly reduces the chance of dying from secondary infections that result from viruses. The main component of erythromycin is something called a polyketide. Polyketides are useful not only in antibiotics but also as anticancer agents, antifungals, antiparasitics, immunosuppressive agents and cholesterol-lowering agents such as lovastatin.[65] The one used to make erythromycin comes from a bacterium that lives in soil. Unfortunately, making polyketides from their natural bacterial source is a long, hard process. So researchers at Stanford University in California took the necessary genes out of the bacterium that originally produced the erythromycin polyketide and inserted them into the workhorse of biotech pharmaceuticals, *E. coli.* The original bacterium, *S. erythraea,* grows slowly, taking four hours to double its population. But *E. coli* doubles every twenty minutes, producing over two thousand times as much product in the same amount of time.[66] "It's not difficult anymore" to make the polyketides, said Chaitan Khosla, the chemical engineer at Stanford who did the gene work. "It opens the door for harnessing the enormous power of molecular biology in *E. coli.*"[67]

Unfortunately, early on in the age of antibiotics we discovered that bacteria can develop ways to become resistant or immune to them. (Resistance to penicillin, for instance, developed within a few years of its widespread use.)[68] And for various reasons, the problem is growing.[69] The World Health Organization estimates that fourteen thousand Americans die each year from drug-resistant germs in hospitals, while antibiotic resistance is an even greater problem in much of the rest of the world. I've never been among those who say (sometimes quite vociferously) that we're "losing the war" against bacterial resistance.[70] My reasoning is simple: We have brains; bacteria do not. And we have computers that increase in power by the month; the ability of bacteria to evolve defenses will never increase. Ultimately, we have to win.

But science realizes that this is a war and one way to wage it is to keep fresh drugs in the pipeline to combat these resistant strains. While screening organisms found in soil and elsewhere for antimicrobial activity will be the mainstay of antibiotic development for some years to come, biotech allows completely different ways of killing the bugs that can add powerful weapons to our antibacterial arsenal. Ultimately, it may be possible to design a drug that attacks a part of the bacterium that cannot mutate, hence rendering it incapable of developing resistance. The first step toward this is to sequence the bugs genetically, and as of April 2003, researchers worldwide had sequenced over 180 microbial genomes, including some of the most deadly and medicine-resistant.[71] Indeed, in that very month, public health authorities around the world realized the need to map the genome of the germ suspected of causing a pandemic called severe acute respiratory syndrome or SARS. With scientists from several nations working on it and trading information via the Internet, it took a mere ten days to complete the sequencing. This was immediately posted to the Web, where medical researchers all over the world could use the information to develop diagnostic tests, screen it against currently available drugs, and begin work on new drugs and vaccines.[72]

The reason a bacterium develops resistance is that at some point a mutant offspring will have the right characteristics to survive the antibiotic attack. If it keeps replicating, eventually this mutation becomes dominant and the bug becomes less vulnerable or even impervious to the drug. Using an amazing technique called "directed evolution," scientists have forced and observed changes in the lab in weeks that might take bacteria decades in nature. This allows scientists to see what might otherwise happen years down the road—something like capturing an enemy's battle plan. It can allow researchers to develop new mechanisms of killing germs even before they're needed.[73] Directed evolution originally meant using chemicals to try to make a gene mutate.[74] Nowadays, researchers are able to select specific genes and cause them to mutate anywhere from thousands to even trillions of times in the lab.[75]

One of the most promising approaches to killing both resistant and nonresistant bacteria involves turning one microbe against another. The attack microbe is a small virus known as a bacteriophage (or phage), which attaches to the surface of the bacterium and then inserts its DNA so that it can reproduce. So many new viruses are generated that the bacterium bursts. This finding provides a new approach to designing a whole new class of antibiotics to combat many serious bacterial diseases, including *E. coli*, pneumonia, staph infection, ear infections, Lyme disease and cholera in humans, as well as bacterial diseases in pets, livestock and even crops, according to researchers at Texas A&M University.[76] The discovery of this bacteria-killing mechanism is a "milestone in our quest for antibiotics," said Dr. Sankar Adhya, chief of developmental genetics at the National Cancer Institute in Bethesda, Maryland. He adds that the simplicity of the mechanism suggests a quicker route to designing new antibiotics that can be effective against bacterial resistance.[77]

Phages "let the cell commit suicide by dividing without making a new cell wall," explains Dr. Ryland Young, a biochemist at Texas A&M.[78] "Ideally," said his co-researcher, genetics professor Dr. Douglas Struck, "the small bit of protein responsible could be mimicked by a pharmaceutical company, and a drug could be made to be general against many bacteria, or specific against a certain pathogen, and even better, it could easily be changed to overcome resistance."[79] The beta virus is not of the type that infects humans, animals or plants. Rather, it is an essentially dormant bundle of DNA or RNA until it comes into contact with bacteria. It then goes into action, replicating within the bacterial cell and, after only a few minutes, exploding it. Relief for sufferers can be as quick as it is effective.

One type of phage is being expressly designed to combat bioterror, specifically anthrax attacks. (For the type of anthrax used in the mailings following 9/11 there were plenty of effective antibiotics available, although Cipro was mentioned most often.) These new phages are being developed to counter newly engineered strains against which the current drugs might have little effect. "Any sophisticated terrorist could easily engineer strains to be resistant to antibiotics," said anthrax researcher Stephen Leppla at the National Institutes of Health.[80] The new drug dissolves a vital component of the bacterium's cell wall, so that it would be virtually impossible to create strains that could withstand it. A protein from the phage, called PlyG lysine, can kill 50 million bacteria in two minutes, making it more powerful than many antibiotics. A single injection saved three-fourths of mice given an infection that would otherwise kill them, and it's believed a higher dose could perhaps cure 100 percent.

Beyond that, the phages can even be used to make a quick handheld detector for checking contaminated sites for all strains of anthrax.

Suspected spores are dropped into a filter and treated so that they sprout and take on a green glow. Within minutes, as tiny an amount of anthrax as a hundred spores can be identified.[81] In the anthrax attacks following 9/11, vital government buildings had to be shut down for weeks.[82] With this advance, that could be shortened to the length of a lunch break.[83]

Antibiotics have been and still are miracle medicines. But as the case of Lauren Doby shows, there are times when antibiotics are not enough.

A sophomore at Bellaire High School in suburban Houston, Doby contracted meningococcal septicemia shortly after the start of the school year in 2000.[84] Bacteria had entered her bloodstream, found their way into her spinal fluid, and inflamed the membranes that protect the spinal cord and the brain. Her mother, Kathy Doby, remembers initially feeling guilty for wanting Lauren to live, knowing that if she did survive she could be retarded, deaf or blind, or could lose a limb. But Kathy discovered that Texas Children's Hospital was testing a new medicine for severe sepsis.[85] Seeing little to lose, she gave the okay for her daughter to receive it. The drug worked, and Lauren is attending class again. "Lauren lost the tip of one finger, but you can't even tell it," her mother told the *Houston Chronicle*.[86]

What probably saved Lauren's life and health was one of the most remarkable drugs approved in 2001, a product named Xigris produced by Eli Lilly. The research was carried out on 1,690 patients with severe sepsis at 164 sites in eleven countries. Doctors gave patients either a four-day infusion with Xigris or a placebo. Deaths were 6.1 percent lower in those receiving the Lilly drug than in those on placebos.[87] This reduction in death came even though more than 70 percent of the patients treated were already in shock and three-fourths were on mechanical ventilation.[88] "That's just a striking reduction in mortality," said Dr. William Macias, medical director for the Xigris product team.[89]

Xigris is not an antimeningococcal medicine per se, and it's not an antibiotic at all. Instead, it's meant to be used when antibiotics fail or can't do their job fast enough. It acts when sepsis (sometimes used interchangeably with "septicemia") strikes. When this happens, the bacteria send out a mass of endotoxins, poisons that infect the body through disease, wounds or burns, or during surgery. It's like "proteins cascading out of control, like a nuclear bomb explosion," said Dr. Robert

H. Rubin, chief of surgical and transplant infectious diseases at Massachusetts General Hospital in Boston.[90] But it is the body's overreaction to these toxins that really does the damage. The resulting massive inflammation, accompanied by blood clots in small vessels, wreaks havoc on tissues and organs and dangerously lowers blood pressure. In its most severe form, called septic shock, it shuts down vital organs.[91] The rate of death from severe sepsis ranges from 30 to 50 percent despite advances in critical care.[92] It kills 1,400 people around the world every day, or about half a million per year. Even in a nation with a highly advanced health care system, such as the United States, it kills nearly one-third of its 750,000 victims each year.[93]

These grim statistics are not for lack of effort on the part of researchers. According to Lilly, about twenty major products for sepsis have failed to gain federal marketing approval or have proven ineffective in patient tests.[94]

Like so many other biotech drugs, including Lilly's own recombinant insulin products, Humalog and Humulin, Xigris is a version of a chemical the body produces naturally.[95] According to Rubin, "This is the first significant advance we've seen in 20 years."[96] The chemical in this case is activated protein C. In studies beginning in the mid-1980s comparing patients who died of sepsis with those who survived, it was found that those who succumbed had abnormally low levels of the protein.[97] Xigris turns the tables by giving recipients abnormally *high* levels. Genes from the protein are spliced into an "immortalized" human kidney cell line, meaning the cells never stop dividing. These cells then become factories for more protein C.[98] Xigris is the first drug made using a human cell line—difficult and expensive, but a necessity in this case and a true leap forward in the ability to make recombinant proteins.[99]

Further, like so many biotech drugs, the active ingredient of Xigris has also shown indications of being effective against at least one completely different disease from that for which it was developed. Activated protein C has been found to greatly reduce neural damage in lab animals when administered after a stroke. In fact, in mice given strokes, some 65 percent of the cells that would have died after the stroke survived when the mice were treated with the protein.[100]

In addition to saving lives, the new biotech drugs sometimes affect these lives in marvelous ways. Consider the case of young Bryon Vouga, whose kidneys began failing at age sixteen. Soon he was one of 200,000

Americans kept alive through dialysis. That was unpleasant enough, but the treatment also made him so anemic that he could no longer attend class. Today, in his thirties, he's a high school teacher in southern California. His prospects changed in 1989, when biotech pharmaceutical pioneer Amgen of Thousand Oaks, California, began mass-producing Epogen.[101]

Healthy kidneys produce a hormone that stimulates production of red blood cells, which then carry oxygen to muscles. But when kidneys are destroyed by infection, as Vouga's were, they stop clearing waste from the blood and stop producing a hormone, erythropoietin, that is vital in making red blood cells. The result is anemia, often requiring blood transfusions. Epogen, made by inserting human erythropoietin genes into hamster cell lines, replaces that hormone.[102] Vouga is still on dialysis because two attempted kidney transplants failed. But in 1999 he completed a 2,700-mile bicycle journey sponsored by the National Kidney Foundation.[103]

"First and foremost, I'm doing this to reach my goal," Vouga said. "It's always been a dream. Second, it's a fund-raiser for the National Kidney Foundation. Third, I hope to inspire other people to reach their dreams and also to exercise."[104] Vouga is clear on what has enabled him to do all this. "Epogen has saved a lot of lives," he said. "Patients used to have to get blood transfusions all the time. Now, you almost never get one."[105] Yet far from resting on its laurels, Amgen is trying to make those cross-country bike rides even easier. While Epogen requires three injections a week, the company has developed a successor, Aranesp, that requires only one injection per week or less, and is also useful for patients suffering anemia from cancer and chemotherapy treatment.[106]

Miracle Milk

Milk is an antibody medium with tremendous promise. In fact, the purpose of creating Dolly, the original cloned mammal, was to generate a line of sheep that would carry in their milk a human gene for factor IX, a clotting protein that people with hemophilia type B lack. Once a sheep had been made that could produce this protein, the cloning technique used to make Dolly could build herds of living medicinal factories.[107] One of Dolly's creators, PPL Therapeutics, is also making sheep whose milk will provide alpha-1-antitrypsin (AAT). (PPL's partner is the Roslin

Institute; both are just outside Edinburgh.)[108] AAT is a human blood protein that seeks out and digests an enzyme called neutrophil elastase, which appears to contribute to cystic fibrosis, chronic obstructive pulmonary disease generally, and congenital emphysema.[109] There are 150,000–200,000 sufferers of severe AAT deficiency in the United States and Europe.[110]

Another PPL sheep milk product undergoing human testing is a naturally occurring human enzyme known as bile-salt-stimulated lipase (BSSL). This enzyme breaks down fats for digestion, and is normally produced in the pancreas and secreted in breast milk. PPL hopes to use BSSL to treat conditions where the pancreas does not produce enough fat-digesting enzymes, such as cystic fibrosis and chronic pancreatitis, or for pre-term babies, who need a high-protein diet but have difficulty digesting fats. According to the company, there are currently 250,000 potential adult patients and 400,000 pre-term babies each year who would benefit from its protein.[111] The milk would not be consumed directly; rather, the recombinant proteins would be extracted from the whey, using techniques common in the dairy industry.[112] They would then be injected or administered in some other way.

The largest transgenic animals being used as pharmaceutical factories are cows. One cow can produce as many as 2,600 gallons of milk per year, from which a ton of medical proteins can be purified.[113] Scientists at AgResearch of New Zealand are attempting to use proteins introduced into cow's milk to combat multiple sclerosis.[114] This nervous system disease is believed to result from a breakdown in the myelin sheaths, the thin layers of fatty cells that wrap around and insulate nerve fibers in the brain and spinal cord.[115] By introducing the human gene that codes for these proteins into cows, the scientists hope they can make them reproduce the proteins in their milk.[116]

The Dutch company Pharming, now a subsidiary of Genzyme, is using cows on a farm in Wisconsin to produce lactoferrin, a natural protein with both antibiotic and probiotic qualities (discouraging harmful bacteria while encouraging helpful ones) as well as anti-inflammatory properties.[117] Its first clinical trials were for effectiveness in restoring normal coagulation in the blood of open-heart surgery patients and reducing bone swelling in persons with arthritis.[118] Pharming has also begun collaborating with a British company to develop lactoferrin to treat the hepatitis C virus, and is developing two types of collagen for tissue repair and an oral drug for rheumatoid arthritis, along with two blood products.[119] These are fibrinogen, a blood-clotting protein, and factor VIII for hemophilia A.[120]

Another industry leader in extracting proteins from transgenic animals is Genzyme Transgenics Corporation of Framingham, Massachusetts, which has found more than sixty-five such proteins.[121] In September 1996, Genzyme became the first company to begin human clinical trials with a drug collected from the milk of an animal, in this case goats. Unfortunately, development was halted, though not because the experiment wasn't working; rather, it was because of the cost of additional clinical trials compared with the market size.[122] Now Genzyme is working with cows to produce HSA, both because cows produce so much protein and because the market for HSA is so large.[123] PPL is also working to develop transgenic HSA in cow's milk.[124]

Goats are still the protein producer of choice for the Canadian company Nexia Biotechnology, which became the first lab to clone goats in 1999.[125] Nexia is working on one of the strangest projects in all of biotechdom: to mass-produce silk from goats that have had spider genes inserted into them. The milk of a single goat supplies more silk protein than could be extracted from a great mass of spiders—which wouldn't work anyway, because spiders are cannibals. ("Put a bunch of them together and soon you end up with one big, fat, happy spider. It's like trying to farm tigers," remarked Jeffrey Turner, a molecular geneticist and chief executive of Nexia.)[126] This silk isn't for scarves or neckties; instead, it constitutes the strongest fiber known. A woven cable as thick as your thumb can bear the weight of a Boeing 747 airliner.[127] The fiber is slated for use in wound healing, tissue repair, artificial tendons and ligaments, prostheses, and superthin, biodegradable sutures for ocular surgery or neurosurgery.[128]

Nexia's "BioSteel" could also play a role in the war on terrorism. Ever wonder why American soldiers seem to take so few casualties compared with the number they inflict? Weaponry certainly has much to do with it, but the role of body armor is greatly underrated. As the *Washington Post* noted shortly after the second Iraq war, "The vast majority of American soldiers who suffered life-threatening wounds in combat in Iraq were hit in the limbs, not the torso, suggesting that the body armor now worn by all soldiers is remarkably effective."[129] Currently the lightweight armor is made from DuPont's Kevlar, which has large pockets into which soldiers slide ceramic plates with much greater stopping power. But BioSteel is expected to be three times as bullet-resistant for the thickness, either reducing the need for the ceramic plates or supplementing them to provide even greater protection.[130] Nexia silk was first produced in early 2002 in mammalian cell lines, but the company is building up a herd of goats to commercialize the process.[131]

Cows are useful because they produce so much milk, while goats and sheep have the advantage of breeding much faster. But since using milk to create proteins is limited by the delay before female sexual maturity and the onset of first lactation as well as the intermittent nature of lactation itself, other animal-produced liquids are also being examined. One, oddly, is semen. Researchers have created transgenic mice that secrete human growth hormone in their ejaculate. No, you're not going to get much product from even the randiest rodent. But the idea is to transfer the technology to animals that produce larger amounts. One such is the boar, which can produce a prodigious half-liter of semen per ejaculation.[132]

Britney Lays an Egg

When it comes to fast, efficient and ultimately cheap protein production, though, few things beat the egg. So it was with some fanfare that first one British newspaper and then over two dozen newswires and newspapers around the world announced that the Scottish Roslin Institute had produced "Britney"—presumably in honor of the young entertainer, Miss Spears—a cute little chick intended to produce human antibodies in its egg whites.[133] As it happens, "There is no 'Britney,'" Roslin Institute assistant director Harry Griffin told me. "This fictional bird was the invention of one of the Sunday newspapers in the U.K.," specifically London's *Mail on Sunday.*[134]

What Roslin *did* do was announce a collaboration with a Florida biotech company, Viragen, which plans to use egg proteins to make a cancer vaccine.[135] Proteins in egg whites are produced according to instructions encoded in the hen's genes. Altering the genetic material in a cell's nucleus can lead to a chicken that will lay eggs full of proteins that can be used in drug production. Such a hen would have the potential to produce large amounts of medicinal proteins in a small space at a very low cost. A chicken lays about 250 eggs a year, and can produce about 100 milligrams of drugs in each egg.[136]

Although Roslin's nonexistent chicken grabbed the headlines, several American companies are far along in the feathered pharmaceutical race. One of these, AviGenics of Athens, Georgia, provides on its website a table comparing potential costs of production of a monoclonal antibody made from cell cultures, goat's milk (which like cow's and sheep's milk is still in an experimental stage), and chicken eggs.[137] While a hen lays an egg every day or two, a randy rooster can produce two thousand offspring per month. So all you need is a handful of roosters and a roomful of hens to produce protein at commercial volumes. And that drops costs dramatically. With cell cultures, the cost comes out to $100 per gram of product; with the goat's milk it's an estimated $2–$20; with chicken eggs it's a mere 10–25 cents.[138] AviGenics' cofounder Robert Ivarie refers to these special birds as "hen oviduct bioreactor technology" (which sounds rather like something that can propel chickens into the next solar system). Ultimately, he says, they will produce biopharmaceuticals, blood factors, monoclonal antibodies, and enzymes to help animals digest and process their food. For now, though, the company will focus on making "crude proteins" for pharmaceutical companies.[139] Another firm, Gene Works of Ann Arbor, Michigan, already has deals to use the eggs of transgenic chickens to produce fourteen proteins for six drug companies around the world.[140]

TranXenoGen, based in Shrewsbury, Massachusetts, is crowing about its roosters that produce transgenic offspring. Its first product will probably be insulin, but TranXenoGen has no plans to put all its eggs in one basket; it says it will also begin producing monoclonal antibodies soon. Later on, when other researchers come up with the right proteins, they hope to turn chicken farms into factories for drugs treating Alzheimer's, Parkinson's and Huntington's disease.[141]

Groups opposed to genetic engineering are bent on making biotech birds a rallying point. The British organization Compassion in World Farming claims that engineering animals to produce medicine is unnecessary because there are alternative ways to do it.[142] Obviously this is true. The issue is not producing medicine but producing it in quantities large enough and cheap enough to be widely available. Spokesperson Joyce da Silva obviously thought she was scoring points when she said, "It's as if we are determined to develop a sub-class of animals which have been tampered with so we can extract things which might possibly be of benefit to man."[143] But that's essentially what a chicken is—a bird that has been radically domesticated into a flightless provider of flesh, eggs and feathers for human benefits. Meanwhile, a spokesman for Friends of the Earth claimed, "Genetically modifying animals so they become drug factories raises serious ethical questions. The technology is well ahead of the debate."[144] Although we eat chickens and their eggs by the zillions, we should debate whether it's okay to use their eggs to treat horrible, hitherto incurable diseases? Apparently he's ignorant that we began using chickens as vaccine factories decades ago.[145]

FOUR

"Plantibodies"

It has long been known that the effectiveness of a vaccine won't ensure its use. Indeed, the Salk polio vaccine won out over Sabin's because it could be delivered to children with a tasty sugar cube instead of a painful injection.[1] Yet another way of painlessly and easily delivering immunizations would be through consumption of transgenically modified fruits, vegetables and grains. Nicknamed "plantibodies," they are being made from a wide array of foods which themselves fall under the category of "biopharming."

Basically, three types of pharmaceutical proteins are being developed in plants. One type makes antibodies that can be administered to help the immune system fight a disease that has already taken hold. Another is designed to prime the immune system so that a disease can't take hold in the first place; it works like a traditional vaccine. The third type comprises a wide range of biopharmaceuticals such as anticoagulants, anemia treatments and hypertension drugs.[2]

Some of these plantibodies will be delivered directly, by eating of the fruit or vegetable. Others will be grown in plants that won't be eaten but will provide stock from which material can be extracted, similar to the milk taken from transgenic animals.[3] Plants with high protein contents, such as tobacco and corn, are best for this.

Using food vaccinations was the brainchild of Charles Arntzen, now with the Department of Plant Biology at Arizona State University in Tempe.[4] The original inspiration came in the early 1990s, when the World Health Organization called for inexpensive oral vaccines that needed no refrigeration. Over 500 million injections against several potentially fatal diseases are administered in developing countries each year through the WHO's Expanded Program on Immunization (EPI) alone.[5] The WHO had set lofty goals, including the eradication of polio, and it appeared to be reaching those goals. Then, because of funding cuts and social chaos resulting from regional warfare, vaccination rates began a dramatic and deadly drop. Nigeria's overall coverage fell from 80 percent in 1990 to 27 percent in 1998; the Congo's went from 46 percent to 25 percent; Togo's 100 percent rate was cut almost in half; and the Central African Republic's coverage dropped from 93 percent to 53 percent. The result of this decline in protection, according to the WHO, is the death of an estimated three million people a year from polio, mostly children.[6]

At the same time, though, vaccinations with needles can themselves be deadly in poor countries because needles and syringes are usually reused and often not properly cleaned and sterilized. In 1999, the WHO estimated that unsafe injections in the underdeveloped world each year cause an incredible 8–16 million hepatitis B infections, 2.3–4.7 million hepatitis C infections, and 80,000–160,000 HIV infections. Indeed, according to the organization, as many as one-fifth of all new hepatitis B infections in developing countries might be spread in this way.[7]

Shortly after the WHO had made its plea for cheap oral vaccines, Arntzen traveled to Bangkok. While there, he saw a mother soothe her crying baby with a piece of banana. This was his "Aha!" moment. He knew that crops were being engineered through transgenics to fight off plant disease, and it struck him that perhaps human disease-fighting genes could be introduced into food. Once developed, the plants could be grown locally as cheaply as a normal banana or potato.[8] Nothing would have to be shipped in, no refrigeration would be required, and no needles would be necessary.

Super Spuds and Other Edible Vaccines

"What we are doing is oddball stuff in terms of what people think of as traditional research," Arntzen admitted back in 1993. "We are on an idealistic quest. We have to come up with a way to deliver genetically engineered vaccines to places like Bangladesh, India, China and Thailand, in a production system that is appropriate for the Third

World." He added, "The simplest thing is to give them a seed. Let them plant it, grow it and harvest it."[9]

Idealistic? Maybe, but it's hardly quixotic. Within two years, Arntzen had published a study in the journal *Science* showing that it was possible to immunize mice by having them eat pieces of raw potatoes into which genes from disease-causing antigens had been spliced.[10] In three more years he had demonstrated the same in humans.[11] Of eleven volunteers who ate the potato vaccinations transgenically made to combat the effects of the virulent form of *E. coli,* ten had immune responses that "exceeded the level we anticipated," said Arntzen. As for the controls, the volunteers who ate placebo potatoes, they spent a lot of time "quite sick" with diarrhea.[12]

In the same paper in *Science,* Arntzen and his colleagues revealed that they had also found that mice that ate the potatoes expressed what's called a "mucosal immune response."[13] Mucous membranes line bodily cavities and canals that lead to the outside, primarily the respiratory, digestive and urogenital tracts. These include the inside of the nose and mouth, throat, lungs, digestive tract, anus, urethra, vagina and penis.[14] Injected vaccines initially bypass mucous membranes, going directly into the bloodstream from the injection. This results in weak mucosal immunity.[15] But edible vaccines make contact with the lining of the digestive tract, which could include the esophagus and throat. In theory, this could activate not only systemic immunity, but immunity where you most want it: right where the germs enter.[16] It's always best to keep the enemy outside the castle gate rather than have to fight them in the courtyard. You also want to have a fully stimulated gastric system because the intestinal lining alone is estimated to contain more lymphoid cells (those that carry germs off to the lymph glands to be destroyed) and produce more antibodies than any other organ in the body.[17] An eaten vaccine could also accomplish this.

One problem with Arntzen's potatoes, though, is that the vaccine protein could not survive cooking and people don't care much for raw potatoes. So a process was developed at Loma Linda Medical Center in Riverside, California, in which a vaccine (against cholera) was put in a potato so that it could survive baking, boiling, even frying.[18] When Arntzen's technology was introduced into tomatoes, the vaccine remained stable even if the tomatoes were processed or dried. Thirty original plants, producing fifteen pounds of tomatoes a week, were enough to produce thousands of doses of effective vaccine at one cent per four doses.[19] The vaccine cannot yet go into production, however, in part because the FDA has no licensing mechanism for plant vaccines. When it does get approved, Arntzen believes, it will be a moment that will shake the world. "We have rid the world of smallpox through vaccination; we are close with polio. Now I believe we can do it with hepatitis B too," he said.[20]

Worldwide, eighteen companies at the beginning of 2001 were working on producing pharmaceuticals in plants, according to the Bowditch Group, a Boston-based agbiotech consulting firm.[21] In addition to potatoes and tomatoes, carrots are being experimented with to produce vaccines for hepatitis B.[22] Potatoes are also being tested against foodborne *E. coli*, and against the cause of type I diabetes, cholera and other illnesses.[23] In July 2000, Arntzen and other Cornell University researchers announced they were able to use potatoes to induce immunity in 19 out of 20 volunteers in a study against the Norwalk virus, an illness transmitted by food, causing nausea, vomiting, stomach cramps and diarrhea.[24] According to the federal Centers for Disease Control and Prevention, an amazing 23 million people in the United States alone are infected annually by the Norwalk virus or by similar viruses—compared with only 79,000 cases resulting from *E. coli* contamination, 2,500 cases of *listeria* and 1.4 million cases of illness from *Salmonella*.[25]

Potatoes make a great plant vaccine for the Americas and Europe, but they're not high on the menu in other parts of the world. That's why bananas are also being developed to prevent diarrhea, which kills some two to three million children in underdeveloped nations every year.[26] Bananas are also being engineered to prevent hepatitis B. It's expected that these "inoculations" would cost merely pennies a dose.[27] Most recently it was announced that Japanese scientists have succeeded in transferring the hepatitis B antibody into rice plant genes, cultivating the plants, and confirming the rice's ability to attach to and kill the virus.[28]

Tomatoes and spinach are being developed as vaccines against rabies, a terrible problem in some parts of the world.[29] More recently, Indian researchers have developed a rabies vaccine in muskmelon.[30] Tomatoes have also passed the first stages of testing as a vaccine against respiratory syncytial virus, a plague of infants and older people in nursing homes.[31] Specially engineered soybeans appear effective against genital herpes and corn is being used to produce an anti-herpes gel.[32] Australians are making an edible vaccine for measles, while the lowly turnip is being tested as a way of producing interferon for treating hepatitis B and C.[33]

An AIDS vaccine is being developed in corn. ProdiGene in College Station, Texas, received a $300,000 grant from the National Institutes of Health to work on this project after demonstrating that it had a good handle on the technology.[34] (It's also working on corn that would vaccinate against hepatitis B and so-called "traveler's diarrhea.")[35] There's no reason to expect that ProdiGene's HIV vaccine will be the first such vaccine implemented, assuming it works at all; many HIV vaccines are much further along in development. But ProdiGene's could well become the cheapest. Vaccinating millions of Westerners against HIV, even with an expensive drug, is quite feasible. But the same injection that may be affordable for an American could be a year's wages for somebody in Nigeria, Chad or China. That's where a product that can be produced cheaply in tremendous volumes and administered perhaps just by eating it could really pay off.

In 2002, ProdiGene became the first company to begin producing recombinant medicinal proteins in plants in commercial amounts.[36] One is trypsin, an enzyme that's critical in the manufacture of insulin. It is produced in the pancreas of all animals and was originally extracted from cows, though worries about pathogens, especially bovine spongiform encephalopathy (BSE), have pushed the market to recombinant versions.[37] Another trypsin application is in the wound care markets as an oral treatment for inflammatory edema, hematoma and pain associated with a wide variety of internal and external wounds.[38] It even has uses in the industrial enzyme market, such as softening leather.[39] It's estimated that the worldwide demand for trypsin will increase five-fold in the next five years, in part due to a tremendous increase in diabetes, so ProdiGene's supply will come just in time.[40] The other ProdiGene drug is aprotinin, a protease inhibitor used in cardiac surgery, for healing wounds, and as a component in making pharmaceuticals. Traditionally aprotinin has been extracted from cow lungs.[41]

Plant proteins could be even cheaper if we made them do their own extracting. That's what Ilya Raskin of the Biotech Center at Rutgers University in New Jersey has been working on for the last several years. The basis for the idea is that during the night, when leaves lose less moisture by evaporation, pressure builds up inside them, thereby squeezing out fluid, a process known as guttation. The fluid contains a bit of protein, which Raskin figured must come from the fluid in the spaces between cells. So Raskin and some of his colleagues engineered tobacco plants to produce three foreign proteins in this intercellular fluid, including a fluorescent green one from jellyfish. As they had hoped, the new proteins showed up in the dew on the leaves. Now that they know it works, the next step will be to splice medicinal proteins into the plant's genes. The "sweat" drops could be taken off the leaves each morning, processed and purified.[42]

"It would provide a system for obtaining fluid that is already purified and concentrated," according to Hugh Mason of Cornell University, who works on vaccine expression in potatoes. The amount of protein that Raskin and his colleagues have been able to get from guttation—about 2.8 percent of all the protein in the fluid—is comparable to what other people have been able to extract from the plant itself, he said.[43]

But Raskin doesn't limit himself to using the parts of the plant we can see. Indeed, he thinks the real mother lode of medicinal proteins may be what lies beneath. Root networks can be gigantic compared with the plant parts that poke forth into the sunlight; according to one estimate, 2.5 acres of rye can grow 87 million miles of roots, enough to stretch from the earth to the sun.[44] Because roots can't use antipredator devices like thorns or bark or waxy coatings, they rely entirely on producing chemicals to ward off everything from microbes to weeds to animals. So it occurred to Raskin that they might make ideal factories for chemicals, including some that could be useful medicines, if he could fool them into believing they were under attack. It worked. With the right triggers, or "elicitors," he found that the roots produced an incredible spectrum of chemicals. He has now set up a company, Phytomedics, to exploit what he calls "rhizosecretion."[45] Raskin calculates that making proteins using mammalian cells costs about $5,000 per gram, while using bacteria, yeast or fungi is perhaps $500 per gram. But when you include the cost of purification, these figures can double or triple, he says. "We think we can produce around 100 kilograms (220 pounds) of unpurified protein for each acre of greenhouse space each year, which works out at around $5 per gram."[46] Even after purification, the final cost may be little over $100 per gram.

One company focusing on infectious disease antibodies in corn and rice is Epicyte Pharmaceutical of San Diego, founded in 1996.[47] Among its current projects are medicines to protect against the transmission of HIV, genital and oral herpes, and human papillomavirus. Epicyte is also developing a contraceptive that grabs sperm and holds onto it like flypaper.[48] While many plantibodies are meant to be consumed directly, in this case the grain will just provide stock from which the antibodies are extracted. "We like the idea of using grain because it can be stored very inexpensively for years," said Epicyte's plant antibody project leader at the time, Kris Briggs. Acknowledging that tobacco also makes a good plant into which to engineer antibodies because it has such a high protein content, she points out that "you have to harvest and purify it right away" to remove the antibodies, whereas grain can be stored for years. Most of Epicyte's products are intended to be put in gels or films that can be spread directly onto sexual organs. "We're also working on a vaginal ring that would release antibodies over a month-long period," she said.[49]

While Epicyte's focus is sexually transmitted disease, it also seeks to grow antibodies against infections that plague both the very young and the old. The first is the most common cause of lower respiratory tract disease in infants and young children worldwide.[50] Called respiratory syncytial virus (RSV), it causes both pneumonia and bronchiolitis.[51] Almost 70 percent of infants have primary RSV infection during the first year of life, and more than 95 percent experience infection by age two.[52] Each year, over 90,000 infants are hospitalized with the virus and 4,500 of them die.[53] The second infection, *Clostridium difficile,* is usually acquired in hospitals or nursing homes. It's usually a problem with elderly persons who have undergone antibiotic treatment and are showing evidence of resistance to conventional antibiotics.[54] Briggs expects antibodies against these infections to be on the market in 2006.[55]

Plant-derived vaccines would also benefit farm animals. Transmissible gastroenteritis (TGEV), a swine virus, is one of the top five diseases in the hog industry.[56] According to Bruce Lawhorn, a swine veterinarian at the Texas Agriculture Extension Service, "Once an outbreak starts it's nearly impossible to stop it from going through the whole operation. It's very contagious. Nearly 100 percent of the animals get sick with diarrhea and most baby pigs die."[57] Stauffer Seeds, in conjunction with ProdiGene, plans to sell corn shown effective against TGEV.[58] Edible vaccines are certainly easier to administer than injections, since no handling of the animals is required, and they could significantly reduce costs as well. Dekalb Genetics, a division of Monsanto based in Dekalb, Illinois, is also developing corn to fight virus infection in poultry.[59] Corn spliced with an interferon gene may become an alternative or adjunct to conventional vaccines.[60] "Interferon is a naturally-occurring chemical produced by animals to help the animal's immune system fend off viruses," said Monsanto researcher Alan Kriz.[61] He expects such corn will be available as early as 2004.

Monoclonal Antibodies

Monoclonal antibodies are already among the most versatile biotech drugs, with that versatility expanding like a supernova.[62] Originally conceived of as cancer fighters, they were actually first applied in a wide range of diagnostics. As of late 2002, almost ninety different monoclonals were in various stages of human testing, the majority aimed at treating cancer or immune system diseases.[63]

As an example of what's in the pipeline, the Irish pharmaceutical company Elan, working with Biogen of Cambridge, Massachusetts, has a drug named Antegren in Phase III trials to treat both multiple sclerosis and Crohn's disease.[64] Like rheumatoid arthritis, Crohn's is an autoimmune illness. It affects the intestines and commonly causes abdominal pain and diarrhea, as well as rectal bleeding, weight loss and fever. Bleeding may be serious and persistent, leading to anemia. Children with Crohn's suffer delayed development and stunted growth. In severe cases, ulcers eat through the intestines, causing open holes ("fistulae") that link the bowel to the anus, vagina or surface of the skin. Sections of diseased bowel can be removed surgically, but symptoms often return. Crohn's is believed to affect as many as a million Americans, while perhaps 300,000 Americans and Europeans have the severe form.[65] Crohn's patients receiving Antegren in Phase II studies had about twice the remission rate of those receiving a placebo. There were also serious side effects.[66]

Like gene therapy, monoclonals were terribly slow in getting off the ground. One article about this therapy in the *St. Louis Post-Dispatch* began: "Its development won its creators a Nobel prize. Its introduction was accompanied by talk about a 'magic bullet' for cancer. But its progress in drug therapy has been slower than scientists and financiers had expected."[67] Yet this article appeared in 1989; the Nobel Prize was awarded back in 1984; and monoclonal antibodies had been discovered nine years before that.[68]

What is a monoclonal antibody? Antibodies are a certain type of white blood cell that the immune system produces to fight off infectious agents such as viruses or bacteria, or invading bodies such as tumors. The antibodies bind to proteins on the invaders called antigens. For each antigen, a specific antibody is created. That's why we never become immune to the common cold: while we do get antibodies to one strain of the cold virus, we don't get them to the vast number of other cold viruses.

The response of the immune system to any antigen, even the simplest, is polyclonal, meaning that the system makes a great range of antibodies. A *monoclonal* antibody, conversely, isolates the specific antigen to which we want it to react. Antibody-producing white blood cells are fused with a benign (noncancerous) tumor cell. This gives the monoclonal the characteristics of a natural antibody while having the rapid reproduction capability of a tumor.

Most monoclonal antibodies currently used are made from white blood cells removed from the spleens of mice that have had human antigens inserted into them.[69] Thus, you may see monoclonals referred to as "chimeric," a term taken from the creature in Greek mythology built from parts of several animals. You may also see them called "murine," the adjectival form of "mouse." The special mice in which these monoclonal chimeric murine antibodies are created "give us a degree of continuity that is not available in other animals," according to Monsanto biologist Kevin Glenn. Further, he said, "There is little genetic variability because they have been in-bred for a long time."[70]

One limitation of these murine antibodies is the possibility that the human immune system will react to the mouse part, causing side effects like fever and rash and reducing the effectiveness of the antibodies. Thus, some companies and academic institutions are experimenting with growing them in plants, while still others have developed techniques for creating more humanized forms and have widely succeeded in replacing about nine-tenths of the mouse portion.[71] But nothing will substitute for fully humanized monoclonals, which finally arrived with the FDA approval, on the last day of 2002, of Humira, a rheumatoid arthritis drug from Abbott Laboratories. Other companies such as Biosite of San Diego and Genmab of Copenhagen, Denmark, are developing fully humanized monoclonals through use of transgenic mice.[72]

While therapeutic uses for monoclonal antibodies have been long in coming, tremendously useful diagnostic tests using monoclonals arrived early and in rapid succession. Every day, innumerable women use them to tell whether the stork will soon be making a delivery, by detecting a pregnancy-associated hormone in samples of urine.[73] Monoclonal antibodies are used for blood-typing, tests for heart disease and home ovulation tests (to increase or decrease your chance of that stork coming), and they have become a standard for detecting the presence of many types of cancer.[74] The best known is the PSA, or prostate specific antigen, used for prostate cancer. It was also through the use of monoclonals that doctors recognized in 1981 a bizarre and apparently new virus that would soon spread disease and death around the world. We now call it AIDS.[75]

The first monoclonal antibody product approved by the FDA for therapeutic use was Johnson & Johnson's Orthoclone OKT3, designed to block the immune system's T cells from rejecting transplanted organs like kidneys.[76] Two others may constitute the best weapons we'll have against type 1 diabetes. More remarkably, one may be part of a cure for the disease and not just a treatment replacing insulin injections, while the other may prevent the disease from ever developing in the first place.

Diabetes is one of America's top killers, causing about 65,000 deaths a year.[77] Type 1 involves a lack of insulin, a chemical secreted by the pancreas, a finger-sized organ behind the stomach. Insulin is needed to efficiently transport and store glucose (the main sugar absorbed from food) in the liver and other tissues. Lack of insulin causes glucose levels in the blood to rise abnormally, and also disturbs fat and protein metabolism. With type 2, often brought on by obesity, insulin is produced but not enough, because the body has become resistant to it.[78] Lifestyle changes can prevent or control this type of diabetes.[79]

Type 1 accounts for only about 1 million of the 16 million diabetics in the United States, but it's the most serious variety.[80] For reasons not yet understood, the body's own immune system attacks the pancreas and renders it worthless. To survive the ravages of this disease—including blindness, loss of circulation in the limbs leading to amputation, and even death—the type I diabetic must adhere to a strict diet and exercise regimen. He must constantly monitor his sugar level by jabbing himself to draw blood and testing it, and give himself repeated insulin injections. Even then, no matter how often the diabetic tests and injects, nothing can replace the exquisite ability of a healthy pancreas to inject just the right amount of insulin at the right time.[81]

Pancreas transplants are one way of treating the disease. Whole pancreas transplants succeed about 85 percent of the time, but this is major surgery and about one in twenty patients dies.[82] For a quarter of a century, doctors tried to take only the insulin-producing cells, called islet cells, from cadaver pancreases and place them into type I diabetics. The results were dismal: the transplanted islet cells kept being killed off by their recipients. But that changed when researchers in Edmonton, Canada, used a new combination of immunosupressants including Zenapax (daclizumab), a monoclonal from Swiss-owned Hoffman-LaRoche.[83]

Armed with this new drug combination, the researchers removed islet cells from the cadavers and then gave seven patients a sedative and a local anesthetic. Using X-rays for guidance, they threaded a long, thin plastic tube through the skin of the recipients' abdomens and into the portal vein, which carries blood to the liver.[84] The cells were then squirted into the portal vein, after which the islets lodged in small veins in the liver. Follow-up studies have shown that surviving cells will probably secrete insulin permanently. "The liver is essentially functioning like a pancreas," said James Shapiro, the research team leader.[85] Seven persons received new islets in the study; all seven soon found they were able to throw away their jabbers, needles and insulin. The daclizumab combination also appeared to suppress the body's immune system reaction that originally destroyed the patient's own cells.[86]

Unfortunately, there aren't nearly enough donor islet cells to go around. "Right now, even if we could use every single pancreas, we could transplant less than one percent of the patients that might benefit," said Camillo Ricordi, director of the Diabetes Research Institute at the University of Miami.[87] The answer to the shortage lies in cloning the needed cells in the laboratory. University of Florida researchers have cloned large amounts and have apparently cured diabetes in mice with them.[88]

All of this is wonderful news not just for diabetes sufferers, but for many others as well. "When you look at the Edmonton success," said Dr. Stephen Rose, head of the transplantation and immunobiology branch of the National Institute on Allergy and Infectious Diseases, "it could provide insights into how we can manipulate the immune system and achieve success in all these other areas" of human transplants.[89]

Monoclonals have also shown tremendous promise in stopping the progression of type I diabetes. A June 2002 study at Columbia University in New York and the University of California at San Francisco showing that a dozen young people recently diagnosed with the disease kept it in check for more than a year by taking a two-week course of infusions from an immune-suppressive monoclonal that protected besieged pancreases. There was no decline in their insulin production beyond what they had already lost prior to the trial, and side effects during the transfusion period were quite mild.[90] The group has since been expanded to forty-six people and "the results look as good as, if not better than, for the first group of patients," said one of the study's lead researchers, Kevan Herold of Columbia.[91] FDA approval of the treatment could come in 2005, he added.

Another use of monoclonal antibodies relies on their ability to locate and bind to a specific target. Monoclonal antibodies themselves are usually too weak to destroy a target, but by linking a toxin or radioisotope to the antibody, you get a guided missile that zeroes in on the enemy.[92]

Monoclonal antibodies have also been effective against heart disease and in preventing reclogging of the coronary arteries in patients who have undergone angioplasty.[93] One such approved a long time ago (by biotech standards) is ReoPro, produced by the Johnson & Johnson subsidiary Centocor of Malvern, Pennsylvania, and co-promoted by Eli Lilly.[94] It entered the market in 1995. Trials with ReoPro on heart patients showed that it tremendously reduces the chance of dying or having another heart attack, and decreases the need for repeat medical procedures.[95] More recent work (February 2003) indicates that administering ReoPro as much as six hours after an ischemic stroke may increase recovery of normal or near-normal function and reduce the chance of death.[96] Ischemic strokes, meaning complete blockage of a blood vessel carrying oxygen to the brain, constitute 80 percent of all strokes.[97]

Another Centocor drug, Remicade, has been approved by the FDA for use against both rheumatoid arthritis and Crohn's disease.[98] When it was approved in August 1998, Remicade was the *only* FDA-sanctioned drug for Crohn's.[99] Scientists believe that Remicade reduces intestinal inflammation in patients with Crohn's disease by binding to and neutralizing a protein known as tumor necrosis factor-alpha (TNF-alpha) and by destroying cells that produce the chemical.[100] In the primary study used to get FDA approval for Remicade, all of the fistulae in more than half of the patients closed up after treatment with the drug.[101]

Like Crohn's, rheumatoid arthritis is an autoimmune illness. It causes severe joint inflammation, pain and stiffness, and afflicts more than two million Americans.[102] (It shouldn't be confused with osteoarthritis, a much more common affliction generally considered to be caused by wear and tear on the joints and their protective cartilage.) Since Remicade was approved for one autoimmune illness, Crohn's, some physicians began prescribing it for rheumatoid arthritis and found that it often works.[103] The FDA gave its formal approval for use of the drug for rheumatoid arthritis in late 1999.[104] It has been a godsend for sufferers of the disease. While all the previous anti-inflammatory medicines temporarily reduced pain and swelling even as the disease progressed, X-rays of patients using Remicade in combination with another drug have shown that together they can actually stop progression of the disease.[105]

A whole new method of vaccination using monoclonal antibodies is also being studied that would allow the drug to be absorbed through the skin as creams or gels or be taken as tablets. As we have discussed, a traditional vaccine works by fooling the body into thinking it's infected, prompting it to produce antibodies to fight the infection. Then when the disease comes along, the immune system is already primed and ready to go into action. But this new technique would introduce antibodies directly. "Since one of the things a vaccine will do is stimulate antibody production against a pathogen, let's just skip that step," said Larry Zeitlin, a research scientist at ReProtect, a biotechnology company in Baltimore.[106]

Zeitlin and some colleagues from Johns Hopkins University in Baltimore described the potential of their method in a 1999 issue of *Emerging Infectious Diseases.* Most infections enter the body through mucous membranes, which happen to be places where vaccines have problems constructing a readymade immune response. But monoclonal antibodies wouldn't rely on the body's prepared defenses; they would *be* the defense. "As soon as you apply them, you're protected," Zeitlin said.[107] The protection wouldn't last nearly as long as a vaccine, but the antibodies could be reapplied as often as needed. One trial under way uses toothpaste to deliver antibodies that prevent tooth decay.[108]

But what of cancer, the original target for monoclonals? The first such drug, which finally received FDA approval in 1997, was designed to fight non-Hodgkin's lymphoma (NHL), a disease of the lymphatic system. Named Rituxan, it came from a collaboration of Genentech and the IDEC Pharmaceuticals Corporation, based in San Diego.[109] According to the American Cancer Society, about fifty thousand people in the United States were diagnosed with NHL in 2000; it is the fifth most common cancer in the country, excluding nonmelanoma skin cancers.[110] Of those who contract NHL, about one-half have what's called "low-grade" or "indolent" lymphoma.[111] Standard treatment consists of radiation or chemotherapy, or both. These are usually helpful, but patients normally have several relapses and die after six or seven years. These are the persons for whom Rituxan is intended.[112]

Rituxan works by binding to an antigen on the surface of healthy lymphocytes (white blood cells that combat infection) as well as malignant cells, stimulating an immune response. Administration is easy, with just four outpatient infusions over three weeks.[113] Side effects are relatively mild, as well, consisting primarily of flu-like symptoms such as fever and chills during infusion, and usually occurring only during the first session.[114] About half of recipients appear to show some improvement, according to the latest studies, while 6 percent have gone into complete remission.[115] But because it's well tolerated, it has great promise as part of a combination therapy. In one small study, researchers adding Rituxan to a chemotherapy treatment found that all patients with low-grade NHL improved, and in almost two-thirds the tumor completely disappeared. Then it was tried on persons with the high-grade, more aggressive (but paradoxically more treatable) form of the disease. Here the results were almost as good: 97 percent of patients improved while 61 percent had their tumors vanish altogether.[116]

Monoclonals can also sometimes be combined with radiation therapy. One such drug is Bexxar, from Seattle-based Corixa Corporation and GlaxoSmithKline of Philadelphia.[117] Bexxar uses radioactive iodine-131 to attach to cancerous cells; another drug, Zevalin, hits its target with yttrium-90. Results from late-stage trials in 2000 showed that Bexxar was effective in 70 percent of patients, with 40 percent entering remission, according to researchers.[118] In 2002, Corixa had collected data from five different clinical trials showing that 70 percent of advanced non-Hodgkin's lymphoma patients who had a complete response to Bexxar were still alive and disease-free after more than seven years.[119] Bexxar is available in Europe and received FDA approval in 2003.[120]

Monoclonal antibody medications offer hope where conventional therapies once offered almost none. Such is the case with a certain ravenous form of breast cancer that affects about one-third of breast cancer victims. These women have a gene that overproduces a protein called HER2/neu, which prompts cells to reproduce like crazy, spreading rapidly throughout the body.[121]

Genentech's Herceptin, approved in 1998, is a monoclonal antibody that blocks HER2's protein production, often shrinking and sometimes eliminating the malignancy. Again, it's no magic bullet. But in clinical trials, about one-fourth of patients who received Herceptin plus chemotherapy had no signs of tumor growth after one year, while this was true in only one-eighth of those who received chemotherapy only.[122] Herceptin is useful only in a certain type of breast cancer, which accounts for 25 to 30 percent of all cases. But patients can be tested beforehand to see if they are in that 25 to 30 percent.[123]

Making Sense of Antisense

Yet another—and completely different—biotech approach to fighting cancer is with "antisense" technology, which in effect screws up the ability of a cancer-promoting gene to do its promoting.

All cancers spring from genetic errors. One such error causes genes to start producing dangerous mutant proteins. In another, the cell pumps out the normal protein but just doesn't know when to quit. Most current drugs are designed to interact with specific proteins each gene controls, but antisense compounds intercede to prevent protein production altogether. How? Once the gene sends out the production order, the assembly process begins with the making of a "message." The DNA creates an exact copy of a small portion of one of its two strands, called messenger RNA (ribonucleic acid) or mRNA. That mRNA molecule goes into the cell, which decodes the message and produces the protein. When the antisense molecule encounters that mRNA, "it goes up to it and zips up just like a zipper," as one specialist put it.[124] This interrupts the "sense" that the cell is trying to make of the mRNA's message, hence the term "antisense." No message read, no cancer-causing protein produced.[125]

At least that was the idea. But what can be described can't always be implemented in something as complex as the human body. A decade of research produced nothing but one heartbreaking failure after another. That was until 1998, when Novartis got approval for Vitravene as a treatment for cytomegalovirus (CMV) retinitis in AIDS patients.[126] Although CMV is generally harmless in persons with healthy immune systems and, indeed, most adults carry it, it used to be a major cause of blindness in AIDS patients.[127]

Then in 2000, researchers from Genta of Lexington, Massachusetts, reported success against a gene that conspired in causing malignant melanoma, the deadliest form of skin cancer.[128] The disease afflicts about 48,000 Americans a year and kills about 7,700.[129] In about 90 percent of these cases, a gene pumps out a protein called Bcl-2 that protects the tumor from chemotherapy that could otherwise kill the malignant cells. It produces the proteins at such a terrific rate that doctors simply can't kill cancerous cells fast enough. Genta's Genasense slows this protein production so that chemotherapy can work. In the first trial one of the patients, a ninety-year-old woman given but a few months to live, had her cancer completely disappear. In two other cases, cancer growth stopped and more than half the tumor went away, while three other patients had lesser responses but still remained alive.[130] Genasense itself is now in Phase II or Phase III trials for chronic lymphocytic leukemia, multiple myeloma, non-small-cell lung cancer, acute myeloid leukemia, chronic lymphocytic leukemia, mantle cell lymphoma and prostate cancer.[131] There are other antisense drugs under development for other viruses and respiratory disease.[132] The skeptics are beginning to turn. "In my wildest dreams I didn't think we were going to get this far," remarked Columbia University professor Cy Stein.

This has been a mere sampling of the many new biotech approaches that are specifically tailored to exploit weaknesses in cancer cells. The attitude of ordinary chemotherapy toward dividing cells is "Kill them all; let God sort them out!" It's an approach that can cause a variety of terrible side effects—including death. Or it can make the patient so sick that chemotherapy must be halted, and then death follows. These new techniques zero right in on the troublemaker, allowing chemotherapy to do its job with few complications.

FIVE

The Book of Angiogenesis

Almost three decades after it was declared by Richard Nixon, the "War on Cancer" was looking an awful lot like the bogged-down trench fighting of World War I. Lots of lives were lost, lots of hopes were raised and dashed, and massive amounts of money were spent—with no end in sight. Incremental advances were made in the 1970s and 1980s, but most of the progress resulted from the new emphasis on early detection. Oncologists could bring their weapons into action earlier, but the weapons themselves weren't changing much. Standard therapy for cancer remained what it had been for decades: so-called "slash, burn and poison." Surgery and radiation therapy techniques have certainly improved; so has the use of chemotherapy. But these traditional approaches to cancer were for the most part where they were years earlier. One cancer drug still in wide use dates back to the 1950s. Others, such as cyclophosphamide, have been in the oncologists' arsenal for decades.[1]

Yet now the battle front is changing. We're beginning to push the disease back not a trench at a time, but rather in great sweeping medical flanking movements. The main reason is that we finally have a good idea of how cancer works. Dr. Samuel Waksal, a former National Cancer Institute researcher and former CEO of ImClone Systems who unfortunately is now infamous for his part in the Martha Stewart insider trading scandal, provided an apt metaphor.[2] "For years in oncology, you looked at a cancer cell and it's like *Mission: Impossible* when you open up the bomb and you see a green wire, a red wire, a blue wire and all these lights," he said. "You'd take a sledge hammer and try to smash it. If you were lucky, you could destroy the tumor without blowing up the patient. Now we actually know what the blue wire does and what the green wire does. Now we can clip one wire, and do it in a very specific fashion."[3] More and more often, that new wire clipper uses biotechnology.

We've already seen the role played by vaccines. Another new attack on cancer involves "angiogenesis." In normal tissue, new blood vessels are formed during growth and repair and during pregnancy, but rarely otherwise. That's not true for malignant tumors. Once they've reached the size of a small pea, they send out messages to the body to grow new blood vessels to supply the tumor with oxygen and other nutrients necessary for survival and growth.[4] If you look at a large enough tumor, you can see these vessels, a little like the appendages of a crab, even with the naked eye. (That's why the crab has come to represent the fourth sign of the zodiac in astrology, the sign of cancer, and why the German word for cancer, *Krebs,* is the same as the word for crab.)

The purpose of these vessels has been recognized for over a century, but only in the 1960s did a Harvard doctor named Moses Judah Folkman postulate the basic theory of angiogenesis, which holds that these tumors need those blood vessels to survive and grow. He was laughed at and considered a fool by many.[5] Folkman is now said to be on the short list for a Nobel Prize, and he became the subject of a fascinating biography in 2001.[6]

Folkman's work focuses on how the endothelial cells create the walls of these new nutrient-carrying blood vessels. These cells have a remarkable ability to divide and migrate. When signals from the tumor activate the cells, they secrete enzymes that degrade the surrounding tissue, known as the extracellular matrix. The cells can now invade the matrix and begin dividing. Eventually, strings of new endothelial cells organize into hollow tubes, creating new networks of blood vessels.[7]

Endothelial cells are always there, but most of the time they lie dormant. When the body calls upon them for tissue repair, they allow short bursts of blood vessel growth in a small area. Normally, this growth is tightly controlled by a finely tuned balance between factors that activate endothelial cell growth and those that inhibit it. But cancer destroys this balance. Numerous proteins are known to activate endothelial cell growth and movement, including angiogenin, epidermal growth factor, estrogen and vascular endothelial growth factor (VEGF). Many chemicals are being tested that inhibit or block one or more of them; these are potential antiangiogenic drugs.[8] A tumor that has already grown beyond the 100–300 cell point, even far beyond, can still be choked off by an antiangiogenic drug. The tumor may grow more slowly, stop growing, or even recede and disappear.[9] Just stopping the growth is more important than it sounds, because a fully contained tumor can't kill you.

Yet those same blood vessels that provide tumor nourishment also play another diabolical role. They allow cells from that tumor to enter the bloodstream, whence they travel to the rest of the body, a process called metastasis.[10] The immune system kills most of the free-floating mutant cells before they can plant themselves among normal cells and begin to grow. But given time, a metastatic cell will take hold somewhere and grow into new tumors.[11] Antiangiogenic chemicals fight this by shutting down the vessels coming from the original tumor and by inhibiting growth of the mutant cells that do escape from the tumor.[12] "So angiogenesis is an important factor in preventing metastasis as well," said Dr. William Li, president and medical director of the Angiogenesis Foundation in Cambridge, Massachusetts.[13] "And it's usually not the primary tumor that kills, it's the metastasis."[14]

Antiangiogenesis is exciting because it singles out cancer as the enemy and attacks only newly growing blood vessels.[15] Chemotherapy, by contrast, carpet-bombs dividing cells (including those in bone marrow) indiscriminately. This often makes the patient sick and limits both the dose and duration of chemotherapy treatment.[16] Even more disturbing is that many of the agents used in chemotherapy are themselves carcinogens, meaning they may cause cancer years down the road.[17] Still another problem with chemotherapy is that tumors often shrink under the initial chemical attack, but because cancer cells are constantly mutating, some of the mutations will become resistant to the chemotherapy. These become predominant and the tumor comes roaring back.[18] But the blood vessels that tumors create for nourishment and mobility do not mutate, so it appears they can be treated with antiangiogenic drugs indefinitely.[19]

True, sometimes the body does need new vessels, primarily during tissue repair after an injury but also in women as they build the lining of the uterus each month before menstruation and form the placenta after fertilization. But the blood vessels of tumors are abnormal. They are, as two angiogenesis experts noted, "tortuous, dilated and leaky," and further, "the cells that compose them display certain molecules on their surfaces from a class known as integrins that are absent or barely detectable in mature vessels. Biologists have recently produced small proteins, called RGD peptides, which preferentially recognize the integrins on tumor vessels. These peptides can be linked to cell-killing drugs to target such therapeutic agents to tumors without damaging other tissues."[20]

Another advantage of antiangiogenic agents is that they appear to work against a tremendously broad array of cancers. A drug that's initially tested against and approved for a certain tumor, perhaps that of the breast, could actually turn out to be even more effective for prostate cancer. Maybe one that's slightly effective against skin cancer could devastate a brain tumor. "We now know that angiogenesis is a common denominator in diseases long thought unrelated, including over thirty types of cancer," said Li.[21] These aren't just "solid tumors" cancers like those of the breast, prostate, lung or colon, he points out, but "liquid" tumors as well, such as leukemia.[22] Even these will respond because leukemia needs to form new blood vessels in the bone marrow. The tumors grow on these vessels, said Folkman, like "berries on a bush."[23]

There are more than sixty antiangiogenic agents in various stages of oncology clinical trials, including more than a dozen agents that have reached advanced Phase III studies, according to the Angiogenesis Foundation.[24] If you look at the chart of prospective cancer treatments posted on the website of the Pharmaceutical Research and Manufacturers of America, you'll find that almost all of the cancer drugs are listed as having promise against this or that specific type of cancer, while those labeled "anti-angiogenesis" do not.[25] That's because all of them are seen as having antineoplastic activity against a wide number of tumors. One of these drugs is combretastatin, part of a collaboration between OXiGENE, based in Boston and Stockholm, and Bristol-Myers Squibb.[26] When researchers injected the drug into mice, they found that within only minutes 93 percent of the blood flow to the rodents' tumors had been cut off. Within twenty-four hours, 95 percent of the cancer cells were dead.[27] At this writing, combretastatin has also been successful in the first phase of human clinical trials.[28] Although all of the biotech antiangiogenesis drugs are in early phases of human testing, some antiangiogenic nonbiotech drugs (that is, discovered through traditional screening and made from chemical compounds) appear to be effective in blocking cancer.[29] This is what's known as "proof of concept."

One such is the antiarthritis capsule Celebrex from Pfizer. Originally FDA-approved to relieve the aches of arthritis, Celebrex later received a second approval for treatment of precancerous polyps of the colon.[30] Dr. Karen Siebert, who directed the colon cancer research that resulted in the second approval, told the *Chicago Tribune*, "There's no doubt in my mind that we have an absolute effect on angiogenesis and that we have an absolute effect on tumor development."[31] Her published research backs her up.[32] The way Celebrex appears to work is to inhibit the production of an enzyme called cyclooxygenase-2, or COX-2, which causes joint inflammation and pain. (Other so-called nonsteroidal anti-inflammatories such as aspirin, ibuprofen and naproxyn sodium also block COX-2 production, but unfortunately do likewise with COX-1 which helps protect the stomach from gastric acids.) It happens that COX-2 also encourages blood vessel growth associated with tumors, so drugs that block the arthritic swelling of COX-2 also block vessel growth.[33]

Another COX-2 inhibitor named Vioxx, from Merck, works in a way similar to Celebrex and therefore is antiangiogenic as well.[34] The over-the-counter drugs that block both COX-1 and COX-2 have also been found to be antiangiogenic.[35]

Another antiangiogenic chemical was once the most notorious pharmaceutical ever: thalidomide. It was originally prescribed to Europeans as a sleeping pill and as a morning sickness treatment for expecting mothers. It didn't hurt the mothers, but before it was yanked from the market it led to over ten thousand babies with horrific birth defects.[36] Decades later, however, when Folkman and other scientists analyzed *why* it did this, the answer seemed to be that it prevented blood vessel formation.[37] Now it's being tested, and showing great promise, against many different cancers, including those of the prostate, breast and brain, along with Kaposi's sarcoma. Thalidomide is commonly prescribed for a deadly blood cancer, multiple myeloma.[38]

One of the many advantages of antiangiogenesis drugs is that while they may be used instead of the traditional anticancer weapons of radiation, surgery and chemotherapy, they could also complement their use.[39] Amazingly, angiogenesis inhibitors appear to "normalize" tumor vessels before killing them. This can help anticancer agents reach tumors more effectively.[40] "While the first generation [of antiangiostatins] in the pipeline appears to have promise as monotherapy," said Dr. William Li, "most experts believe they will be most useful when combined with standard therapies. The gold standard of oncology is to combine effective medicines together."[41]

It's also possible, indeed probable, that antiangiogenic drugs will be used to prevent cancers before they even have a chance to form. What makes that possible is not so much the effective side of "safe and effective," as the safety side. You can't prescribe a drug to prevent a problem that doesn't exist and may never exist unless that drug is extremely safe and remains so over a period of many years. Many antiangiogenic drugs seem to have that quality, and indeed the idea of using them as preventatives goes way back to the 1970s, although with so little research having been done, they were then only ideas.[42]

Bypassing the tumor to attack its roots? Attacking cancerous growths while leaving healthy cells alone and not causing the wicked side effects of chemotherapy such as fatigue and hair loss? Like so many things in biotech, it sounds rather too good to be true, but it's not. Antiangiogenesis is "the single most exciting thing on the horizon," said National Cancer Institute director Dr. Richard Klausner.[43]

Yet it may surprise you to hear that there are also many biotech drugs under experimentation that are *pro-angiogenesis.* Like so many body functions discussed in this book, blood vessel generation isn't bad in itself; it's only bad under certain circumstances. Indeed, the whole purpose of coronary bypass is to insert surrogate vessels to circumvent clogged ones. But what if we could entirely bypass bypass surgery by getting the body to generate its own vessels? That's the idea of GenVec of Gaithersburg, Maryland. Its BioBypass uses gene therapy to deliver a form of vascular endothelial growth factor to the heart so that new blood vessels are formed in blood-starved tissue.[44] A Phase II trial announced in late 2002 showed significant improvements in total time to chest pain on a treadmill and improvements in quality of life as measured by the patients' ability to get dressed, to resume play with their children or grandchildren, and to reduce their need for cardiovascular medications. As the company's press release put it, "The magnitude of improvement seen was similar to what has been observed after angioplasty."[45] Already, "keyhole" surgery is beginning to decrease the pain and the recovery time now associated with open-heart surgery (sawing open and separating the breastbones).[46] BioBypass or some other pro-angiogenic agent may make even keyhole surgery unnecessary.

SIX

Medicinal Matchmaking

"One size sits all." How often have you bought something with that claim, only to find it didn't fit *you?* Traditionally it has been that way with medicine. Sure, we distinguish between children's and adults' doses, we look at possible interactions with other drugs, and sometimes gender or pregnancy is a factor. But for the most part, drugs are prescribed simply on the basis of what has worked best for most people. The problem with this lies within our genes.

For all the tremendous fuss over the Celera and the Human Genome Project sequencing efforts, there's actually no such thing as "*the* human genome." Each of us is genetically slightly different from everybody else, unless you're an identical twin. In that sense, there are actually more than six billion human genomes. The Celera genome came from five people, but primarily from its own colorful president, J. Craig Venter.[1] The genes in the Human Genome Project came from genes of individuals sequenced in labs all over the world.[2] But you can't get an "average" genome just by sequencing more people, any more than you can get an "average" color by mixing together as many colors as possible. It *is* fair to say our genomes are about 99.9 percent identical to those of other humans. It *is* fair to say that each of our genomes is more like any other human's than it is like that of any other creature. But when it comes to such things as pharmaceutical applications, it's that tiny bit of I percent left that makes all the difference. The science that's taking into account those tiny but important differences is called pharmacogenomics, a simple combination of "pharmacology" and "genomics."[3]

Pharmacogenomics is often equated with "designer drugs," but a better term is "individualized medicine." You don't necessarily need a drug tailored to your genome; off-the-rack ones may do just as well so long as it's been determined that the fit is a good one. That's much of what pharmacogenomics is about.

Whether inherited or caused by external agents such as radiation or viruses, genetic variants may be harmful, helpful or neither. A variant can make us more or less disposed to disease, more or less disposed to reacting favorably to drugs. What we're concerned with here are the inherited forms, which are also called "polymorphisms." If a polymorphism occurs in at least 1 percent of the population, it's called a single-nucleotide polymorphism.[4] This is abbreviated to SNP, pronounced as "snip."[5]

A SNP can be within the gene itself or located on uncoded DNA that lies on the chromosome between genes. The average person has perhaps 50,000–100,000 SNPs, though the number may be as high as 150,000.[6] Most are probably inconsequential. But the search is on for those SNPs that do play an important role. Identifying all such SNPs in the entire human race will be an incredible undertaking. Still, it's essentially a mathematical process that will get faster as computers become more powerful. The more difficult problem is matching SNPs to diseases and disease therapies. One method that's been under way for some time is studying families, ethnic groups or isolated villages with a high rate of a particular disease. This works well when a defect in a single gene is to blame.

The best-known example of this process is the discovery of the BRCA (breast-cancer-associated) gene variants. It was clear to researchers that many breast and ovarian tumors are genetic, because they so often run in families. Beyond that, little was known—such as whether it was one gene variant or multiple ones, or whether the gene or genes in question were tumor promoters or broken tumor suppressors. But now we know that at least three variants in two tumor-suppressor genes are involved. Two of the variants are on the BRCA-1 gene, with the third on BRCA-2.[7]

To find these variants, researchers looked first for distinctive gene traits in individual families with high rates of breast cancer, and then at the entire ethnic group that was suffering high rates.[8] In this case, the group was Ashkenazi Jews.[9] Having found these variants, researchers then started looking for and often finding them in other ethnic groups that had high rates of breast and ovarian cancer.[10] The knowledge was also quickly translated into a test to let women know if they are at increased risk of disease, and then into specific drugs (such as Herceptin) that specifically target the trait.[11]

Researchers have long known that people react differently to different drugs based on their genetic make-up. During the Korean War in the early 1950s, for example, scientists found that about 10 percent of black servicemen became anemic after receiving an antimalarial drug that rarely caused problems in white soldiers. The anemia was ultimately traced to a variant of the G6PD gene, which is disproportionately found among people with roots in Africa, the Middle East and Southeast Asia.[12]

Sometimes these variants can make a common drug deadly. It's believed that about 7 percent of Caucasians carry a variant of the 2D6 gene that causes carriers to metabolize about half of the one hundred most-used common drugs, including most antidepressants, much more slowly than do other people.[13] Others have a variant of 2D6 that makes them fast metabolizers; their problem isn't overdosing, but rather needing more of a medicine than the average person would for it to be effective. If they don't know they have these variants, they'll simply reject a medicine as worthless to them when in fact the dose simply wasn't high enough.[14] One team of researchers pored over several years' worth of studies on adverse drug reactions and metabolization differences due to gene variants, finding twenty-seven drugs frequently cited as causing bad side effects.[15]

Orchid BioSciences of Princeton, New Jersey, is focusing on five enzymes known to interact with three hundred of the most commonly prescribed medicines.[16] The idea is to develop a set of tests to determine how quickly a patient metabolizes a drug, and use that to set the dose. Orchid hopes to have the tests ready in 2003.[17] The Mayo Clinic in Rochester, Minnesota, uses a routine blood test that can identify childhood leukemia patients whose SNPs make them unable to metabolize the normal dose of a chemotherapy drug.[18] If they test positive, they're given one-tenth the dose.[19]

The discovery of the BRCA genes and the subsequent development of tests to detect them in individual women and drugs to treat those women are a wonderful portent of things to come. Unfortunately, finding other variants related to disease and its treatment will take a lot more work, because it's thought that most diseases are associated with multiple variants, not to mention numerous environmental factors. Locating these genes might require thousands of patients. "This is not going to happen in some esoteric studies with a couple of families here and there," said Mihael Polymeropoulos, a vice president of Novartis Pharmaceuticals in East Hanover, New Jersey.[20] "We're going to need lots of people."[21]

That's why Estonia is planning a genetic database of all of its 1.4 million citizens, as is the Polynesian island kingdom of Tonga.[22] But beating them both is DeCode Genetics based in Reykjavik, Iceland, which is taking genealogical as well as genetic information from that entire island nation.[23] Iceland was considered ideal for several reasons: it has a relatively small population of about 275,000; detailed individual medical records have been maintained by the public health services since 1915; and almost 80 percent of its citizens can trace their family trees to the ancestor who immigrated to the island. Often these go all the way back to the Norwegian Viking Ingolfur Arnason and the original settlers. One cluster of 104 Icelandic asthma patients turned out to be descended from a single common ancestor born eleven generations back, in 1710, according to Dr. Hakon Hakonarson, a DeCode physician who searches for asthma genes.[24]

The Icelandic database has been controversial because of concern over potential privacy violations.[25] Further, it turns out that the population is not nearly so homegeneous as was previously thought, with apparently a much more diverse population of immigrants than anybody knew until genetic testing began.[26] Yet the DeCode work is already yielding important information. In mid-2002, DeCode told the *New York Times* that it had already mapped the general location of errant genes for 20 of the 50 common diseases on its list, including Alzheimer's, anxiety, asthma, hypertension, Parkinson's disease, macular degeneration and rheumatoid arthritis.[27] DeCode also said it had found a region holding a longevity gene. In late 2000, researchers announced that the database had established that late-onset Parkinson's disease has a genetic component as well as an environmental one.[28] Shortly thereafter the Swiss drug giant Hoffman-LaRoche said it was going to develop a schizophrenia drug using information gathered from the Icelandic gene pool.[29]

Other genome-scouring groups are using other techniques to snoop out SNPs and their functions. One company recently established in part by double helix co-discoverer James Watson is California-based DNA Sciences. Among its numerous ventures, including a very instructive website on genetic disease (www.DNA.com), is what it calls "The Gene Trust." On that same website people can fill out confidential health and family profiles, after which they are asked for blood samples if they fit a research project that's planned or in progress. These samples become a part of DNA Sciences' database, where they are stored by anonymous codes.[30] The first areas of study are type 2 diabetes, asthma, coronary artery disease, Parkinson's, breast cancer and multiple sclerosis.[31] "The response has been phenomenal," chief business officer Steven Lehrer told *Fortune* magazine. "Our typical volunteer is someone who's watched a sister go through a lot of terrible treatments for breast cancer and wants to do something to help save others from having to go through the same thing."[32]

London-based GlaxoSmithKline says its DNA recruiting program has already found genes associated with migraines, diabetes, psoriasis and Parkinson's disease.[33] Genset, a French company specializing in studies on diverse populations, says it has found genes linked to prostate cancer and schizophrenia.[34]

One problem with this approach is that it can be used to find only the more common diseases—you've got to get enough "dots" in your defined population to connect them. But if you think about it, other than the seriousness of a disease, what better way to set your priorities than by the *frequency* of that disease? Moreover, we are talking about what we can do with the tools that are at our disposal here and now. As we learn more about how specific genes function, as computers get faster, even the rarest of diseases with genetic associations will be revealed.

In April 1999 a group of fourteen companies, including essentially all the pharmaceutical giants, established the SNP Consortium to identify these associations and add the knowledge to the public domain, rather than patent it for commercial use.[35] It sounds like a noble thing to do, but really it made good economic sense, since developing biotech drugs often entails a nightmare of patent licensing. Often it's better for everybody to let everybody share in the discoveries from which to draw their research.[36] Moreover, if each company went its own way, the costs would be outrageous and perhaps even prohibitory. The initial "kick-in" was $45 million between all the companies; they wouldn't do this if they didn't think their investment would pay off.

SNPs on Chips

Even when scientists have found all the important SNPs, it won't matter to us unless we can identify *our* personal SNPs. Further, this must be achievable at a reasonable cost and in a reasonable amount of time. The best path at this time seems to be the DNA chip. Also known as gene chips, DNA microarrays, or—my favorite—"SNPs on chips," these are similar to a computer chip but are imbedded with DNA molecules instead of electronic circuitry. They're designed to allow the study of vast numbers of genes at one time using densely packed samples of known sequences of DNA nucleotides. These are typically arranged on a piece of glass (like a microscope slide) or a piece of silicon that's a bit bigger than a postage stamp. The slides carry a DNA configuration to screen for certain genes. The double-helix structure of DNA means that one half of the strand can be unzipped and used to hunt for its mate, since the two sides always join up the same way.[37]

Gene chip technology has exploded in the last few years, putting to shame the speed with which computer microprocessors have been doubling in capacity. The pioneer and current leader in the field is Affymetrix of Santa Clara, California.[38] By May of 2001, Affymetrix's chips were looking at up to 400,000 DNA sequences simultaneously, but the company is now working with a single wafer that can analyze 60 million sequences.[39] More computational power, miniaturization and automation are all driving down the cost and increasing the speed at which genetic data can be read, according to Robert Lipshutz, vice president for corporate development at Affymetrix. Shortly after the publication of the Celera/Human Genome Project's so-called "draft" sequence, Lipshutz said his company had completed screening chromosome 21 in fourteen individuals. The process involved looking at 3.5 billion of the chemical base pairs that are the building blocks of DNA. That's more than the 3.1 billion sequenced in the Human Genome Project, yet it took Affymetrix just ten weeks and cost only $4.5 million.[40]

In early 2002, the Mayo Clinic announced it was working to become the first medical center or hospital anywhere to warehouse electronically the medical records of every one of its patients. Ultimately the plan is to have not only medical records but detailed genetic information as well.[41] It will not only help the Mayo's own patients but serve as a database for researchers, doctors and patients around the world seeking to connect symptoms to genomes. At Mayo, many researchers are simply excited about the prospect of taking a leading role in genomic medicine. "If you look at this as a ladder, then we're only on the first one or two steps," said Dr. Hugh Smith, chairman of the clinic's board of directors. "But boy is it worth climbing."[42]

"The entire genome of individuals will eventually be sequenced at birth routinely," predicts Darrin Disley, who is responsible for drug discovery technologies at Britain's Generics Group.[43] The information could then be stored in any number of ways, such as on a DVD or a magnetic hard drive, but the medium of choice right now is those chips that Affymetrix pioneered. Sequencing an entire individual genome could be done in a day with technology that should be available in less than five years, according to George Weinstock, co-director of the Baylor College of Medicine Human Genome Sequencing Project in Houston.[44] Obviously there are privacy concerns here, but they are no greater than those we currently have with medical records. Perhaps they're even less. Would you prefer that it be leaked that you have a gene on chromosome 8 that indicates you have a 50/50 chance of developing Alzheimer's by age 45, or leaked that you have HIV?

Aiding AIDS Patients

There's already a device that can select individualized therapy for people with HIV or full-blown AIDS, based on which strain or strains of the virus they have. Long gone is the time when a doctor couldn't tell a person with AIDS much more than "get your affairs in order." There's now a long list of therapeutic drugs, and it's getting longer all the time.[45] The problem is that many mutational strains of HIV have developed. If you have a certain mutation, the same drug that otherwise might knock down the amount of virus in your blood to an undetectable level will prove utterly ineffective. Experts estimate that about 60 percent of patients have a strain that is resistant to at least one drug.[46] "You've got a huge medicine cabinet to pick from and no good way to pick," said Richard Daly, chief executive for Toronto-based Visible Genetics.[47]

That's why his company invented the Trugene HIV genotyping test, which received FDA approval in September 2001.[48] The device replaces a $300,000 gene sequencer the size of a refrigerator with one that fits easily on a desk. It decodes the HIV genes in a patient's blood, identifying all the genetic mutations. Software programs match those mutations to a list of more than seventy mutations linked to resistance in specific drugs. This allows doctors to avoid prescribing drugs that won't work.[49]

"It's made an enormous impact on how we select drugs for patients," a physician and managing director of a California medical group told the *New York Times.* "Before resistance testing, we used what I lovingly referred to as the eeny, meeny, miney mo method."[50] Another advantage of the Trugene tests is that they cost from $300 to $500, roughly half the price of the tests they replace. They also require less time to get the results.[51] HIV resistance testing "is a model for what's going to happen in the future," said Jorge Leon, vice president for applied genomics at Quest Diagnostics, a New Jersey–based lab that does testing for doctors and hospitals.[52] "You'll see the same thing happen with cancer, Alzheimer's, diabetes and cardiovascular disease."

The discovery and testing of SNPs may help end the problem of drugs being yanked from the market even though they are helpful for most users and harmless to *almost* all. In recent years, many such prominent pharmaceuticals have been pulled because they caused extreme side effects in a handful of takers. A few really bad reactions and the medicine gets the hook, no matter how many people it was helping. Among these have been Seldane, for allergies; Lotronex, for irritable bowel syndrome in women; the diabetes pill Rezulin; and most recently Raplon, an injectable muscle-relaxing anesthetic. Such drug withdrawals occurred eleven times during 1997–2001.[53] Some advocacy groups claim this is because the FDA doesn't monitor the drug companies' tests closely enough, but a better explanation is that SNPs are uncommon enough that they may not show up even in large clinical trials of several thousand people, yet sufficiently common to show up when the drug is approved and subsequently used by hundreds of thousands or millions of people.[54]

In the case of Raplon, for example, the maker said it received five reports of death from its product. Five deaths are a tragedy. But it's also a tragedy to withdraw a drug that had safely helped over a million patients.[55] As a *New York Times* headline poignantly put it after a number of such valuable drugs had been withdrawn, "FDA Pulls a Drug, and Patients Despair."[56]

The flip side of the problem is that by luck of the draw, a few people with SNPs that react poorly to an otherwise wonderful new drug can doom it in a clinical trial. Pharmacogenomics can help prevent both these types of tragedies. First, people can be tested before they take the drug to find out if they will be among the tiny proportion that reacts badly (or, for that matter, won't react positively). Second, during the drug development process, the drug can be tested to see if it might cause extreme side effects in a portion of the population that would probably be too small to show up in clinical trials involving perhaps a few thousand people.

SNP-testing could also help track a drug's safety after approval. In one proposal, hundreds of thousands of patients who received the medicine would have blood spots taken and stored on filter papers. When rare, seriously adverse events occurred, DNA from patients who suffered them would be extracted and compared with DNA from control patients who received the drug but experienced no problems. This would allow SNP profiles for patients susceptible to the adverse event to be determined.[57] "Imagine the benefit to the development of new therapies if drugs entering clinical trials are almost ensured to be well tolerated in the body and to have the desired effect," writes Chris Sander, chief information science officer for Millennium Pharmaceuticals, in *Science* magazine. "Or imagine relatively short clinical trials, [as] confirmatory final tests to guarantee that drugs and diagnostics are safe and effective."[58] Now drug companies are indeed routinely collecting such information, though they are still limited in the extent to which they're able to use it.[59]

Designer Therapy

Often when we hear about the promise of pharmacogenomics, it's about "designer drugs" of the future, tailored to persons with specific SNPs. Drugs like Herceptin, designed for women with BRCA variants, are already being made. Yet pharmacogenomics can also be applied to drugs currently available. It's just a matter of finding out which drugs react well or poorly to those with particular SNPs. This could more aptly be termed "designer therapy."

One fascinating ongoing research project is that of Dr. Stephen Liggett, a lung specialist and molecular geneticist at the University of Cincinnati College of Medicine.[60] Liggett knew from the research of others that a variant of a specific gene determined to a great extent how responsive congestive heart failure patients would be to medication. Beyond that, however, the science was murky. So Liggett and his colleagues recruited one hundred healthy people and analyzed the beta-2 adrenergic receptor gene in each one, eventually finding three variants.[61] Then they studied over 250 sick people being evaluated for heart transplants. They found that patients with one variant of the receptor gene responded far worse to medication than those with the other two variants. Within a hundred days after they arrived at the hospital, about 60 percent of the patients with the least responsive variant had died or required a heart transplant; of those with the most responsive variant, about 95 percent were still alive.[62] Now, said Liggett, "We test everyone who has congestive heart failure," and those with the worst variant "should be put on the transplant list early and they should have more aggressive drug therapy."[63]

Pharmacogenomics is also being used to tailor treatments with current medicines to specific tumors. "Traditionally, cancer treatments have been selected on the basis of tumor type, pathological features, clinical stage, the patient's age and performance status, and other non-molecular considerations," editorialized Dr. John Weinstein of the National Cancer Institute in the *New England Journal of Medicine.* "We have generally accepted with a certain fatalism that some patients pigeonholed into a given category will have a response to a particular therapy, whereas others will not. The difference is often viewed as a matter of luck, like the result of a coin toss."[64]

But the one-size-fits-all approach is changing, in great part because of a company named Response Genetics, which types tumor genes.[65] It has long been known that people with the same type of tumor often responded quite differently to the same chemotherapy drug. Now we know why: tumors have distinctive gene sets, too. Response Genetics has already identified two genes in gastrointestinal tumors that can determine which line of chemotherapy should be tried first. This is vital, because some patients may not get a second chance. The company is now expanding to the more common cancers, such as those of the lung, prostate, and breast.[66]

SEVEN

Ingenious Gene Therapy

"Twenty years from now gene therapy will have revolutionized the practice of medicine," predicted Dr. W. French Anderson in 1999. Anderson is director of gene therapy at the Keck School of Medicine at the University of Southern California in Los Angeles, and the most prominent pioneer in the field.[1] "Virtually every disease will have gene therapy as one of its treatments."[2] But that is what people were saying ten years earlier with almost nothing to show for it. Now gene therapy is the "comeback kid" of biotech, with over six hundred clinical trials involving almost 3,500 patients by late 2002.[3]

All of the early gene therapy experiments took advantage of the infectious power of viruses, which by nature burrow into the nucleus of living creatures and set up housekeeping there. Viruses comprise a tiny section of DNA or RNA stuffed inside a protein envelope. Researchers render the virus benign by deleting the harmful gene or genes; then they splice the therapeutic gene into the remaining genetic material and mix it with human cells. The altered virus, called a carrier or a vector, can deliver the therapeutic gene into the nucleus.[4]

The first real breakthrough in gene therapy came in April 2000, involving two infants born with a life-threatening illness called "severe combined immunodeficiency disorder" that had forced them to live inside protective sterile "bubbles."[5] French researchers first removed millions of stem cells from each infant's marrow. (Stem cells are those that haven't yet matured into a final cell type. They'll be discussed at length later on.) Then the researchers employed genetically altered viruses to deliver to those cells healthy copies of the gene the children lacked, and reinfused the altered cells into the children.[6] Over two years later they reported that the children were still healthy and that they had successfully treated another two children as well.[7] In June 2002 came the announcement that two more children were apparently cured of the disease through gene therapy. In a truly international effort, Italian and Israeli researchers treated a Colombian child and a child from an Arab community in Israel.[8] "We've known it ought to work, and fortunately it did," said W. French Anderson. "This tells us that if you can get a high enough percentage of cells fixed, gene therapy will cure you."[9]

Gene therapy holds great promise for correcting hereditary disorders. (Five percent of children worldwide are born with congenital or hereditary problems and nearly 40 percent of adults are thought to have some genetic predisposition to common illnesses, ranging from minor ailments to killers like breast cancer and sickle cell anemia.)[10] Originally, the purpose of gene therapy was to do precisely what the French researchers did: treat a disease caused by a single defective gene by replacing it with a good one. Then scientists made a discovery that changed all that: they found that adding genes that caused the right proteins to be produced could potentially alleviate any number of disorders.

At virtually the same time the world heard about the "bubble boy" successes, the *Wall Street Journal* reported the case of a woman with terminal lung cancer who had a new gene inserted to replace the defective one, the p53, that had allowed the tumor to form and grow. After insertion of the new gene, her tumor "shrank into a little ball, and surgeons handily removed it." In a male patient, the *Journal* reported, treatment appeared to have wiped out a tumor the size of a large lemon. Genetic therapy involving repair of p53 is now in advanced human trials for head and neck cancer, ovarian cancer and lung cancer in combination with radiation.[11] In fact, while most people still equate gene therapy with fixing hereditary problems that strike children, almost two-thirds of clinical trials are now directed against cancer.[12]

Some results have been simply stunning. "Malignant glioma (a cancer of the brain) is almost always fatal, due to very poor response to surgery, radiation and chemotherapy," declared Matthias Gromeier of Duke University at the annual meeting of the American Society for Microbiology in May of 2001.[13] But, he went on to say, "We have developed a new form of therapy against malignant glioma based on viruses that can cause brain infections in humans."[14]

The virus is essentially the same one that causes crippling and often fatal poliomyelitis. But Gromeier's team tamed it by inserting a small piece of genetic material from a cold virus into the poliovirus genome, effectively rendering it "completely unable" to cause polio in healthy neuronal cells. They injected the hybrid virus into mice with malignant gliomas and found that tumors were "eliminated by the replicating virus within days," and none of the mice contracted polio.[15] And in a related FDA-directed study, none of fourteen monkeys injected with the disabled virus "showed any signs of weakness or symptoms associated with poliomyelitis," according to Gromeier.[16] This was no chance finding. Gromeier's team had focused on the genetics of malignant gliomas, and found that a specific molecule, CD155, abnormally expressed in glioma cells contributes to their explosive growth. It was a "Eureka!" moment because viruses like polio appear to be specifically attracted to CD155, using it to target and destroy cells.[17] The findings are "a dream come true for people who work on brain tumors," Gromeier said, adding that his people had found active CD155 in "every single brain tumor—and more specifically, malignant gliomas—that we have so far analyzed."[18]

Special Delivery

"There are only three problems in gene therapy," according to Inder Verma, a professor at the Salk Institute for Biological Studies in La Jolla, California.[19] "Delivery, delivery, and delivery. It isn't going to be a problem to make gene therapy work—if we have an appropriate set of tools to deliver the genes."[20]

The need for better gene delivery, however, is leading to the development of some ingenious techniques. University of Pittsburgh scientists have trimmed a gene down to size to be able to insert it.[21] No, this wasn't the "fat gene" that some people invoke to explain why they're overweight; it's actually a gene that's fat in size. It codes for a dystrophin protein that holds great promise for curing sufferers of Duchenne muscular dystrophy, an inherited disease for which there is no real treatment and which cripples and eventually kills about one in 3,500 males.[22] The protein vexed researchers because it's 640 times larger than the virus into which it would have to be inserted to be useful as gene therapy. So Dr. Xiao Xiao and his colleagues at Pitt essentially stripped away any part of the gene that didn't appear vital to its function. The result was three different "mini-genes," all of which were inserted into the virus and then injected into the calf muscles of mice lacking natural dystrophin protein. Two of the slimmed-down genes stimulated the manufacture of the protein in 90 percent of treated tissue. Dystrophin continued to be pumped out for at least a year—the duration of the experiments.[23]

Another way of inserting larger genes employs bacteria rather than viruses. Virginia Waters of the University of California, San Diego, has found that bacteria can "have sex" with mammalian cells.[24] The bacteria she used had been genetically engineered to contain pieces of DNA called plasmids. These readily transfer between two bacteria and are actually a vital tool in crop biotechnology. But mammalian cells aren't known to accept bacterial DNA. Perhaps it was just something that occurred rarely and had never been observed.

To find out, Waters inserted a gene into one of the plasmids that would show up under a special light. She then laid the bacteria on top of a layer of mammalian cells, and after eight hours used the light to determine that the desired hanky-panky had indeed transpired.[25] This "bacterial conjugation," as she described it, could conceivably deliver even the largest genes to a variety of sites in humans.[26]

Unfortunately, size isn't the only problem with viral vectors. Viruses can have immunological side effects. In 1999, the first known death clearly attributable to gene therapy occurred at the University of Pennsylvania's Institute for Human Gene Therapy.[27] Eighteen-year-old Jesse Gelsinger, who'd had a gene that had been engineered into a weakened adenovirus (a cold virus) injected into his liver to treat a rare metabolic disorder, subsequently died of a severe immune reaction. The tragedy was that Gelsinger's disorder was not fatal.[28] His death prompted the FDA to suspend all human gene therapy trials at the institute.[29] It appears that Gelsinger died from proteins on the surface of the coat of the virus, which had not been removed.[30] But the possibility of harm from using viruses was not lost on the medical community. This concern was reinforced three years later when a three-year-old "bubble boy" was diagnosed with a disease very similar to leukemia, followed by another such child in early 2003. Both children had been part of experiments to build their immune systems using retroviruses, which researchers had long thought in exceptional cases could cause cancer if they lodged in or near a cancer-causing gene. That appears to be what happened with both children. But hundreds of humans and far more lab animals had been given such viruses without harm. Indeed, the researchers who originally treated the boys had success with nine out of eleven, so one could make the utilitarian argument that the children were given a treatable albeit *possibly* terminal disease in place of a *certainly* lethal one. Nonetheless, "do no harm" demanded that related types of research be suspended, putting an even higher premium on finding nonviral vectors.[31]

Researchers at Friedrich Schiller University in Jena, Germany, are using infrared lasers to cut temporary holes in mammal cells and insert DNA.[32] The genetic material is encoded so that it glows green, which has enabled the scientists to see that only the cells they aim for are hit, that they are undamaged, and that they go on to grow and divide naturally.[33]

Scientists in Britain are experimenting with a combination of ultrasound and "microbubbles," which are gas bubbles measuring only about 3 microns across that are filled with new DNA and then injected intravenously. When ultrasound is applied to microbubbles, they pop and this can cause small perforations in the target cells that permit the DNA to enter. This has been used as a gene therapy technique only on mice so far, although microbubbles are already commonly employed to improve what the doctor can see with ultrasound scans. The first gene therapy application of this technique on humans will probably be for children suffering from muscular dystrophy.[34]

Bringing New Life to Older Blood Vessels

Gene therapy also holds tremendous promise for treating heart disease by replacing or strengthening blood vessels. Dr. Keith Channon of Oxford University and his colleagues are using modified viruses to put new DNA into veins, making them stronger.[35] "It has great potential as this is an extremely large problem in modern medicine. It could save lives and improve outcomes from surgery," Channon said.[36] Others have successfully used synthetic blood vessels, but these tend to clog up and cause dangerous blood clots, especially when they are thin and implanted in parts of the body where blood flow is sluggish.

If you have clogged arteries, you may get to bypass coronary bypass surgery with angioplasty, a medical procedure in which a doctor inserts a small tube (a catheter) with a balloon on the end into the affected artery. The doctor will inflate the balloon after it is placed in the narrowed area, stretching out the artery and improving blood flow.[37] But how can you keep the artery from constricting again after the balloon is deflated and removed? One way is with stents, tiny metallic scaffolds that are a miniaturized version of the way tunnelers keep the walls and ceiling in place while they dig further. But the metaphor ends there, because usually your body reacts to the little scaffolds as intruders, and cells soon repopulate and block the artery. That's what happened to Vice President Dick Cheney in February of 2001, when doctors discovered that an artery into which they had inserted a stent just three months earlier was already 90 percent blocked.[38] Doctors at Children's Hospital of Philadelphia, however, may have found a way around this problem, by coating the stents with genetic material that would inhibit cell production.[39] Dr. Robert Levy and his team, using live pigs, found that about 1 percent of the cells from the walls of coronary arteries that had received stents coated with the polymer-DNA mixture expressed the gene within a week. More than that, gene expression was largely restricted to the coronary arteries, spreading only slightly to other organs. Levy said gene-therapy stents could be used in humans by 2003.[40]

Therapy for Blindness

The video showed a scene that was a bit bizarre and perhaps seemingly cruel. Three dogs were walking around a room, occasionally bumping into objects. Curiously, though, they bumped into objects on one side only. That's because this breed of dogs, Briards, though congenitally blind because of long-term breeding that went awry, could now see out of one eye treated with gene therapy. The dogs had functional genes injected into their eyes piggybacked aboard a harmless adenovirus. About ten thousand Americans now alive were born with essentially the same disease the dogs had, Leber's congenital amaurosis (LCA). Children with LCA are born with little or no vision, and whatever they may have at birth they lose.[41] The illness is one of several incurable forms of blindness collectively known as retinitis pigmentosa, which afflicts more than 100,000 Americans by destroying specific nerve cells in the retina.[42] To sufferers of LCA, these dogs represent hope for the miracle of sight.

This study, led by researchers at the University of Pennsylvania, marked the first time that congenital blindness has been cured in anything larger than a mouse.[43] But "exactly the same approach could be used with humans," said lead researcher Dr. Jean Bennett. "The protocol we used to deliver the therapeutic reagent [to the Briards] is exactly the same that is used every day by retinal surgeons to remove fluid under the retina, [and to] remove abnormally growing cells."[44]

The scientists first gave the dogs injections into their left eyes, but in an area far from the retina. As expected, this had no effect. But into the right eyes, they injected corrected genes directly behind the retina, very close to the retinal pigment epithelial cells.[45] The videos, along with electrical measurements of the retina and measured pupil responses, clearly showed that the animals had attained a relatively high degree of vision. "We have to be careful not to fill people with false expectations or false hopes," said Albert Maguire, an ophthalmologist at the University of Pennsylvania's Scheie Eye Institute who wasn't involved in the study.[46] "But that said, it's hard not to get very excited about this, because it's a very dramatic result. I mean, basically these dogs were blind and now they are not blind anymore."[47]

The Pennsylvania researchers believe that similar gene therapy treatments may prove to be a cure for as many as 150,000 Americans suffering from retinal diseases.[48] Further, if the gene therapy works in LCA patients, it could be the vanguard for treating a broad array of hereditary vision diseases that strike the retina. "It should open the floodgates," said Dr. Gerald Chader, chief scientific officer of the Foundation Fighting Blindness in Owings Mills, Maryland.[49]

Hemophilia

In 2000, kids at Children's Hospital in Philadelphia with hemophilia B were successfully treated with genes that helped their bodies make a blood-clotting protein they lacked, called factor IX.[50] About five thousand Americans suffer from hemophilia B, a hereditary disease in which the body produces insufficient levels of this protein. As a result, blood leaks into their joints. Treatment for the disease involves injections of purified factor IX, which itself is now bioengineered. Traditional transfusion treatment improves symptoms but is no cure. Many patients are disabled by age thirty, and some still die of complications.[51]

The new gene's effect was "modest, but measurable," said Mark Kay, the study leader and a member of the Department of Genetics at the Stanford University School of Medicine.[52] "It changes from severe disease to moderate, which really increases the quality of life for the individual."[53] Jeanne Lusher, a hematologist at Children's Hospital of Michigan in Detroit and chief medical advisor to the National Hemophilia Foundation, said the findings would have to be demonstrated in much larger numbers of people, "But the fact that they are getting reductions in bleeding and some increase in factor IX is really very exciting," she added.[54] "All of us in the hemophilia community believe that the best hope for a cure is through gene therapy."[55]

Tremors

We've already discussed efforts to use biotech to fight Parkinson's disease. It's estimated that about 1.5 million Americans suffer from this progressive neurodegenerative disorder. While it's rarely fatal, it can be terribly incapacitating. Symptoms include tremors, slowness of movement, stiffness of limbs, and problems with gait or balance.[56] It's generally a disease of older persons, but as actor Michael J. Fox's struggle with Parkinson's reminds us, that is not always the case.[57] Although no one knows what causes the disease, it involves a gradual loss of neurons in one region of the brain. These neurons specialize in producing the signaling chemical dopamine, and as dopamine production declines there is a gradual loss of muscular control.[58] There are various treatments for the symptoms, the major ones designed to add dopamine back in.[59] One that seemed promising over a decade ago was called glial-derived neurotrophic factor (GDNF), but doctors tried repeatedly to inject it directly into the brain, without success. Until recently, that is. A gene therapy study in late 2000 was "able to completely reverse the changes" that lead to Parkinson's symptoms in aged Rhesus monkeys. In young monkeys given a Parkinson's-like disease, "We halted the disease process, and in fact reversed it," said neuroscientist Jeffrey Kordower of Rush Presbyterian–St. Luke's Medical Center in Chicago.[60]

Looking ahead, many gene therapists think the next big advance will be a mechanism in the vector that won't just implant a gene but will regulate it as well. "Most diseases and most drugs require modifying the dose," explained James Wilson, director of the Institute for

Human Gene Therapy at the University of Pennsylvania. "But the genes carried into cells by currently used vectors are either on or off." Wilson said the mechanism he envisions "will be like a genetic rheostat. The gene will not work until you take a pill, and the more pills you take, the more the gene will be expressed—and if you want to cut off the supply, you simply stop taking the pill."[61]

Don't expect any widespread applications of gene therapy in the next five years. (Even putting a drug on the so-called "fast track" usually speeds up the process by perhaps only six months to a year.) Individuals in trials will continue to benefit, with some completely cured, but there are just too many kinks to be worked out now for broader use. When official approvals come, however, they will bring sweeping changes, as they already have to a lucky few in clinics.

EIGHT

The Genomic Generation

For the most part, drug development today is merely a more sophisticated version of what scientists, doctors and even folk medicine practitioners have always done: treat sickness by trial and error. In its earliest form, this meant directly observing the benefit—or harm—conferred on a patient. The mostly widely used synthetic drug, aspirin, got its start this way. Clay tablets from the Sumerian period describe the use of willow leaves to treat rheumatoid arthritis, while the Egyptians were also aware of the pain-relieving effects of potions made from myrtle or willow leaves. Willow bark was found to have the highest concentrations of the pain-relieving anti-inflammatory known as salicylic acid. Eventually a German scientist discovered the compound's chemical structure and made it artificially in 1859, allowing it to be produced for one-tenth the cost of bark extract.[1]

Today drug development usually means trying to find novel compounds by sending somebody to the steppes of Mongolia or elsewhere to scoop up soil samples containing microorganisms or clip leaves from plants. It may also involve searching the ocean, which promises a cornucopia of potential medicines.[2]

This material is then brought back to the lab and screened for properties that seem as if they might be therapeutic. After the candidates are analyzed in test tubes and Petri dishes, the most promising move on to animal testing. The winners there go on to human testing, and about one time in five thousand a drug candidate becomes a drug.[3] Computers have replaced notebooks and robots have replaced drudge labor by lab workers, but the system remains terribly time-consuming, inefficient and expensive because when we pay for that successful drug we must pay not only for its development but for research on the 4,999 that fell by the wayside.[4]

Another basic problem with this method is that one of the best ways of finding useful new drugs is to find compounds that are somewhat similar to ones that have already proven useful. In this sense, it's rather like the "profiling" that law enforcement agencies use. Profiling compounds saves time and money, but it tends to help in simply discovering better drugs for the same old targets, as opposed to finding completely new drugs for completely new targets. A result is that out of perhaps 3,000 to 5,000 potential drug targets, we've hit only about 500, said Francis Collins, director of the National Institutes of Health's Human Genome Research Institute, and "We've beaten those targets to death."[5] Consider how many remedies there are for headaches or heartburn on the one hand, and how many for the insidious autoimmune disease scleroderma on the other. (None.) Others estimate that perhaps ten thousand genes could provide therapeutic targets.[6]

Genomic-based drug discovery will change all of this. In traditional drug discovery, diseases are like locks and therapies are like keys. Hanging on your belt is the most massive key ring in the world. One or more of those keys may fit that lock, but you have no idea which. Or you may not have the key at all. Or maybe you'll find a key that fits but not particularly well. This means that for some humans you'll open the lock to a cure and for others you won't. Or for some humans you'll open the lock while causing no side effects, while for others the complications of inserting that key into that lock are so bad that the patient dies.

Now let's fine-tune the metaphor. The locks are still there, but now you can take out a delicate instrument and measure the tumblers perfectly. Then, instead of trying one key after another, you go to a computer database where the exact dimensions of each key you own are stored, and you run a search to find out which key fits. "If you understand the genetic basis of a disease, then you can predict what protein it produces and set about developing a drug to block it," said Collins.[7] By identifying genes, their proteins, and the functions of those proteins, we can more quickly and more cheaply match up existing drugs or existing compounds to a particular disease. That's genomic-based medicine.

Dr. C. Kent Osborne, director of the Breast Cancer Center at Baylor College of Medicine in Houston, told delegates at the European Breast Cancer Conference in Brussels in September 2000 that within just a few years doctors should be able to determine how all the genes within the tumor are acting, which receptors are responding to what stimuli, how much they are responding and exactly which proteins are functioning as co-activators or co-repressors of receptors.[8] Rational-based design requires designing those proteins rather than simply screening lots of molecules to see which will bind and how well they bind.[9]

The result is what Dr. Osborne calls a complete DNA "photofit" of the tumor that will ultimately be able to determine what treatment, in what amount, and for what duration will work best for that woman.

A baby step in this direction was the aforementioned Herceptin. From the beginning, it was intended specifically to deal with tumors that are resistant to normal therapy because of overproduction of protein from a specific gene. "Herceptin proved the concept that if we show what is broken in the cell we can improve therapies," said Dr. Dennis Slamon, director of the Revlon/UCLA Women's Cancer Research Program at UCLA's School of Medicine and one of the scientists most responsible for Herceptin's development.[10] "The big news is that this is the first time a product has been designed against an underlying defect in a cancer cell," said Susan Hellmann, senior vice president and chief medical officer for Genentech. "It really seemed to bring to fruition the promise of biotechnology." Noting that the company is also looking into whether overexpression of the breast-cancer-promoting protein HER2 is common in ovarian, prostate, lung and colon cancers, she added, "We are beginning to move toward the day when we treat a cancer based on its genetics rather than the organ it arises in."[11]

Gleevec

If Herceptin was one small step for women, Gleevec was a giant leap for mankind. It's taken as a pill, has few side effects, and is extremely effective against at least one form of cancer and perhaps many more. Developed by Swiss-based Novartis Pharmaceuticals, Gleevec ("Glivic" outside the United States) received its first approval for CML, chronic myelogenous leukemia.[12] This is a slow-growing but ultimately fatal cancer of the blood in which white blood cells proliferate like ants on a sugar pile. The disease constitutes 15 to 20 percent of all adult leukemias and afflicts about five thousand Americans and five thousand Europeans a year.[13]

Gleevec works by blocking so-called "oncoproteins" (cancer proteins) called BCR-ABL, pumped out by a mutant gene. This mutation causes the body to go into hyperdrive making white blood cells. Sadly, scientists knew about this mutant gene's damaging effects for almost four decades before they could finally do anything about it.[14] But it was clear very early on that there was something special about Gleevec. Its Phase I trial included thirty-one CML sufferers who had all failed to respond to standard interferon therapy. They were dying. In Phase I a drug isn't really being tested for effectiveness, but rather for safety, so doses are typically kept quite low. That's what stirred up all the excitement about those little pills: even at low doses, every single patient went into remission. Some patients saw their white blood cell counts drop from over 300,000 to the normal range of 5,000 to 7,000.[15] "We were amazed by how well it worked," said Dr. Brian Druker, one of the developers of the medicine.[16] Subsequent trials did nothing to dampen the initial enthusiasm.[17] "I don't know that in medicine we're going to very often, if ever, do better than this with a single agent," said the National Cancer Institute's Richard Kaplan.[18]

In May 2001, the FDA approved Gleevec for leukemia after less than three months of consideration, "an all-time record for a cancer drug," according to Health and Human Services director Tommy Thompson.[19] Just ten months later the drug received approval for gastrointestinal stromal tumors, a form of stomach cancer.[20] Because many cancers use the same mechanism that Gleevec blocks, it's in numerous clinical trials for other types of malignancies as well.[21]

What's most important about Gleevec is that it represents a whole new paradigm in drug discovery, relying on discoveries about the genome. "This is not some leaf from the Amazon," said Dr. Marc Lippman, director of the Lombardi Cancer Center at Georgetown University. "This is no longer just grinding up tumors and throwing in a little eye of newt, and hoping for the patient to throw away their crutches and walk. This is, 'Let's figure out what causes CML; let's now rationally design a drug that goes after that.' And it works."[22]

Gleevec is what's known as a tyrosine kinase inhibitor. Kinases are enzymes that play a key role in transmitting signals between and within cells, and drive a wide variety of important cellular functions and responses. There are more than five hundred known kinases in the kinase protein family, and they are implicated in most major diseases including cancer, autoimmune and inflammatory disease, cardiovascular disease, metabolic disease and neurological disease.[23] The existence of kinases and their role in causing malignancies is hardly news, having been discovered more than a quarter of a century ago. But many researchers wrote off the idea of making drugs that targeted tyrosine kinases because it was thought that inhibiting them necessarily meant blocking the adenosine triphosphate (ATP) binding site of the kinase. Since many noncancerous enzymes also bind to this site, it was believed that this would be like unplugging your computer to stop spam e-mail. You'd be keeping away *everybody;* or in physiological terms, you'd cause lots of side effects.

"Fortunately the pioneers have forged ahead and we have inhibitors coming through, and the concerns are beginning to disappear," said Paul Workman, director of the Cancer Research Campaign Center for Cancer Therapeutics in Surrey, England.[24] Thanks to genomics there are myriad medicines in the pipeline to counteract them, and scientists know that what works (or doesn't) to shrink a breast tumor may be effective in fighting heart disease or combating autoimmune illness.

Another promising tyrosine kinase inhibitor is Iressa, from AstraZeneca.[25] The drug targets a single kinase, the epidermal growth factor (EGF) receptor. EGF is found in many tumors including those of the kidney, breast and lung. Moreover, high levels of this enzyme usually mean a particularly aggressive tumor—and therefore a particularly deadly cancer.[26] As with Gleevec, Iressa is taken as a pill and appears to have limited side effects. Although the original tumor targeted by this drug was non-small-cell lung carcinoma (NSLC), persons with colorectal and head and neck cancers are now in clinical trials and Iressa has shown success against human tumors implanted in mice, including those of the prostate and the vulva, the external parts of the female genitals.[27] In the lab, it's been effective against cancer of the breast, ovaries and colon.[28] In May 2003 the FDA gave Iressa its first approval, for NSLC—the most common form of lung cancer in the United States.[29]

While it's obvious that there are real advantages in a single drug that can target several cancers and even diseases beyond cancer, some of the applications might not occur to you at first. Consider these: Once they've proven that the drug is safe, researchers no longer have to repeat safety trials and can focus on efficacy. Something approved for one type of tumor can quickly be approved for several other types. Another advantage is for cancers that have metastasized. The pill you take for the primary cancer may help knock out the secondary. Often the secondary cancer is found first (this is common with liver cancer, which is usually metastasized from another site), so the pill you take also goes after the primary cancer.

It's common to read that genomic-based medicine is being driven by the sequencing of the human genome by Celera Genomics and the Human Genome Consortium. In fact, while the projects have been great publicity for the participants, they actually contributed little more than hot air. The purpose of the sequencing was to identify the "base pairs" or the "letters" in the alphabet that spell out the language of our genes.

In February 2001, the media blasted it around the world that the genome had been mapped, with headlines like "Reading the Book of Life," "Recipe of Life Revealed" and "Road Map to the Core of Mankind."[30] Never mind that over two years later, in April 2003, scientists were still only describing the project as "essentially" done, prompting the wry *New York Times* headline: "Once Again, Scientists Say Human Genome Is Complete."[31] The metaphors of map and recipe actually obscured the fact that the two competitors hadn't done what most people thought. They had not completely mapped the genome nor even prepared what in any meaningful way could be described as what many dubbed it: a "draft." (Were it a draft, it would be like an architect telling you he knows the basic size of your new house and the location of some of the rooms, but not much else.) Which didn't stop newspapers from running headlines like "Scientists Crack Code of Life; the Human Genome Project Breakthrough to Aid Battle against Almost All Diseases; May Cut Cancer Deaths to Zero."[32] Getting just a bit ahead of ourselves, aren't we?

Even when the genetic sequencing was essentially completed in 2003, it was of limited use because what our genes do is far more important than where they are located.[33] No therapies have been developed as a result of the sequencing efforts, which is essentially why J. Craig Venter was forced to step down as president of Celera in January 2002. Under his leadership, the company he founded repeatedly grabbed headlines and magazine covers but never produced anything sellable.[34] The hard task of identifying drug targets against which to screen compounds has fallen to smaller organizations such as Incyte Genomics of Palo Alto, California.[35] Incyte sequences DNA to map out genes and the proteins they express, then clones them, then makes those clones available for licensing. You'll never buy a drug produced by Incyte, but you may be buying drugs that wouldn't exist without it.

Instead of trying to sniff out genes from the entire sequence, Incyte and others in the commercial sector look for the specific function of a particular gene to see how it codes for proteins.[36] "We can usually get the protein pattern, as well," said Incyte CEO Roy Whitfield. "At that point the drug companies or even individual researchers can get online and see what Incyte has to offer. When somebody sees something interesting, they license it from us." Over thirty thousand such licenses have already been purchased from Incyte. "Then from that clone they get a protein that works." Whitfield said that over 50 percent of the proteins obtained by the pharmaceutical giant AstraZeneca have come from Incyte.

But it's not just the big boys doing the ordering anymore, thanks to the Internet. "We used to ship thirty-three CD-ROMs a month and the computer power necessary to use those required investing tens of thousands of dollars," said Whitfield. Now, "the lone medical researcher can have access to the same powerful genomic and protein database as the largest drug companies in the world. We used to call it clone by phone; now we call it clone by click."[37]

Stalking Killer Genes

Before the Human Genome Project ever began, scientists were looking for specific genes that either contributed to illness or, as with tumor suppressors like "p53" (to be discussed later), helped to prevent it. At the very least, the information they find about genes can be used to put somebody on notice that they have an extraordinary chance of contracting a certain disease and if that disease is treatable, to be on the lookout for it. At best, finding the gene leads to techniques or medicines to counter its dangerous activity.

Before proceeding, let's be clear on something. As Matt Ridley points out in *Genome,* "To define genes by the diseases they cause is about as absurd as defining organs of the body by the diseases they get: livers are there to cause cirrhosis, hearts to cause heart attacks, and brains to cause strokes." Thus, he says, it's misleading when we hear of a "New Gene for Mental Illness" or "Gene for Kidney Cancer Isolated."[38] This makes it sound as if only a single gene is involved in a disease. In some cases that's true. But it appears that in the vast majority of cases, a combination of multiple genes *plus* environmental factors such as diet or cigarette smoking is responsible.

Moreover, a gene can have several functions. Theoretically, you could eliminate a gene that causes hair to go white, only to find that it played a powerful role in preventing cancer. That's hardly an acceptable trade-off. We know, for instance, that one gene variant that appears to make the carrier much more resistant to HIV also appears to make the person far more *susceptible* to hepatitis C.[39]

Other genetic diseases aren't the result of *having* a gene, but rather of lacking one. Wolf-Hirschhorn disease, which causes numerous birth defects, is caused by a chromosome that has a whole chunk missing from it.[40]

Sad to say, but at this early point in the biotech revolution, the discovery of a gene or gene combination or mutation that causes a problem often leads to no immediate advantages. For example, in late 2000 researchers discovered a genetic link between ALS (amyotrophic lateral sclerosis, a progressive neuromuscular illness better known as Lou Gehrig's disease) and a form of dementia in which nerve cells die.[41] Such discoveries usually get lots of play in the general media, but probably should be relegated to professional journals where they will not raise false hopes. As one reporter explaining this particular finding put it, "Don't look for any breakthroughs in fighting this disease, but this research adds to the store of knowledge."[42]

Malignant Melanoma

So we know not to confuse "association" with "causation," and not to match one gene to one illness. But when hunting for ways to prevent or cure illnesses, it's also important to understand that finding genes that combat disease may be more vital than finding those that contribute to it. The most important of these is p53. While it's often called "the p53 cancer gene," it doesn't cause cancer. It would be more accurate to say that the *lack* of a properly functioning p53 *allows* cancer. The job of p53 is to tell a "runaway" cell to stop replicating or, barring that, to commit suicide. This mechanism is called apoptosis, or programmed cell death. Normally, the cell obeys, faithfully doing itself in; but if something goes awry with p53, cells can and will go out of control, eventually becoming malignancies.[43] "Mutation in the p53 gene is almost the defining feature of a lethal cancer; in 55 percent of all human cancers, p53 is broken," explains Matt Ridley.[44]

What about that other half? As you might guess, cancers are now being linked with tumor-suppressor genes other than p53 that, for whatever reason, aren't doing their job.[45]

One such cancer is malignant melanoma. It's been known for some time that those who have this deadly form of skin cancer usually have a p53 that's working just fine.[46] Yet the cancer is somehow evading p53's order to commit suicide and is able to ward off the chemotherapy drug of choice, called Adriamycin.[47]

In early 2001, researchers reported in *Nature* magazine that they thought they'd found at least a good part of the answer: it's not to p53 that the body assigns the responsibility of reining in malignant melanomas, but to another gene or genes. The culprit they nailed causes the manufacture of a protein called Apaf-1 (short for "apoptosis activation factor-1"). Apaf-1 was inactivated in 42 percent of human skin cancer tumors they studied in cell lines. Apparently, if the Apaf-1 gene isn't functioning, the malignant cells get no suicide order.[48] As to why the Adriamycin is blocked, many researchers have demonstrated in the laboratory that numerous chemotherapeutic drugs work by inducing apoptosis.[49] So if Apaf-1 can't do the job on its own, neither can it be helped by drugs.[50]

"The study is significant because it demonstrates that accurate diagnosis and treatment of malignant melanoma, and perhaps other cancers, should include an assessment of the status of Apaf-1," said principal study author María Soengas, now with the University of Michigan Medical School.[51]

"But what makes this study doubly interesting is that the inactivated gene—Apaf-1—is merely switched off, instead of being completely lost or mutated" as is the case with p53-related cancers, according to Peter Jones of the University of Southern California, who wrote an accompanying editorial in *Nature.*[52] Testing for Apaf-1 would allow advanced notice of who is at high risk for malignant melanoma. This in turn can lead to tumors being caught early, before metastasis, when the cure rate is still about 90 percent.[53]

But the Apaf-1 discovery can help even with a tumor that has already metastasized. Knowing that the gene is turned off will tell you that you need to turn it on. And when will medical science be able to turn *that* trick? It already has, at least in human cells in Petri dishes where they are treated with something called 5-aza-2'-deoxycytidine (5aza2dC), a compound that's been known for decades to be effective in suppressing various tumors, especially leukemia.[54] Gene therapy—inserting a switched-on Apaf-1—would also do the trick, the researchers wrote in *Nature.*[55] Hopefully, 5aza2dC will work as well in the body as it does in the lab. But in any case, there could be any number of other compounds already in companies' inventories that might also be able to flip that Apaf-1 switch. Yet another alternative, said Soengas, would be to "find drugs or protein factors that work like Apaf-1."[56]

It isn't hard to imagine the next step: screening people regularly for Apaf-1. If they test positive for a switched-off one, it could be switched back on before they even have a chance to develop melanoma.

Small Molecules; Big Results

Another goal of genomic-based medicine is to make more small-molecule medicines. All of the early recombinant proteins were large molecules. To some this may sound rather oxymoronic—any molecule you drop into your pocket is going to get lost. It also doesn't help the imagination that the reference is actually to molecular *weight*—to most of us, all molecules are quite light. But everything is relative and so there are large molecules and small molecules. And to the pharmaceutical world, small is beautiful.

Both large and small molecules may fight disease in a similar manner. For example, by fitting into the contours of a receptor on the surface of a cell like a key specific to a lock, they can trigger a chain of biochemical reactions. Only a small part of the larger molecules actually go into the "keyhole," with the rest—as much as 95 percent—merely providing support. To keep the analogy going, it's like a big piece of wood that a gas station attaches to the bathroom key to make sure you don't accidentally pocket it and drive off. It *does* serve a purpose, but it's a lot more convenient if the key is just big enough to spring the lock.

Why? In part, because a small molecule can be absorbed into the bloodstream through the stomach or intestinal wall. Large molecules when taken by mouth are broken down and digested; many break down even when injected. For example, you can toss back insulin by the mugful, but your body rapidly breaks it down into useless components. Thus, while small molecules can be taken as pills, larger ones must be injected somehow. Further, making small-molecule drugs is considerably cheaper.[57] Since large-molecule proteins are the stock-in-trade for biotechnological pharmaceuticals now, some have compared biotech companies trying to make small-molecule drugs to "Microsoft making televisions or typewriters."[58]

Emisphere Technologies of Tarrytown, New York, has finished Phase I trials for insulin in pills.[59] Once inside the cell, these stealth molecules spring back into proper shape and go to work. The drug has sparkled in early human testing and holds the promise of completely eliminating insulin injections for diabetics.[60] Emisphere's *raison d'être* is making large-molecule drugs available orally, be it as pills, syrups or inhalants (nose drops).

Other companies are also looking to package their proteins into small molecules. Genentech, for one, plans to market an injectable drug that attacks immune cells involved in causing psoriasis, but now has a "rather advanced small-molecule program that blocks the same interaction," according to Dr. Dennis Henner, senior vice president of research.[61]

The Proteomic Promise

You don't hear much about the proteome, but it's there just the same. The proteome is often loosely defined as the collection of all human proteins. Just as genomics is the study of genes and their function, proteomics concerns what proteins the cells express and what those proteins do.[62] Especially if you're an investor in biotech enterprises, you're likely to hear that proteomics is "the next big thing" or that it "will leave genomics in the dust." Actually, proteomics is simply the next evolutionary step in our learning process. Genomics wasn't possible without the development of certain knowledge and blazing-fast computers; likewise, proteomics requires far more amassed knowledge and incredibly faster computers. That's why IBM is working on a 2004 roll-out of Blue Gene, which should be about a hundred times faster than the current fastest supercomputer. Its purpose will be to figure out how proteins fold, which in turn determines their function.[63]

Why might proteomics be superior to genomics? Genes are essentially the instruction manuals for what we're most interested in, the proteins themselves. "You don't raise an antibody [or other therapies] against a gene or against thin air," explained Michael Kranda, the CEO at UK-based Oxford GlycoSciences.[64] "You raise an antibody against a protein."[65]

Almost all existing drugs either are proteins or influence the production of proteins in the body. Genomics companies routinely hunt for drug candidates by comparing which genes are turned on in healthy and diseased tissues or cells. Thus, if a company finds an overactive gene in prostate tissue from men suffering prostate cancer, it might develop a drug that targets the protein that gene codes for. But here's the problem: a gene's level of activity may have little relationship to the amount of protein it makes.

"Looking broadly, there's no correlation," said Scott Patterson, who leads the proteomics project at Celera Genomics. "Some of the correlation is negative, some of it's positive, and some we don't understand."[66] Likewise, Christopher Parkes, chief executive of the British company Proteome Sciences, noted a recent discovery of one of the gene defects that "cause" breast cancer.[67] "Great news, but less than 4 percent of breast cancer patients actually carry that gene," he said. "It drove home the point that the relationship between genes and disease is far more complex than was once thought."

In fact, some genes code for many different types of protein, which helps explain why the link between genes and illness can be tenuous. The production of proteins depends not only on the DNA code but also on the chemical environment of cells.

Proteomics enthusiasts argue that genomics provides only rough clues about the workings of the body. They also question the many scientists who have tried to bind the deciphering of the human genome tightly to drug discovery, including Patterson's own company, Celera Genomics. "To some extent, they've sold the public a bill of goods in genomics," said N. Leigh Anderson, founder and CEO of the Plasma Proteome Institute in Washington, D.C.[68]

So protein discovery has the advantage of being far more specific than gene discovery. But gene discovery has its own advantages, one of which is that it's a lot easier to do. "You can take DNA from anything—yourself, bananas, barnacles—and put it through a machine," explained Sydney Brenner, founder of the Molecular Sciences Institute in Berkeley, California.[69] "That's because it's all the same stuff. [But] there are no good techniques to try and handle proteins."[70] More important is the sheer volume of proteins. There are 5 to 20 times more proteins than genes in the human biological system. Further, disease can result from the collusion of as many as 100 different proteins. Some researchers have estimated that the number of possible protein interactions tops 10 billion.[71]

That's why, for all the talk about "mapping the proteome" or launching a "Human Proteome Project"—and there *has* been a lot of such talk—the smart proteomics companies are focusing on specific proteins, just as the smart genomics companies have focused on the functions of individual genes.[72] "We're going disease by disease," said Michael Kranda. "If we know all the proteins involved in Alzheimer's, that has vital implications for diagnosis and treatment of the disease."[73]

Celera too, in spite of its grandstanding with the genome, has developed more realistic goals than finding all of the proteins from every tissue, organ and cell. Instead, the company is studying tissues and cells from people with specific diseases, investigating proteins found in the membranes surrounding cells that have a great chance of being susceptible to drugs. "We're looking in a very targeted way for potential therapeutics," said Patterson.[74]

This is much more realistic than trying to map out a human proteome, but the task is still daunting nonetheless—like "searching for one needle in millions of haystacks," as one writer put it. To a great extent, the ability to do this will depend on raw computing power.[75] While the speed of the latest desktop computers is far greater than most of us can ever use, quite the opposite is true concerning supercomputers and most areas of biotechnology. Scientists in these fields could readily apply computers a thousand or even ten thousand times faster than those they currently use. They're making do with what they have now, but there's a powerful connection between raw computing power and gene, protein and drug discoveries. With the speed of computers increasing geometrically, so too will the speed of discoveries.

One invaluable tool already in use is the protein chip, a variant on the gene chip. "This is where people will get answers about how disease develops, how drugs work, and how to find new drugs," said Peter Wagner, chief technical officer of Zyomyx in Hayward, California.[76] Each protein chip contains thousands of protein samples that can be analyzed quickly and cheaply. These chips have tiny grids of protein fragments or molecules like antibodies that can trap proteins. For example, you might take spinal fluid from somebody with Alzheimer's and from somebody without the disease and spread it over the chip. Most of the molecules in the fluid will simply wash off, but some will stick. They can be identified by markers such as fluorescent dyes visible when hit by a laser. Molecules that remain do so because they're binding with proteins. By identifying those that bind in the sick person, you can zero in on the differences between the Alzheimer's and non-Alzheimer's patient. Similarly, the chips can be used to find antibodies that bind tightly to protein molecules from sick persons. Because it's that ability to bind that makes pharmaceuticals effective, this could be a shortcut to finding new drugs. The chips can also help diagnose illness. If a specific protein is associated with any medical condition, this could be a quick way of identifying that condition long before it would show up on any other type of test.[77]

Lots of Billion-Dollar Molecules

A stunning example of the future of fast-paced genomic medicinal development is what's happening at Vertex Pharmaceuticals of Cambridge, Massachusetts, a company that became famous as a result of Barry Werth's book, *The Billion-Dollar Molecule.*[78] The book traces the ordeal of Joshua Boger, a brilliant chemist who gives up his comfortable career path at a huge pharmaceutical company to start a scrappy little outfit that would live or die by its success in developing a single drug molecule.[79] Now Boger and Vertex have gone in exactly the opposite direction, not developing one molecule for one disease but seeking countless molecules for countless diseases.

This makes sense from many perspectives. Because small biotech companies are so heavily dependent on the price of their stock or partnerships with large pharmaceutical companies to fund their development programs, the fewer drugs they're working on, the more vulnerable they are to being devastated by a single setback. It's far better for them to spread their risk over several drugs. It's also smart to develop drugs along parallel paths because something that doesn't work for one disease may prove effective against another, or for that matter, a drug that proves successful against one disease may be therapeutic for others. You may find that out several years down the road, as when the prostate drug finasteride turned out to grow hair on balding men. In any case, the sooner a company finds multiple applications, the wider its revenue stream.

So why didn't people do things this way before? Simple—the technology wasn't there. That's why Thomas Edison could invent the light bulb but not the transistor. But now, between the exponential growth in knowledge of the genome and the incredible increase in computer power, massive parallel drug development is workable. Vertex has dubbed its process "chemogenomics," defined as "the identification of all of the potential drugs that interact with all potential protein drug targets." A broader term for it is "combinatorial chemistry," which means testing hundreds of thousands of potential drugs at one time.

Vertex's lab starts with an organizing principle: that the genes and the proteins they code for are grouped as families or superfamilies, related by their atomic structures. These families can have as few members as about ten, and as many as a thousand or more. Vertex uses the genetic information in its databanks to group these proteins together. They literally map out every atom in a target protein and put the information into their computers as a 3-D model. Just as a military commander would make a model of a fortress he wanted to assault in order to probe its weaknesses, Vertex uses these protein models to identify the areas that are vulnerable to attack by a known molecule. If there appear to be no such molecules available, this information may be used to try to design one.

As a fascinating animated "tour" on the Vertex website explains, "Using the atomic map we have created of our target, we use proprietary software to screen billions of chemical compounds per second to see which ones 'fit' best in the vulnerable underbelly of our protein target. With the computer's rankings, we then choose only the best 100 or so compounds for synthesis and screening in the lab." This allows the company to "quickly generate not one but potentially dozens of lead classes of drug candidates." Many that are active against the initial target will also prove active against other targets in the same family, thereby allowing Vertex to "quickly establish a large number of lead classes of drug candidates directed at closely related targets."[80]

Moreover, because these atomically related targets may be involved in different disease processes, the lead classes discovered for any given gene family will probably produce useful drugs for a range of diseases that would seem to most of us to be completely unrelated. For example, one drug that Vertex has in Phase II trials is pralnacasan. While it's being specifically tested against rheumatoid arthritis, the company believes it may also combat osteoarthritis, heart failure and stroke.[81] As for the company's strategy of developing numerous drug candidates at one time, the numbers speak for themselves. Although it's still a relatively small company, as of October 2002 it had a total of eight different drugs in human trials. The disease targets have an equally remarkable range, including AIDS, hepatitis C, autoimmune illness, cardiac disease and diabetic nerve damage.[82] This little laboratory near the banks of the Charles River has turned itself into a pharmaceutical powerhouse.

On the other coast of the United States, Exelixis of South San Francisco has established a high-throughput screening lab to support its efforts in cancer and angiogenesis drug discovery.[83] Just during 2000, the company built a collection of an amazing one million compounds by both internal synthesis and acquisitions from outside sources. It has established collaborations with Bayer, Pharmacia Corporation and Bristol-Myers Squibb, as well as with U.S. government agencies and academic centers worldwide. A few years ago, acquiring a library of compounds this large would have been considerably less worthwhile than obtaining a haystack to search for a needle. Exelixis has different ways of screening, the fastest being a pooling of compound mixtures that allows an incredible 500,000 compounds to be looked at in a single day.[84]

NINE

Gene Testing

Gene tests at the beginning of 2001 were already available for more than eight hundred genes or genetic variants that cause or increase the probability of disease, with hundreds more under development.[1] One company, GeneTests of Seattle, has an amazing set of search engines on its website that allow you to choose tests by disease name, body part, gene name, laboratory or other options.[2]

There is much value to such testing. But there is also tremendous room for abuse and ignorance. An excellent piece by *Washington Post* ace biotech reporter Rick Weiss illustrated how access to such tests is easy but both laypersons and even medical professionals are often grossly ignorant as to what the results mean. For example, one genetics counselor told Weiss, "People come in saying, 'Oh my baby's going to have spina bifida!' when in fact the test had only found that they were at a moderately increased risk, like 5 percent."[3] Spina bifida (the type we normally think of) is a disorder involving incomplete development of the brain, spinal cord and their protective coverings caused by the failure of the fetuss spine to close properly during the first month of pregnancy. It occurs in about one birth per thousand.[4] Thus the positive result means the odds have really only gone up from 100 in 100,000 to 105 in 100,000. But since the condition is also quite serious (though not necessarily fatal), people who don't know this are likely to have an abortion they wouldn't even dream of getting if they understood what the test results meant.

All too many people confuse genetic predisposition with genetic destiny. Some gene tests do indeed say that you *will* get a disease or your unborn child *will* have a disease; one such test is for Down syndrome, which is indicated by an extra chromosome.[5] But most tests simply indicate that your odds for some kind of malady are increased, especially in the presence of outside environmental factors like smoking or bad dietary habits. These are analogous to high blood pressure carrying an increased risk of cardiovascular disease. Such a condition is called a "complex trait"; it is an inclination or influence but not destiny.[6]

Researchers tend to break down genetic influences into two categories: high-penetrance and low-penetrance. There's no bright line separating these categories, but "high" means genetic influence plays a major role in the development of disease. Diseases caused by high-penetrance genes include muscular dystrophy, sickle cell anemia and cystic fibrosis. They are rare and appear to occur independently of outside factors.[7] Generally speaking, about 5 to 10 percent of common cancers are thought to result from highly penetrant genes.[8] Low-penetrance means the gene plays *some* role, but outside factors make a much greater difference.

Alas, far too few laypeople and even health writers know this. Hence we hear repeatedly of this or that "cancer-causing gene," when often—indeed, *usually*—the issue involves low-penetrance genes. As a result, some people will avoid taking a test that could provide useful information in preventing that cancer because they wrongly see a potential positive test as a death sentence. Many women in particular think that if they test positive for a BRCA variant, they are destined to get breast or ovarian cancer. Rick Weiss wrote in the *Post* about a woman who tested positive for the BRCA-2 variant and began a virtual epidemic among her female relatives of whacking off breasts and cutting out ovaries.[9]

But most women with either of these BRCA variations probably won't actually get ovarian cancer; they simply have a higher risk for it. (And for that matter, many who don't have those genetic variants will get cancer.) Moreover, an elevated risk means different things to different women. For example, considering that cure rates for breast and ovarian cancer have been going up, the significance of a positive test for a BRCA variant should be interpreted rather differently for a fourteen-year-old girl than for a fifty-year-old woman.

Even health care providers may be woefully ignorant of what test results mean. Sometimes this stems from a mere misunderstanding of basic statistics. If a doctor gets a result indicating that a man has five times the chance of getting a fatal disease as those without a certain genetic variation, should he tell the man he has terrible news and that he ought to get his affairs in order? What if the disease strikes only one in 100 million, *and* never strikes before age 75, *and* this man is only 20? Then a positive result doesn't mean much, does it?

Properly used, genetic testing can be a powerful tool and the tool is getting better all the time. As medicine progresses, patients will be offered far more than relief at finally knowing their odds; they'll be offered treatments and cures. Already there are prophylactic drugs in use against breast cancer such as tamoxifen or Eli Lilly's raloxifene (Evista).[10] But tamoxifen increases the risk of endometrial cancer.[11] It must be given only to women at high risk of breast cancer because only then is the tradeoff justified. Gene tests make that possible.

There are also some very low-risk and even over-the-counter therapies for inherited diseases. Recall the University of Washington findings that people with hypertension who have a certain genetic variant can dramatically reduce their risk of heart attack and stroke through use of a common diuretic.[12] Cheap anti-inflammatory drugs like aspirin, ibuprofen and naproxen sodium have all shown powerful preventative abilities against cancer.[13] Celebrex and Vioxx, newer prescription anti-inflammatories, may be even better at preventing cancer. Celebrex has already received FDA approval as helpful in preventing precancerous polyps in the colon. There's reason to think its prime competitor Vioxx will likewise be effective.[14] These drugs are showing the ability even to prevent such dread cancers as that of the pancreas.[15] Regular use of anti-inflammatories (though possibly not aspirin) also appears to slash the risk of developing Alzheimer's, while a baby aspirin a day can keep heart attacks and stroke away.[16] Everybody knows that exercise can stave off a host of ills, but fewer know there's evidence that women with BRCA-1 or BRCA-2 mutations can delay the onset of cancer by six to nine years with regular exercise.[17]

Finding that you have an increased risk can also be helpful for monitoring purposes. For the most part, the sooner a malignancy is found, the better.[18] Sometimes there's even a nongenetic treatment that basically amounts to a cure. Consider the revelation in the *New England Journal of Medicine* that mutations in women's genes can cause late-term miscarriages. Such miscarriages have often been associated with abnormalities in the placenta that cause the fetus to get too little blood, but it wasn't known *why.* Believing it might have a genetic link, researchers at the University of Milan in Italy tested sixty-seven women with unexplained late miscarriages.[19] They found women with mutations in their genes for a blood-clotting factor called prothrombin were over three times as likely to have late miscarriages as women without the mutations.[20] So what can a woman with this mutation do? Not have children? Wait for genetic therapy that can fix the mutation? Sure, if she wants to. But the researchers said that preventing the placental abnormalities and miscarriages could be as simple as giving the women blood thinners while they are pregnant.[21]

Therapy will one day be the best answer to the ethical issue of employers or insurers testing employees for genes that predispose one to a particular disease. For now, many governmental bodies have taken the simple route of outlawing such testing. In 2000, President Clinton issued an executive order prohibiting federal employers from requiring or requesting genetic tests as a condition of being hired or receiving benefits. It also prevents them from using genetic information to classify workers in a way that deprives them of opportunities for advancement, and from obtaining or disclosing genetic information about workers or prospective employees. The president of the Biotechnology Industry Organization, an umbrella information and lobbying group for biotech, supported the move.[22] Further, according to the Council for Responsible Genetics, as of August 2000, forty-two states "have enacted laws that provide some level of protection against genetic discrimination in health insurance," while only "three states have *no* legislative protections against genetic discrimination in any setting."[23]

Currently, the problem is probably overstated. It's one thing, for example, to find out that a potential employee could get sick any time and then quickly die, as a positive HIV test once meant. It's another to find out that your potential employee is 50 percent more likely to die of a heart attack within the next three decades. That said, an ignorant employer could give a test result a meaning it doesn't merit. That's just what happened in early 2001 when the news came out that the federal Equal Employment Opportunity Commission had filed a suit accusing the Fort Worth–based Burlington Northern Santa Fe Railway Company of violating the Americans with Disabilities Act. The railroad was administering gene tests to employees who filed claims for carpal tunnel syndrome, a wrist condition believed to be caused by repetitive hand motions.[24] The tests looked for a genetic trait known as a chromosome 17 deletion. Some studies have indicated that a person with that trait is more likely to develop some forms of carpal tunnel syndrome.[25] But since nobody has suggested that this genetic trait by itself causes carpal tunnel, the company could hardly claim it didn't have liability on that basis. "It sounds like really dumb corporate behavior," said Stanford law professor Henry Greely.[26] It was. But it was the first time the EEOC had challenged genetic testing and the American Management Association reported in 2001 that only about 7 of 2,100 companies surveyed in 2000 had conducted genetic testing of their employees.[27]

But as testing improves, it will become more accurate. At some point, insurance nondiscrimination laws will become problematic because they will allow the insured to find out his status while keeping the insurance company in the dark. This is especially a problem with life insurance companies, since extremely large sums of money may be involved. Insurance companies make money by spreading risk over as large a population as possible, calculating that the healthy will pay for the sick or that those who die later will balance out those who die sooner. Unless prohibited by state law, they can discriminate with higher premiums based on a person's medical record—whether he has suffered a heart attack, for example. Let's say a test comes along that can determine that a person has a very high chance of dying of a certain disease in the next few years, and that national or state or local governments pass legislation limiting the ability of insurance companies to use the results of those tests in determining rates or insurability. Then let's say the people with a positive test take advantage of this to buy huge life insurance policies, and businesses then spring up that buy those policies from them before they die.

This is not a hypothetical example. Back in the late 1980s, many states and cities limited the ability of insurance companies to use HIV tests. Many people who knew they were HIV-positive bought tremendous amounts of life insurance and then sold their policies to "viatical" businesses. These companies deal in "death futures"—that is, they speculate that the holder of the policy will die sooner rather than later. They buy the policy at a discount, then cash it in for the whole amount when the insured dies. It's quite legal and it did help a lot of people who expected to die soon to pay their bills, including medical ones, when they were too sick to work. But in providing such a powerful incentive to lie to insurance companies, it was also a mighty incentive to rip them off. In 1999, the North American Securities Administrators Association cited viatical scams as one of the country's top ten financial cons of the century.[28]

Often the insurance companies shared the blame because they didn't take full advantage of what the law did allow in conducting their own investigations. But the worst of it was when governments got involved and allowed AIDS patients to stop insurance companies legally from even seeking important medical information. Washington, D.C., passed the most restrictive law, prohibiting insurers from requiring HIV tests for either health or life insurance, so most insurance companies stopped writing policies entirely in the city. Suddenly, residents of the District found they couldn't get insurance or had to pay outrageous rates. Ultimately, the city council was forced to relent.[29]

Although people with HIV and AIDS now constitute only half of their customers, the viatical business continues to boom. Imagine what will happen when genetic testing becomes more accurate not only as to *whether* an illness will develop but also *when*. Then imagine that individuals are allowed to find out, but insurers are not. This could devastate insurance companies, forcing them to raise everybody else's rates dramatically. It was perhaps in view of this that life insurers in the U.K. have been given permission to demand the results of genetic tests from people who have voluntarily undergone testing for Huntington's disease.[30] Huntington's, which first caught the public's attention when singer Woodie Guthrie died from it in 1967, is an inherited degenerative neurological disease that leads to dementia and death.[31] It afflicts about thirty thousand Americans at any one time. But why single out Huntington's? Because the gene test for it was one of the first that could indeed pinpoint fairly accurately both whether and when a person will die.[32] The potential for abuse was therefore quite real.

One unpublished yet highly touted study of women who had tested positive for BRCA-1 found that it had little bearing on how much insurance they bought, leading the authors to conclude that testing for other genetically related diseases would bring the same results.[33] But the comparison is silly. For people to try to take advantage of health insurance, the disease in question must be both very likely to occur and likely to occur soon. That was once true if you were positive with HIV and it's still true for Huntington's. But it's not true for BRCA-1. Most of these women will never get breast cancer and those who will have no idea when. So this study was either politically motivated or performed by people who just didn't understand what they were doing.

Still, do we want to have a blanket law saying that a person has to hand the results of all genetic tests over to a life insurer? That may discourage testing for diseases that may be now or in the near future treatable or even curable if caught early enough. It appears there are no easy answers.[34]

DNA, the FBI and MIA

DNA technology has caused one of the most revolutionary changes in law enforcement since the nineteenth century.[35] Everybody has heard of DNA "fingerprinting" in forensics, in which body tissue, blood or semen samples the size of a pinhead can be matched to a specific person with almost 100 percent accuracy. It seems to be a regular occurrence now for somebody to be released from jail or put into jail on the basis of DNA evidence.[36] On a single day one Texas death row inmate lost his temporary reprieve when DNA testing established he had indeed raped and murdered his stepdaughter, while a second Texas prisoner was pardoned from his ninety-nine-year sentence for a similar crime after DNA tests linked another man to the murder.[37] Such evidence can be found in the most unlikely places. In the 1993 World Trade Center bombing case, DNA from a suspect was recovered from saliva on the back of a postage stamp.[38] In another case this evidence was a different bodily fluid from a beret-wearing young woman's blue Gap dress.

DNA tests are proving invaluable outside of the crime lab as well. They can be used for identification of recently deceased persons (as in plane crashes, for example), in paternity suits, to catch poachers, or for identification of human remains.[39] DNA testing of a relative of the ill-fated Romanovs, the last imperial family of Russia, proved that the bodies found in an Ekaterinburg forest were indeed those of Czar Nicholas II and his family, and also showed that the woman who claimed to be Anastasia and to have survived the Bolshevik massacre was a fraud.[40] In one spectacular case, the Vietnam War remains in the Tomb of the Unknown Soldier were exhumed, tested and found to be those of an Air Force lieutenant shot down in 1972.[41] The accuracy of the tests now is such that it will probably be impossible to find the body of another unknown Vietnam serviceman missing in action.[42]

But DNA identification of remains really came into the limelight after the September 11, 2001, terrorist attack on the World Trade Center. Between the incineration of fire and the pancaking of over a hundred floors, only a few of almost three thousand bodies were left intact. Toothbrushes, combs, locks of hair and other personal effects of the missing were sent to the New York State Police Laboratory in Albany. Swabs of cells taken from the cheeks of relatives were also sent.[43] DNA testing will help bring closure to many survivors.

TEN

"Baby by Versace"?

No issue in biotech is more emotional than that of human cloning. That's too bad, because what's needed is less emotion and more facts. From what I've seen, very few of those who write and speak on the subject actually understand what cloning is, much less what it has the potential to do—or, perhaps more importantly, lacks the potential to do. "Human cloning is an enormously troubling development in biotechnology," American delegate Carolyn L. Willson declared at a meeting of the U.N. Committee on an International Convention Against the Reproductive Cloning of Human Beings. She specifically cited humans "being born for spare body parts, and children [being] engineered to fit eugenic specifications."[1] The implication was that a brave new world had arrived in a rush, yet in fact, attempts at eugenic engineering are over a century old, long predating the idea of cloning. And the concept of breeding babies for body parts, as we shall see, is just plain fanciful. None of which is to say that Ms. Willson may not be correct in wanting to discourage human reproductive cloning. She just needs to choose the right reasons.

Our views on cloning are probably far more influenced by Hollywood than by any other source, whether it's older films like *The Boys from Brazil*, with little cloned Adolf Hitlers running all around, or *Star Wars II: Attack of the Clones.*[2] True, in the *Star Wars* film the clones fight on the side of The Force, but it's clear—and I don't think I'm giving anything away here—that they will switch to The Dark Side. The media frenzy over Clonaid also didn't help. This was the bizarre group that claimed—with absolutely no evidence—to have repeatedly cloned human children. They were media darlings for weeks, with the press hanging onto their every word.[3] *Now* you may be forgiven for forgetting who Clonaid is.

That said, it's just a matter of time until somebody does clone a whole human being. In late November 2001, Advanced Cell Technology (ACT) of Worcester, Massachusetts, made headlines around the world when it claimed to have cloned the first human embryo, a claim that turned out to be a bit premature.[4] ACT used two different methods. One is nuclear transfer, the same technique used to produce Dolly the sheep: DNA from the nucleus of an egg is replaced with that from an adult cell and then fused. ACT used nineteen eggs from seven human volunteers, but none grew to more than six cells before dying. The second technique is parthenogenesis, which involves duplicating one set of maternal chromosomes. You get something like an embryo, but it's incapable of leading to a birth since it lacks male chromosomes. ACT used twenty-two eggs for this experiment, but again there was only a bit of cell division before death. Neither method created what could clearly be called an embryo.[5]

The stated purpose of this experiment was "therapeutic cloning," also known as "research cloning." Advocates of embryonic stem cell (ESC) research insist that therapeutic cloning is utterly distinct from "reproductive cloning." It's certainly true that the *purpose* of therapeutic or research cloning is to create a large and unvaried supply of ESCs for tissue regeneration, while the *purpose* of reproductive cloning is to make cloned offspring. But the actual methodology is identical up to a certain point. An embryo created for the one is identical to an embryo created for the other. The only difference is that reproductive cloning goes one step further, to implanting of the embryo into a womb. If it survives, in nine months you have a little clone baby. Therefore you cannot "Ban Reproductive but Not Therapeutic Cloning," as one op-ed title put it; since the two are inextricably entwined. Obviously, the use of nonembyronic (adult) stem cells avoids this problem.[6]

On the other hand, while I see the use of exclusively adult stem cells as a way of avoiding a terrible ethical dilemma over the possibility of eventually producing whole human clones, I'm not as concerned about such clones as most of us seem to be. Why not? Perhaps you should first know that reproductive cloning comes in both artificial and natural flavors. A natural clone is called an identical twin, which is formed when one fertilized egg splits. These twins have identical DNA. (Fraternal twins form when two different eggs are fertilized. Genetically, fraternal twins are no closer than normal siblings, sharing only about 50 percent of their genes.) Although identical twins have the same genotype, or DNA, they have different phenotypes. This means their DNA is expressed in different ways because of factors that have influenced their development other than genetic ones. Thus, schizophrenia is clearly influenced by genes, but if one identical twin has it, the chance that the other also does is only 45 to 50 percent.[7] Even height can vary by as much as four inches between identical twins.[8]

So if identical twins aren't identical, then neither are clones created by mad doctors. The closest match possible would be for a woman to have her DNA inserted into one of her own eggs. That's because a bit of DNA lies outside the nucleus in the mitochondria of the egg, and that becomes incorporated into the clone's genome. Men can't pull off this trick. The closest they can come to being cloned is to use an egg from a female immediate relative.[9]

You can't even re-create your kitty, much less a child.[10] If you look at images of the first cloned cat (CC) and the cat that provided the donor cell (Rainbow), you'll see that while CC is a "Cloned Cat," she's not a "Carbon Copy"; the two animals could hardly have more different markings.[11] How can this be? It's because the womb also has a strong influence on the development of the offspring. Even the same womb at different times has different nutrients, hormones and other factors affecting it. Hence the offspring are different. How much further apart would be the offspring of two different wombs! It appears that a fetus can even be permanently affected by a mother's diet in the short interval after conception and before implantation in the womb.[12]

Further, a clone's personality, intelligence and even looks will be different as a result of being raised in different circumstances. There will never be human cloning as envisioned by Hollywood and most people. You can make people who have similar or, in rare circumstances, identical DNA, but that's it. The very word "clone" may be a misnomer, though it will continue to be used here because everyone else uses it.

That said, when this first human clone is made, as it inevitably will be, hopefully it will be later rather than sooner. That's because the science of cloning mammals is still at a very early stage and failure is far more common than success.[13] Moreover, the current processes of cloning inherently produce large numbers of mutations such that all clones are genetically abnormal.[14] That might be fine for a sheep or a mouse, but for humans it would be horrific. Banning not only the use of taxpayer funds for cloning but also the procedure in general won't prevent it from eventually happening. But if such a ban delays it, so much the better. Fortunately, it also appears that cloning humans will be far more difficult than was previously thought. In fact, there have been over seven hundred failures to clone monkeys.[15]

Forgetting for the moment that cloning is imperfect, we need to ask why somebody would *want to* clone a human. What about making a clone of yourself so that someday when you need it you can rip out its heart or liver to replace your own? Assuming you could find somebody who would do both the cloning and then, many years later, the organ transfer, you'd be as dumb as you are evil. Why go through all this trouble when you could just have a new organ made—as we shall see later in this book. Further, what court is going to rule that killing an identical twin is murder but killing a cloned human is not? If there's any doubt of this, let's pass legislation to make sure.[16]

The most practical application of cloning would be for infertile couples. But even here it's no substitute for normal reproduction, since you're not getting that genetic union of the parents that makes having a child so special.

What about screening embryos to create "designer babies"? It has long been easy to do for sex selection (and for just as long has been considered by most in the developed world to be abhorrent), and it has also now become a simple task to eliminate embryos with a variety of genes that have the potential to cause children to develop certain diseases. Further, something far more drastic than screening of embryos—namely, selective infanticide—is hardly new. It goes at least as far back as ancient Sparta (for hardiness) and continues in modern China (for gender).[17] One could well argue that just as infanticide cheapens the uniqueness or sanctity of human life, so does embryo selection. All pogroms and all massacres have as their theme a central idea: It's okay to kill innocent human beings *en masse* if it's convenient. Some people think embryos are human beings; others consider this idea bizarre. But inarguably they are the offspring of human beings. Allowing them to be eliminated *en masse* because it serves a purpose is not so much a matter of what it does to them, but of what it says about—and does to—us.

Moreover, the very technology that allows us to screen embryos more efficiently for defects is allowing us to repair those defects. Consider the early 2002 announcement that a woman had twenty-three of her eggs removed and screened for a gene variant that causes a rare form of Alzheimer's, such that the child had a 50/50 chance of developing the disease by age forty.[18] But with all the advances against Alzheimer's in just the past few years, it's virtually certain that the disease will be completely preventable long before the child reaches that age.

Redesigning Humans?

What about gene alteration to "improve" the human race, not through negative techniques such as murder or sterilization but through "germ-line" therapy? Currently, gene therapy is directed entirely at those cells in the body, called somatic cells, that do not pass genes on to the next generation. Inevitably, though, people are going to start manipulating germ cells, which produce sperm and ova, passing on information to generations thereafter. Such therapy could remove, replace or change genes that cause inherited disease. But it could also change other genetic attributes, such as intelligence or that distinctive beak-shaped nose. A special report issued in 2000 by the American Association for the Advancement of Science concluded that germ-line therapy is neither safe nor responsible at this time. The report recommends that if a societal decision is made to conduct such research, "a comprehensive oversight mechanism should be put in place to regulate [it] in both the public and private sectors."[19]

While the state of the technology is such that we have a long time to think about this, it's already the subject of hot debate.[20] Johns Hopkins University political science professor Francis Fukuyama devoted many pages to it in *Our Posthuman Future: Consequences of the Biotechnology Revolution,* a book aptly summarized in a *Nature Biotechnology* review as seeking "to arouse fear: fear of science, fear of individual choice, fear of the future."[21] Fukuyama considers germ-line therapy a forbidding specter, and he's hardly alone.[22] Yet most of what he fears has already come to pass through other technologies. Improving your child's looks is as easy as something called "cosmetic surgery." Improving your child's IQ is also quite easy; flip off the TV set and put away the video games. The argument that the rich will have better access to germ-line therapy also falls flat; the rich have better access to everything. Plastic surgery isn't covered by insurance in the United States, and wealth can be used to improve a child's intelligence from preschool all the way through graduate school.

The easiest, cheapest and most efficient way to create superbeings will never be with biotech but rather with bionics—that is, the use of implanted computer chips and electronic or electromechanical devices. It will be a while before we're at the "Six-Million-Dollar Man" or "Bionic Woman" stage, but already monkeys have been implanted with electrodes that enabled them to move a robot arm with brain signals.[23] The same signals that allowed that limb to pick up food morsels could just as easily move a wrecking crane that would put Steve Austin's bionic arm to shame.

Bionics got its start in 1958, fifteen years before gene splicing did, when the first electric pacemakers were implanted. Cochlear implants are starting to become common. These are not hearing *aids,* but rather devices that enable *totally* deaf people to hear—and hear well—through bypassing the ear canal and directly stimulating the auditory nerve.[24] Conservative talk-show host Rush Limbaugh received one.[25] The first deaf Miss America is no longer deaf because of her implant.[26] Similarly, retinal implantable chips are already giving vision (albeit very poor vision so far) to people who were completely blind.[27] Then there are the forthcoming implantable computer brain chips. Some have already been designed with up to ten thousand neurons and have been tested in simulation.[28] They will soon not only help treat brain diseases but actually increase people's range of senses, enhance memory, allow invisible communication with others, and enable consistent and constant access to information where and when it is needed. They will provide 24/7 wireless broadband internet connection, with no annoying pop-up ads.[29]

Attack of the Clones?

How about another science fiction scenario: cloned armies? The time certainly will come when we can genetically alter an embryo to create a bigger, stronger, smarter soldier. But they'll never be able to compete with robotics. Robotic armed aircraft made their first appearance over Afghanistan.[30] It's estimated that by 2015, one-third of the U.S. aircraft arsenal could be robotic, with ground operators perhaps flying several aircraft at once.[31] Meanwhile, the U.S. Army and Omnitech Robotics is developing the Standardized Robotic System (SRS), a kit that will allow remote control of any large vehicle the Army employs, from tanks to bulldozers.[32]

For all the talk about transgenics being "unnatural," biotech—unlike robotics or bionics or computers—is ultimately limited by its very naturalness. You can move genes around, switch them on or off, and correct defects. There are lots of marvelous things you can do with genes. And there are some bad things you could do, if you really wanted to. But when it comes to making truly super people or super warriors, then for good or evil no gene exists or can be altered to do many of the things bionics and robotics already can. Consider this: Someday we might be able to engineer a human who can run a mile in 3.5 minutes, but he still won't be nearly as fast as an ugly orange 1973 AMC Gremlin.

In his 2001 book *The Virtue of Prosperity: Finding Values in an Age of Techno-Affluence,* Dinesh D'Souza writes, "Under certain conditions I will make a case for cloning and genetic engineering." But, "At the same time, there is a very great evil in these technologies," he adds.[33] Wrong. Technologies aren't evil; only people are. "Technology only gives us tools," observes physicist and Templeton Prize recipient Freeman J. Dyson in his book *The Sun, the Genome, and the Internet.* "Human desires and institutions decide how we use them."[34] Was Lizzy Borden's axe "good" when it was cutting firewood, but "evil" when used for the legendary "40 whacks" plus 41 more? Asserting the morality of the *thing* is not only unhelpful, it confuses the issue. The issue is people and their behavior.

Yet no matter how much critics of biotech try to portray the technology as "bad," its boosters must not fall into the trap of treating it as inherently good. If the medical industry and the biotech industry don't shun and openly condemn those who treat humans like a commodity, one of two things is inevitable: Either the occasional Frankenstein among us will be allowed to roam free and we may come to accept those things that presently horrify us, or those who oppose biotechnology for the sake of opposing it are going to succeed in using isolated incidents to straitjacket the entire industry. Either way, the situation cries out for the industry to police itself before somebody else grabs the power to do it.

ELEVEN

Will Biotech Break the Bank?

While we're beholding the wonders that biotech is bringing to improve health and longevity, it seems fair to temper our enthusiasm by asking how much all this is going to cost us. Unfortunately, there's no easy answer. Indeed, the answer depends on how you phrase the question. For example, as we've seen, biotech can make current drugs a lot cheaper by using bacteria, fungi and cell lines as factories. That will be all the more so when therapeutic proteins can be extracted from plants, milk and chicken eggs. So if you asked, "What if we swapped all the non-biotech drugs of today with the biotech drugs that will be strictly replacements?" the answer would be easy. This would save us money. But that's not the way medicine works. Car companies have the incentive to make vehicles cheaper because they deal directly with consumers. With prescription pharmaceuticals, however, third parties such as insurers and the government (for plans such as Medicare and Medicaid) distort the market so that price matters little, while quality is everything.

Thus, some researchers conducted a trial comparing the monoclonal antibody ReoPro to an older, nonrecombinant medicine named Aggrastat (tirofiban) in the expressed hope that Aggrastat would be at least as effective in reducing heart attacks, other complications or death following procedures to inflate and stent closed arteries. To their consternation, they found that ReoPro worked better for each of these outcomes. Why didn't they like this? Because therapy using ReoPro averages about $1,350 for a course of treatment following such a procedure, while Aggrastat therapy costs only $350.[1] Biotech was considerably more expensive, but it was better.

Or here's a confusing conundrum: What if we find a way to diagnose and cure someone in their thirties well before they have any symptoms of Alzheimer's? What if there is a biotech treatment for this that's a one-time therapy costing just $10,000, but there's a limited window of opportunity in which to apply it? Insurers would certainly balk at paying for this, because even though care for someone with advanced Alzheimer's can cost more than that *per year*, chances are the costs would be borne by some future insurer and the individual is not even sick at this point. These things are not going to be easy to sort out.

One thing that probably can be said with certainty is that drug prices in general will continue to outpace the cost of living. There are many reasons for why they have been doing this and will continue to do so; indeed, many books have been written on the subject—often with different conclusions.[2] But one simple reason is that the drug we're buying this year for a given problem is not the same as the one we bought ten years ago. We've seen the same trends with both automobiles and houses. Both have outstripped the Consumer Price Index (CPI), but there's absolutely no comparison in the packages you're getting. Cars are far safer, more fuel-efficient and less polluting. Houses tend to be much larger, with more garage, and have superior heating and air conditioning.[3]

Generics become available for drugs when the patent expires and this knocks down the cost of *that* drug. By the time this happens, however, there may be a newer, superior drug that's still under patent. In some countries, such as the United States, this will mean that most patients will purchase (via insurers) perhaps a cancer drug that gives a 5 percent better chance of survival but costs twice as much. This may not seem socially utilitarian, but if it comes down to your spouse or your child, that extra 5 percent chance might seem significant indeed.

Then there are drugs that can't be compared to earlier ones because they go after diseases that were previously untreatable. Many of the new biotech drugs fall into this category. What price can you put on a cure that was hitherto not possible?

As drugs become available for conditions that were previously untreatable, people will start using more of them. This has been a trend for years and biotech will only accelerate it. For example, there are several forms of cancer that are essentially death sentences right now. You get chemotherapy for a few months and then you die. In the future, you might receive one biotech drug that blocks the receptor of a cancer-causing protein in your system, along with another drug that prevents tumors from sending out blood vessels. Both would probably have to be taken indefinitely, but they would keep you alive and in good health. So instead of paying for a few months of chemotherapy and a headstone, you could now be paying for decades of drug therapy. Nobody better knows how wonderful—and how expensive—such new horizons are than people with AIDS. Within fifteen years the disease has gone from a certain death sentence to one that with proper treatment (meaning expensive drugs) gives people hope of going back to their old jobs and routines and living many years—quite possibly to old age, especially considering that at the very least they're living long enough for better AIDS drugs to be developed.

If we look strictly at the basket of pharmaceuticals we buy each year, then it's precisely because biotech has so much promise that it's going to raise a lot of prices. Pharmacogenomics will save money by allowing drug makers to better structure their tests and not waste time and money chasing false leads. This is known as "failing fast." But in the short term, it cannot outweigh the cost of going after thousands of new drug targets instead of the traditional five hundred or so.

Looking at drug prices in isolation, however, doesn't do the issue justice because, as pharmaceutical manufacturers are wont to point out, drugs can obviate surgery, therapy and special living requirements, and can save employers money by allowing workers to spend more time on the job and less in a sickbed. A study sponsored by the National Institutes of Health found that treating stroke patients promptly with a clot-busting drug netted an average savings of $4,400 per patient by reducing the need for hospitalization, rehabilitation and nursing home care.[4] Another study showed that the drug saved employers $435 a month per employee from a reduction in lost productivity, while the monthly cost of the drug per employee was only $44.[5]

There are also external prices that need to be considered, such as the strain on the environment. Pharmaceuticals can use an incredible amount of resources in terms of water, chemicals and energy. In its 1998 report *Biotechnology for Clean Industrial Products and Processes: Towards Industrial Sustainability*, the Organization for Economic Cooperation and Development compared production of an enzyme that helps the body break down glucose from a recombinant organism with production from the organism in which it was originally discovered.[6] The producer of the native enzyme was a bacterium called *Leuconostoc.* The gene that encoded that protein was then transferred into another bacterium, *E. coli.* The result: Producing the same amount of enzyme from the old bacterium as from the recombinant one (the *E. coli*) took 150 times as much processing water, 500 times as much cooling water, over 50 times as much electricity, 65 times as much ammonium sulphate, 6,000 times as much wastewater, and produced an overall calculated pollution load (factoring in how much one person would use of this enzyme in one day) that was 100,000 times greater.[7] A scientist will tell you that enzymes don't have any color; don't believe it. This one is clearly green.

There are also many up-and-coming biotech therapies that will save on the need for drugs. For example, people with hemophilia B now undergo weekly intravenous injections of clotting factor IX at a cost of $50,000 to $100,000 annually. If their bodies can produce factor IX on their own, they are spared the pain and inconvenience of the injections and they save whoever was paying for them a bundle.[8] As we've observed, many researchers are working on gene therapy treatments to eliminate the need for injections of clotting factor for both type A and type B hemophiliacs.

One avenue that may provide the best of all worlds is cell-based therapies delivered via biologics. Under the current paradigm, a cell is used to produce a therapeutic protein, purify it and then inject it into the patient. Putting the cells directly into the patients, as we saw with the study in which pancreatic cells were planted into livers to produce insulin in diabetics, has many advantages. Implanted cells can deliver a steady stream of the protein, rather than the inevitable spikes from needle injections or an IV. It might be possible to get more of the protein into the proper site. Cell implants would be needed perhaps just once a year rather than every day. And implants would mean great cost-savings as well. One company, BioHybrid Technologies of Shrewsbury, Massachusetts, is developing gelatinous capsules to protect pancreatic islet cells from immune system rejection so they can survive in the body from six months to two years.[9] Other companies are working on long-term, albeit nonpermanent, biologics to treat or prevent disorders of the central nervous system, various forms of hepatitis, cancer and osteoarthritis.[10] All of these hold the promise of not only better but cheaper treatments. Cell-based therapies are a way of offering medical miracles without pushing us to the day when we routinely spend more on our health than we do on our mortgages. It's win-win technologies like these that the biotech and drug companies must pursue.

As we've seen, recombinant proteins can offer a huge cost advantage over some of the natural proteins they replace when the original source was, say, a deadly snake whose venom was extracted at high cost and high risk. Nevertheless, current methods of producing transgenic proteins still tend to be quite expensive. Factories producing pharmaceutical proteins in genetically modified mammalian cells can cost $100 million or more and may produce a few hundred kilograms a year at most. At this stage, it still makes sense for drug developers and their pharmaceutical company partners to try to produce some recombinants any way they can. Lilly's antisepsis drug Xigris is made in human cell lines because that's the only way the company could find to make it. The cost can be justified because Xigris is only used when the patient is at risk of death or permanent injury. But less important drugs may not be worth such expense, nor may be those that require large amounts of protein to be effective. These would include such things as creams that are absorbed through the skin; too much protein is wasted this way.[11] "They cannot make these drugs using the old technologies," said William White of Integrated Protein Technologies, a unit of Monsanto seeking to grow drugs in corn. "It's just not going to be cost effective to do so."[12]

In any event, it's always better to be able to make a product more cheaply. That's why the biotech industry has set the goal of making production more cost-efficient as well as medicinally superior. "There's a lot of excitement about using antibodies as drugs but they can be expensive to produce," said Epicyte's Kris Briggs. "We'd like our products to be inexpensive enough to be used in the developing world as well as here." Another advantage of cheapness, she added, is that "once you can scale up to producing large quantities, you can use proteins for things that hadn't been thought of such as preventatives."[13]

Bacteria, yeast and mammalian cell lines were of necessity the first media for producing recombinant drugs. But while industry is constantly working on ways of producing these less expensively, they are inherently costly.[14] It must be, and indeed is, a goal of pharmaceutical companies not only to produce medicinally superior proteins but also to try to produce them in the cheapest way possible. Industry must continue to move toward far cheaper ways of producing those proteins, such as in milk, eggs or plants, as well as media nobody has even thought of yet.

New Diagnostics

One way in which biotech can make therapy cost less is by developing new types of diagnostics. We've already seen how biotech is responsible for hundreds of gene tests, as well as numerous other medical diagnostic tests, ranging from screening donated blood, to checking various bodily fluids for pathogens or cancer, to providing home pregnancy tests. More specific tests save time, money and lives.

The AIDS epidemic and the collateral tragedy of hemophiliacs and blood transfusion recipients acquiring the virus made us realize how vulnerable our blood supply can be.[15] Since then, both screening of applicants and the use of a variety of tests has made the blood supply of developed countries safer than ever. Unfortunately, this isn't true in other parts of the world where the need for blood is greatest. But Western technology will soon change that. One test developed by the Norwegian company Bionor and approved by the World Health Organization and others will detect the AIDS virus in whole blood, serum, plasma and saliva.[16] At a cost of just $3, it tests for both HIV-I and for HIV-II, a strain rarely found in the West but common in parts of Africa. Results couldn't be simpler to determine: red is positive; yellow is negative. The unit can be powered by either a European or an American car battery, explains Andrew Garnett, international sales and marketing manager for Bionor. The company says the unit itself is as thorough as it is easy to use, that it's unfazed by environmental factors such as water quality and washing devices.[17] "Since it can be performed at ambient temperature, the test is ideal for use in Africa, Southeast Asia, and Latin America," Garnett said.[18]

In July of 2000, the FDA approved the "FastPack," an amazing little unit made by a small biotech company near San Diego named Qualigen.[19] With the FastPack system, physicians can obtain laboratory quality test results in less than fifteen minutes, eliminating the need to send a patient's blood sample to a commercial or hospital lab and wait several days for the results. The first model will perform a total prostate-specific antigen (Total PSA) for diagnosis of prostate cancer.

Additionally, two other prostate-cancer-related diagnostic tests—free prostate-specific antigen (Free PSA) and testosterone—were scheduled for introduction.[20]

The system, resembling something George Jetson's doctor might use, comprises a diagnostic analyzer, single-use disposable test packs with all the required chemicals for each patient's test, and an inkjet that prints directly onto labels that the doctor attaches to a patient's chart. It features "one-button operation," making it suitable for use by non-technical personnel. The operator merely inserts the patient's blood sample into the disposable test pack and loads the pack into the analyzer, then the system automatically completes the test.[21]

But the advantages don't stop there. Qualigen chief executive Michael Poirier said the company had tested the system with five thousand patients by comparing the results on the FastPack with results from conventional laboratory analysis. Qualigen's system, he said, did the work in one-third the time and was more sensitive in detecting small amounts of PSA. That could be important, because patients being treated for prostate cancer must have their PSA checked frequently. A test that spotted an increase in PSA quickly could indicate a relapse, allowing the doctor to take action sooner. Poirier said FastPack is also cheaper than competing systems, about $11,000 compared with $50,000 or more for other systems. That would make it far easier for an individual physician to buy a diagnostic system, and also reduce costs for patients and their insurance companies.[22] FastPack exams cost about $25, compared with $30 to $70 for the standard PSA test. "The concept of being able to get a PSA in 15 minutes is incredibly attractive," said Ralph deVere White, urology department chairman at the University of California at Davis.[23]

But there's more good news about this little device: there's nothing intrinsic about it that makes it useful only for prostate exams. That's just where the company chose to start, because the PSA test is common and becoming more so as men heed warnings to seek early detection. Soon, said Poirier, FastPacks will be able to provide "up to 25 additional blood tests to diagnose multiple other medical conditions such as other types of cancer, thyroid disorders, diabetes, infertility, infectious diseases, pregnancy, myocardial infarction [heart attack] and therapeutic drug monitoring."[24] All of this is coming from a company with a total of two dozen employees.

The bottom line? Biotech may be saving us money and it might not; I'm sure there are computer models that "clearly illustrate" both positions. But there is no argument over the fact that biotech will allow us to have much healthier, happier and longer lives whatever the case. Health is not a commodity like refrigerators. You can have too many refrigerators; you cannot have too much good health. We will always want a better pain reliever, a better chance at surviving cancer, a better way of preventing or treating heart disease. So for the foreseeable future, I believe that drugs in general as well as biotech drugs in particular will continue to cost us a larger portion of our salaries and their prices will rise faster than the CPI. Imagined "miracles" are free or can be bought off drugstore shelves for a few dollars, but they have the drawback of not working. The real drug and therapy miracles cost real money.

PART TWO

The Fountain of Youth

ONE

Turning Back the Clock

Since the dawn of human consciousness, nothing has been more exciting than the possibility of living longer, looking younger, and slowing, stopping or even reversing the natural decay of the body. Today, aside from obesity treatment, there's no bigger health racket—with vast sums of money thrown at books, magazines, potions and pills "guaranteed" to turn back the clock.[1] Presently no book and no amount of money can "stop aging now," much less "turn back the aging process" by so much as a nanosecond. But consider: Until recently, medicinal remedies for baldness were also nothing but charlatanism; yet there are now two drugs approved to do for baldness what so many would like to do for aging—at the very least slow it down, better yet stop it, and best of all even reverse it. We are on the cusp of an era in which revitalization will pass from the sham artist to the doctor.

As a whole we are already living longer than ever. Most of the increase in life expectancy results from fewer of us dying in infancy or youth. After all, a child who would have died at age 2 in a prior era but now lives to be 70 is going to affect the statistics more than dozens of people who would have died at 75 but live to be 76. Old people *are* living to be somewhat older due to a host of factors, especially better health care for seniors and better nutrition.[2] Still, none of this affects what's called "the natural human lifespan." That is, more and more of us are getting closer to the maximum and more importantly we're getting there in better shape. Yet we're repeatedly told that the natural lifespan or outside limit of longevity for humans is fixed at about 120 years.

Or is it?

You might be surprised to find there's no substantiation for this. "Those numbers are out of thin air," declared University of California at Berkeley demographer John Wilmoth, who published a paper documenting increasing lifespans from the 1600s to the present.[3] "There is no scientific basis on which to estimate a fixed upper limit," he said. "Whether 115 or 120 years, it is a legend created by scientists who are quoting each other."[4] So it's possible that simply by curing diseases and replacing worn-out organs, we could eventually get large numbers of people to live beyond 120. In a real sense, this is increasing longevity. But even if there *were* some kind of natural limit, it would be just that. Biotechnology is all about using nature's tools to break natural barriers, whether it's inserting an insecticidal gene from a bacterium into corn or healing severed spinal cords. Through engineering at the cellular level, natural maximum lifespan or no, biotech will inevitably convert what's been an incremental process of extending longevity into a series of quantum leaps.

Not all organisms—even those high up nature's ladder—are predestined to decrepitude and death within 120 years. I remember reading about a 200-year-old lobster pulled from a trap. This creature that had lived when George Washington's troops were freezing their tushes off at Valley Forge now suffered the ignominy of being immediately eaten. Even among mammals, recent evidence indicates incredible longevity. Scientists have dated some bowhead whales harpooned by Alaskan natives, finding the youngest to have been 135 at death and the oldest 211.[5] Keeping in mind that they all came to violent ends, there's no telling how long they would have made it had they escaped the barbed lance of a strong Inuit.

No one knows for sure why our bodies run down and eventually stop like a spring-wound watch. One theory that was fashionable for a while was that our bodies have a built-in self-destruct function. The evolutionary purpose would be to ensure that a species gets a regular infusion of "fresh blood." The problem with this view is that our ancestors (as well as present-day wild animals that age) usually died violently or of starvation long before they reached old age. There was no reason for nature to hard-wire us to croak; it would happen soon enough one way or another.[6] On the other hand, it would make sense for nature to wire us to live long enough to reproduce and raise our children and then not kill us off but simply let us fall apart through weaker immune systems, weaker bones, weaker muscles, and so forth. One recent experiment on mice showed that it's possible that the aging process is sort of a "side effect" of one of the body's mechanisms to kill cancer, that frailty is essentially a trade-off for a much lower cancer rate than we would otherwise have in our youth.[7] Some might think this would put the kibosh on developing anti-aging therapies, if retarding aging is essentially tantamount to encouraging malignancies.[8] But as one prominent gerontologist, Tom Kirkwood of the University of Newcastle upon Tyne in Britain, said, it's really just a matter of affecting certain mechanisms without affecting others. "There's no reason why you shouldn't get greater defence against cancer and greater longevity," he said.[9]

In any event, theorizing over why we age is not the equivalent of counting angels on the head of a pin. If it is true that "death happens," as opposed to a built-in self-destruct mechanism, it will be a lot easier to delay that happening. But whatever the reason for aging, biotech will inevitably affect it. And one of the more heartening aspects of biotech longevity research is how many vastly different avenues are already showing promise.

Any number of medical therapies can be termed "longevity-increasing," depending on the definition of this phrase. Vaccinations have saved the lives of countless children and others, clearly affecting mortality rates. Indeed, nonmedical therapies such as better nutrition tremendously lengthen lives. Proper exercise has long been known to rejuvenate muscle and bone, and evidence now indicates that it does likewise with portions of the brain.[10] So we need distinctions between different types of longevity theories. I divide them into macro and micro. Macro theories affect all or pretty much all of the body's aging mechanism. This is what most of us mean by "stopping" or "turning back" the clock. There are some biotech therapies that, if they pan out, will in fact reverse aging in most of our cells. Others will simply slow down the aging process. Then there are the therapies that work on only one system or part of the body, such as the brain, skeleton, muscles or other individual organs. I'll begin with the macros.

The Telomere Trail

It seems natural to believe that as progress is made against aging, it will go like this: First we'll slow it gradually, then radically, then stop it, and then reverse it. Such a pattern would indicate that age reversal for the body is a long way away. But there's no evidence to suggest that such a gradual progression is required. As one recent paper put it, "reversing mammalian aging is not necessarily any harder than dramatically postponing it."[11]

One approach to reversing aging that's already been demonstrated in lab animals involves telomeres, those protein caps on the ends of our chromosomes. As cells repeatedly divide, the telomere gets shorter until the cell cannot divide anymore.[12] Thus, it may not be so much that our clocks run down, as one popular metaphor puts it, but that our wicks burn down.[13] This was merely a theory until early 2003 when geneticist Richard Cawthon and colleagues at the University of Utah reported they had measured the telomeres in a randomly chosen group of 150 patients age sixty or over.[14] Those with shorter telomeres were eight times as likely to die from an infectious disease and three times as likely to suffer a fatal heart attack.[15] There's no longer any doubt that telomeres play an important role in aging. Yet it remains to be asked: If we re-extended those telomeres by adding the enzyme telomerase discussed earlier, would we be re-extending the wicks of our lives?

We know there are various means of inserting telomerase into the body and we know that telomerase will extend the telomeres. But does this extenstion extend cell life? In other words, just because we know shortened telomeres are linked to cell death, does it necessarily follow that extending them will retard or prevent cell death? In 1998, scientists from Geron Corporation and the University of Texas Southwestern Medical Center in Dallas became the first to show that, at least in a test tube, telomerase could be added to human cells to keep them dividing long after they should have stopped.[16] Cells that would normally divide only 40 to 90 times had done so an additional 40 times or more, and many were still going.[17] "We found that biological aging can be put on hold," said one of the Texas researchers, Jerry Shay.[18]

Then in 2000, Geron published results for their scientists and those at Stanford University demonstrating that telomerase restored the youth of aging human skin tissue that had been attached to the backs of mice. This wasn't as relevant as trying it directly on a human (nor would it be ethical at this early stage), but it was far more telling than an experiment in a Petri dish. Skin comprises two principal cell types: keratinocytes, which form the upper epidermal layer, and fibroblasts, which form the underlying dermal structures. These layers are connected by a tight membrane. The researchers discovered that fibroblasts aged in the laboratory lost the ability to form a strong connection with young human keratinocytes when the cells were put onto the mice's backs. This same condition is seen in the elderly and shows itself by greater frailty of the skin and blistering under the outside layer. But the Geron scientists found that when they added telomerase to the human skin affixed to the mice, it dramatically increased the ability of the cells to divide. It essentially turned old skin young.[19] "This is the first demonstration of a beneficial effect of telomerase activation in human cells in an *in vivo* [in the body] animal model," said Calvin Harley, Geron's chief scientific officer. He added that it brings his company a significant "step closer to a telomerase gene therapy for the treatment of chronic degenerative diseases in the elderly."[20]

According to author and physician Michael Fossel, there are three main schemes for making cells extend their telomeres. One, according to the Michigan State University biologist, would involve inserting a gene that would deliver telomerase throughout the body to any cell that could make use of it. A second approach would be to induce genes to produce their own telomerase. That would probably be the most efficient, but seems to be the farthest away scientifically. That said, researchers for the last several years have already been screening compounds by the bucketful to find just such an "on-switch" compound.[21] Fossel says treatments would probably not be exorbitantly expensive, though they would eventually need to be repeated.

The third method would be via direct injection into the bloodstream of something comparable to telomerase. "Comparable," that is, because the actual stuff would break down far too quickly to be of much help. It would have to be a "supertelomerase" of sorts. But in late 2002, Stanford University scientists announced they had discovered just that, a telomerase mimic. In a published report, they described how they had used their system in test tubes to elongate chromosomes from human kidney cells. They first bound a tiny circle of DNA containing the code for the rebuilding telomeres onto the ends of chromosomes. Then they exposed the chromosomes to a common DNA-duplicating enzyme known as DNA polymerase. For reasons the researchers can't yet explain, this method built telomeres at least ten times as fast as telomerase does. But they do know that they are already routinely elongating telomeres to their full original length. At this writing, they are testing their technique on human bone marrow in test tubes to see if it works in cells.[22]

Nonetheless, telomere extension does raise problems. Probably the biggest worry is cancer, because telomere shortening may be a way the body keeps a cell from running amok.[23] We know that one way cancers become "immortal" is by continually re-extending their telomeres, so that no matter how much the cells divide, the telemores stay long. In this cancer-promotion theory, adding telomerase is likened to pressing a gas pedal, and cutting off the telomerase to hitting the brakes.

But two researchers writing in the journal *Science* explained why this view may be wrong. "Instead of a gas pedal or brake, telomerase and more specifically telomeres may be best viewed as the gasoline tank," they said. "Gasoline is not sufficient to drive or accelerate the car, nor does it affect the brakes, but when the gas is used up the car stops regardless of the status of its brakes or how hard one steps on the gas pedal."[24] So yes, gasoline is required to get a gasoline-powered car to move, but it takes a lot more than that to get a car moving and keep it moving. Cancerous cells usually require telomerase to keep multiplying, but they need a lot more than that to keep a tumor expanding. Further, as we have already seen with Eli Gilboa's study, cancerous cells often find a way of manufacturing their own telomerase in order to keep re-extending their telomeres. If they're going to make as much as they need, there would be no harm in giving them even more.

If telomerase doesn't prove to be a sort of universal age reverser for the entire body, it's nonetheless being used in experiments to treat many diseases associated with aging. That's because you can limit where the telomerase goes. Among these individual areas for which telomerase is being tested is macular degeneration, cirrhosis of the liver and clogged arteries.[25] Researchers also have their eye on AIDS: they're considering telomerase delivery to bone marrow cells as a way to prolong the replacement of immune cells destroyed by HIV.[26]

Geron plays down the lifespan-extension implications of its work. Well that it should. There's been so much hype on this subject for, oh, the past few thousand years that smart people often flip off their hearing aids when you start talking about rejuvenation. It's also wrong to raise false hopes. If you think Geron is going to make it possible for your ninety-year-old father to play tennis with your grandson, you're mistaken.

But the company's CEO, Dr. Thomas Okarma, was being rather cagey when he claimed, "Our objective is to increase health span, not life span," and, "Our hope would be that our children live a great fraction of their life in wellness."[27] The company's founder, Dr. Michael West, now CEO of Advanced Cell Technology, has always seen longer lives as a goal.[28] "When I hear critics saying they don't want to see life span extended, they are thinking about [living longer] in a decrepit state," he said. "That's not what we are talking about doing. As long as people are wanted and happy, I think it's a very noble goal and we should strive for that, and regenerative medicine is one of the means to achieve it."[29]

Methuselah Mice

Even if the telomere trail came to a complete dead end, all would not be lost in the fight against aging. Many more paths are being discovered and explored.

For example, in late 1999 Italian researchers announced they had created what were dubbed "Methuselah mice," after the man in Genesis said to have lived 969 years.[30] These animals lived one-third longer than normal, apparently for no other reason than that one of their genes had been "knocked out." It's too early to say for sure, but it appears that this particular gene tells the body's cells to die. It also appears to play a role in cancer, so eliminating it may independently reduce the risk of contracting this disease.[31] (The mice suffered no side effects other than finding that their life insurance premiums had dropped.) One possible way of controlling the gene would be antisense technology, shutting off the production of the protein that the gene codes for. Another would be to "manipulate the function of this protein by drugs," said Professor Pier Giuseppe Pelicci of the European Institute of Oncology, head of the study.[32] He said that he and his colleagues had already shown that the protein is activated by an enzyme that could be a possible drug target.[33]

Geneticists at the California Institute of Technology in Pasadena, led by Seymour Benzer, discovered that a single mutated gene in the fruit fly (drosophila) gives the insect as much as a 35 percent longer lifespan.[34] (Predictably, it was dubbed the "Methuselah fly.") The researchers also found that the Methuselah gene allowed the flies to withstand higher levels of stress. These insects lived 50 percent longer than normal-variety fruit flies when deprived of food, tolerated heat better, and were more resistant to paraquat, a chemical that can damage cells by generating oxygen-free radicals. This suggests that "if you can resist stresses or better repair damage, then you can increase lifespan," said Benzer.[35] According to Daniel Perry, executive director of the Alliance for Aging Research, "It's a long step from drosophila to healthy ninety-year-olds, but this is all part of a field that is beginning to close in on the aging process and think about ways to intervene in ways that are practical and socially useful."[36]

A millimeter-long fellow called a roundworm has also been found to have a longevity gene, though to the best of my knowledge nobody dubbed it the "Methuselah worm." The California researchers doubled its lifespan with no apparent physiological side effects by silencing a gene at just the right point in a worm's life cycle. "Knocking down the gene after the worms reach adulthood increases their life span without affecting their reproduction," said team leader Cynthia Kenyon of the University of California, San Francisco.[37] The gene, daf-2, is also found in fruit flies and mice, and it may be in humans too.[38] In any case, "it's inescapable that aging is regulated deliberately by genes" said Kenyon, and since "it happens in both worms and fruit flies, you have to be crazy to think it won't happen in vertebrates."[39]

Indeed, such a gene was later discovered in vertebrates. French researchers led by Dr. Martin Holzenberger in late 2002 reported that knocking out a single copy of the IGF-1R gene in mice made them live an average of 25 percent longer than they would otherwise. Further, the modified mice appeared essentially normal in size, appetite, physical activity, fertility and reproduction. As with human aging, the females got the better part of the deal by living about 33 percent longer while the males lived only 16 percent longer.[40]

There's also evidence that a single protein can make a tremendous difference in aging. Scientists writing in *Nature* in late 2002 shook up the longevity field when they revealed that one enzyme can have a tremendous impact on the muscular aging of the roundworm. Even though "middle age" to these little guys is about three days, like humans they appear at that point to begin losing muscle cells in a process known as sarcopenia. But the researchers showed that a single enzyme, called PI3 kinase, has to be present for sarcopenia to occur. That may well also be the case with us.[41] "This kinase probably has a good chance of contributing to muscle decline in humans, which is why you don't have an athlete who's an Olympic champion after age 40," one of the researchers said.[42] Most of those earlier studies have focused on DNA and the genetics of aging, not what happens at the larger, cellular level. The little roundworm contains a grand total of just 959 cells, which greatly facilitated the job. The latest study is "one of the first papers that actually goes after tissue function in aging, so we're pretty excited about this," said Anna McCormick, chief of genetics at the National Institute on Aging.[43]

Meanwhile, many researchers like Dr. Thomas Perls, a gerontologist at Harvard Medical School, suspect what most of us do—that part of the secret to aging both well and long is genetic inheritance. To test that theory, Harvard and Boston's Beth Israel Deaconess Medical Center are funding the New England Centenarian Study, which Perls directs.[44] He has interviewed more than one hundred centenarians, analyzing their family histories as well as their mental and physical health. He has also collected blood samples of six families to search the DNA of siblings who are older than a hundred, all chosen because they look extremely young and are in great shape. "They've got genes that are facilitating the ability to age very slowly and they lack genes associated with aging diseases. It's definitely the two together," Perls said.[45] He sees his mission as identifying these "longevity genes." Then researchers could use them to find treatments to interfere with both the aging process and diseases associated with it.[46]

Perls doesn't downplay the importance of a healthy lifestyle. "What you do now will determine what your health will be like in 20 years," he said. "It's all these bad things we do that give us 10 years less than our genes are capable of. We need to take advantage of the genes we are given."[47] But he's also found that the oldest of the old often had habits in their salad days that included a lot more smoking and drinking than eating salad. "Their diet is all over the place. They just have supergenes that counteract what would normally kill the rest of us."[48]

The Low-Cal Recipe for Longer Living

There is one simple method of greatly extending life that works in a wide array of species and probably works for us, too. But it's not one many of us would choose. Laboratory mice and rats live up to 40 percent longer than usual when restricted to a diet that has perhaps one-third fewer calories than they would normally eat, so long as their food contains all necessary vitamins and other nutrients. The animals even look younger and negotiate mazes with the quickness and ease of young mice, yet they're at an age when you'd think they needed canes and hearing aids.[49] Calorie restriction has similar effects with an incredible variety of creatures, including other rodents and insects.[50] An ongoing experiment with Rhesus monkeys seems to indicate that it works with them as well.[51]

It stands to reason that similar diets would extend youth and life for humans, too. "Okinawans have about 70 percent the calorie intake of the rest of Japan. They eat mainly fish and vegetables," observes Roy Walford of UCLA's School of Medicine, and perhaps the nation's leading longevity expert.[52] "They have less diabetes, tumors and so forth than the rest of Japan."[53] *And,* he says, they have about forty times more centenarians per capita. Unfortunately, in almost all the industrialized countries people are eating *more* calories every year.[54] Some humans have put themselves on extremely low-calorie diets to try to live longer, but they bear a semblance to unwrapped mummies.[55] Who wants to live to be 150 if that's the price? Recent research, however, shows that we may be able to have our cake and longevity, too.

Scientists at the Massachusetts Institute of Technology in late 2000 announced they had discovered how to boost a protein that switches off the part of the human genome responsible for the aging process.[56] The team, headed by Dr. Leonard Guarente, dubbed the gene SIR-2, which stands for "silent information regulator no. 2." It works by making the material surrounding DNA wrap more tightly, keeping cells from getting to the underlying genes. Experimenting on yeast, Guarente's people found that the gene appears to affect the organism's longevity directly. Cells in which the gene had been disrupted died much more quickly than usual. Conversely, those given an extra copy of SIR-2 lived longer than usual.[57] That extra copy keeps "the genome stable," said Guarente. "It is instability that causes aging."[58]

But where does calorie restriction come in?

Yeast cells grown with very little of their food (sugar) live longer than normal. Calorie restriction *does* equal longevity, for yeast as well as for higher forms of life. If the yeast is given more sugar than it really needs, it's working overtime on metabolizing food and this locks the SIR-2 protein out so that it can't do its job.

Great, doc, but we don't like that severe calorie-restriction stuff!

Don't worry; Dr. Guarente knows that. His response is that calorie restriction is just a means to an end. "If we can keep SIR-2 active for longer, we may slow down aging," he said, without giving up breakfast, lunch, and most of dinner.[59] "The aim would be to develop a drug that would give you the benefits without the downside," Guarente said in early 2000. "It's some way down the line, but that's what we are going for."[60] Later he showed that a version of the SIR-2 mechanism has a similar function in the roundworm.[61] "What's true in yeast and worms is probably generally true in all organisms, because these [two] organisms are so divergent," he explained.[62]

There's good evidence that such a drug already exists—at least for fruit flies. In late 2002, researchers at Yale and the University of Connecticut Health Center in Farmington reported in *Science* that they had bred some of the insects to have lower levels of an enzyme called "Rpd3 histone deacetylase," which modulates SIR-2.[63] These lived about one-third to one-half longer than normal fruit flies. Another set of the insects in their lab given an extremely low-calorie diet lived 41 percent longer, so the effect was in the same range. Both had about twice the SIR-2 expression of a normal, untreated fruit fly.[64] "If you decrease the level of enzyme without eating less, you still get life span extension," said Stewart Frankel, the study's senior author.[65] Research published earlier in the year showed that a drug called phenylbutyrate, when given to fruit flies, lowers that enzyme level and extends their lives.[66]

Such findings provide "a starting point in the design of drugs which would have a broad effect on human health, including cancer," according to Dr. Tomas Prolla, a geneticist specializing in aging at the University of Wisconsin at Madison.[67] Prolla also said that using a direct method of manipulating the SIR-2 gene instead of using calorie restriction might lead to even longer life extension than what we've seen in the animal studies. "I don't think a 30 to 40 percent [increase in lifespan] should be considered as some kind of maximum."[68]

It all sounds too good or too facile to think that one gene controls aging, that a single pill or injection could somehow keep death at bay. But genes direct other genes, and Guarente thinks the role of the SIR-2 gene in ensuring a creature's survival could make it something of a master switch. Roy Walford appears to agree. He sees metabolism as a pyramid with a few key changes at the top flowing down to a larger base that affects lifespan. "Probably there's some few changes underlying these multiple changes," he said.[69]

Another approach to combating aging also builds on the calorie restriction factor, and owes its discovery to gene chips. These chips apparently revealed something rather marvelous: not only can calorie restriction retard aging, in some ways it appears to *reverse* it.

Stephen Spindler, a professor of biochemistry at the University of California at Riverside, and his colleagues used Affymetrix gene chips to analyze which of 11,000 different genes coded for proteins in the livers of young and old mice.[70] They also tested the effect of calorie restriction on gene expression. They found 20 genes whose expression increases with age. Several are associated with inflammation, which in the liver can contribute to the development of cirrhosis or cancer. Other genes are involved in protecting and repairing cells. In 14 of the 20 genes, calorie restriction completely or partially prevented the age-related changes. With 26 other genes, protein expression decreases with age. Some of the genes are responsible for preventing runaway cell growth and division. Others aid in the body's detoxification of foreign substances, such as drugs. Of the 26 genes, calorie restriction partially reversed the age-related changes from 13.[71] "Calorie restriction can reverse the majority of the deleterious age-related changes in gene expression that we found," said Spindler.[72]

This shows it's never too late to stop tossing back Big Macs and start eating celery and drinking skim milk. Still, who wants to do that even if they can wait until they're 50 or 60? That's where the gene chips help again. Before they were available, the only way of telling the rate at which animals expressed age was observation and dissection. Now gene chips can be used to screen compounds at a rapid rate, testing anti-aging interventions in mere weeks rather than years and drastically accelerating the search for anti-aging treatments. Said Spindler, "I think the gene expression [test on chips] for the first time provides a method for demonstrating the extent to which a given agent is capable of reproducing the effects of caloric restriction." Subsequent studies, he added, would verify that compounds that pass the screening can prevent the onset of age-related diseases and extend lifespan. "The screening of humans for delaying the onset of age-related diseases with these compounds is a very real possibility."[73]

None of this tosses Guarente's findings with the SIR-2 gene out the window. It may still be that all of the genes that Spindler and his colleagues found are in some way connected to SIR-2, or perhaps not all but many are. In any case, there are now two different approaches being investigated that use the same basic process, calorie restriction, that we know can slow or even reverse the aging process.

A Radical Idea

"Free radical" is a term you've already seen in connection with the long-lived fruit flies. Free radicals are molecules that are missing an electron, and their antics cause oxidation. This is what happens to a piece of metal when it rusts. It's also what turns a sliced apple brown. In a sense, we're all rusting as we get older. The only guaranteed way to prevent this is rather drastic, namely not breathing.

How do free radicals do damage? That missing electron causes molecules to become unstable and then try to steal an electron from other molecules. Free radicals are not subtle thieves. Indeed, they continuously collide with other molecules, like billiard balls bouncing off each other, in order to grab that missing electron without which their lives seem somehow empty and vacant. If they succeed, they stabilize. But what happens to a stable molecule that loses an electron to a free radical? Right; *it* now becomes a free radical. What occurs is a massive chain reaction in which millions of molecules get altered in billionths of a second. The real problem is that these free radicals don't just go bouncing off other oxygen molecules but also those of our DNA, of proteins and cells. Since one electron is like any other electron, these molecules also aren't shy about ripping one out of your cells, causing damage. Or as Bruce Ames put it, "Living is like getting irradiated in some sense."[74]

Ames is a professor of biochemistry and molecular biology at the University of California at Berkeley, a recipient of the National Medal of Science, and widely considered one of the nation's top cancer and longevity researchers.[75] "We have lots of defenses but they are not perfect," he explained. Over time, as organisms age, the damage accumulates. Moreover, production of chemicals that the body creates to combat the free radicals starts to decline.[76]

The part of the cell that suffers most from the free-radical invasion is the mitochondria, bloblike things often called the "powerhouses of the cell." Mitochondria transform low-energy molecules into adenosine triphosphate (ATP), a high-energy molecule that the body uses as a source of energy. ATP powers your muscles, your brain and your biochemistry.[77] "The constant barrage of free radicals pulling electrons out of the mitochondria causes the mitochondria to decay so they don't produce enough energy," said Ames. Therefore, the defense systems don't work as well, and they make more oxidants. It's a vicious cycle that just gets worse.

We know that free-radical damage increases with age. In the first study that could actually measure the effect on DNA, the chief researcher said that "in essentially all cases that we looked at—many strains of male and female rodents—we saw this increase in oxidative damage." Further, he said, "We saw oxidative damage in every tissue that we looked at, in every rodent strain. It varied from a 20 percent increase over life span to almost 200 percent, depending on the tissue that we examined—liver, heart, brain, kidney, skeletal muscle and mitochondrial DNA."[78]

How do we get free radicals in the first place? In part, they're an inevitable byproduct of normal metabolism, required for breaking down food, for example, or fighting off germs. "Every time your white blood cells combat a bacterial infection, they pour out mutagenic oxygen radicals and hypochlorite, which is the same chemical found in Clorox," said Ames. They also create other oxides that cause damage, such as superoxide and nitrogen oxide. "Chronic infection, in fact, causes about 20 percent of cancer, probably due in good part to these oxidants."[79] So free radicals are neither evil nor worthless. You need them.

But we also get free radicals from environmental factors such as radiation (natural or manmade), cigarette smoking, some chemicals, and some of the foods we eat. And while you couldn't get rid of all your free radicals even if you wanted to, what you do want to do—or so the theory goes—is keep the number as close to what you need as possible.

One way eventually may be through gene therapy. Indeed, the aforementioned Dr. Holzenberger said he wasn't sure why mice with only one copy of the receptor for IGF-1 live longer but "For the time being we can only say that part of the additional longevity could be explained by their increased resistance to oxidative stress."[80]

Knocking out genes in humans is years away, though. Might there be something we can do in the meantime to slow the "rusting" process? That's where antioxidants come in. These are chemicals that may prevent or slow the oxidation process by handing free radicals that extra electron they need and telling them to stop acting like jerks. A research psychologist at the Department of Agriculture's Human Nutrition Research Center on Aging at Tufts University found that by feeding extracts of antioxidant-laden blueberries, strawberries and spinach to rats "we can stall or even reverse the effects of aging on the brain, both behaviorally and chemically."[81] Some interesting recent studies on aging dogs found that those given antioxidant-enriched chow bought at the grocery store had significantly improved mental abilities, sometimes in as little as a month.[82]

Some researchers have also found that high doses of vitamin E supplements (2,000 IU per day) delayed the need for institutionalization among moderately to severely demented people, as did another antioxidant, selegiline, though others have found that lesser amounts of neither E nor C seem to delay the onset of dementia.[83]

The research will continue because the brain is an especially good organ for testing antioxidants in that it not only is subject to a tremendous amount of oxidation but also has little in the way of oxidant defense mechanisms.[84]

Many antioxidants are produced by our bodies, but others are available to us only in food or nutritional supplements. Fruits and vegetables contain the most antioxidants. Talk to Bruce Ames about antioxidants and you're guaranteed a lecture on the importance of fruits and vegetables. An improved diet is enough to decrease greatly your chance of getting many types of cancer, and to reduce birth defects (and also quite possibly miscarriages). But to achieve maximum success in soaking up free radicals, and for providing essential nutrients, Ames' research has taken him in the direction of nutritional supplements.[85]

Bruce Ames, who began his oxidation work with cancer and then expanded it to aging, thought that the mitochondria, the "power plants" of every cell, should be the focus of his research on oxidants since they produce the cell's energy by oxidizing (burning) fat and carbohydrate fuel. The oxidant byproducts produced from this process increase mitochondrial disruption or decay that can be caused by aging or a variety of dietary vitamin and mineral deficiencies. But mitochondria also have antioxidant defenses and biochemicals that minimize these oxidant byproducts. One of these is acetyl-l-carnitine, known widely as ALCAR, a chemical normally used to transport fatty acids into the cell's mitochondria to be used as fuel. The other is lipoic acid (LA), another mitochondrial biochemical used in energy production and a true antioxidant. Lipoic acid has been dubbed "the universal antioxidant" because it's both water- and fat-soluble, giving it the ability to go anywhere in the body. Like ALCAR, it also acts synergistically with antioxidant vitamins C and E, enhancing their abilities to soak up free radicals.

In various published studies, Ames' teams sought to measure the effects on rats of ALCAR or LA or a combination of both. They found it reversed many chemical indicators of cellular aging, especially in the brain.[86] For example, malondialdehyde (MDA) is a product of the oxidation of the fatty membranes that surround brain and other cells. "As you get older and the oxidation progresses, the cell gets into trouble," Ames explained. "MDA can cause mutations and it interferes with transport in mitochondria." In Ames' rat studies, the unmedicated geezer rats had by far the highest MDA levels, while those on the combination ALCAR/LA therapy had the same levels as the young rats.[87] Moreover, said Ames, MDA is a marker for other age-related chemical dysfunctions; that which reduces MDA probably reduces many of these others as well.[88] The combination therapy also reversed the dysfunction of carnitine acetyltransferase (CAT), an important mitochondrial enzyme for fuel utilization.[89]

Still, all of these are just markers for brain aging. Could the treated rats actually think better? As demonstrated by the Morris water maze, they could.[90] To understand how it works, you need to know that although rats hate water, paradoxically they are excellent swimmers. When placed in water, they desperately try to get out just as quickly as possible. With this maze, there's a pond and a hidden platform, and the rats have to remember where the platform is. Normally, "the young ones learn quickly and the older ones wander around," said Ames. But the combined supplements, while still not making the old rats as fast as the young ones, did cut their times by about half.[91]

Additionally, Ames' researchers measured "ambulatory activity," referring to the amount and distance of movement of the rodents. Old rats are as fast as young rats, but they don't like to move around much.

ALCAR/LA changed that. The amount of movement and distance traveled more than doubled for the old rats, though it was still less than that of the untreated young rats. Interestingly, even the *young* rats were significantly more ambulatory after being given the combination treatment.[92] "If you feed them the two chemicals together, the rats get up and do the Macarena," Ames said.[93]

Ames admits that none of this proves that what happens in rats will also happen in humans. But rodents aren't the only creatures that seem to age more slowly when given antioxidant supplements. A study in the journal *Science* showed that nematodes (worms)—frequently used to study aging—lived 50 percent longer if given a certain antioxidant.[94] Moreover, there's strong evidence that antioxidants can help combat aging-related human disease. Parkinson's is one such. In October 2002, a sixteen-month study of Parkinson's patients showed that those on the highest daily doses of the antioxidant coenzyme Q10 had an astounding 44 percent less decline in mental function, movement and ability to perform daily living tasks than the placebo group.[95] Previous research had shown that people with neurodegenerative disorders have impaired mitochondrial function and reduced levels of coQ10 in the mitochondria.[96]

So Bruce Ames—and I—have been popping those ALCAR/LA pills for a few years now. (Admittedly I still can't do the Macarena.) In the spirit of full disclosure, it should be noted that Ames owns a partial interest in a company that sells the supplements, Juvenon, but proceeds go directly into his foundation, where the money can be used to fund human studies of the formula along with other related philanthropic concerns.[97]

Numerous epidemiological studies have also shown a relationship between increased antioxidant intake and reduced rates of cancer. That makes sense in part because cancer is partly a disease of aging. Moreover, said Ames, "We think that the mitochondrial decay is going to be a huge factor contributing to all the diseases associated with aging," including not just cancer but Alzheimer's, diabetes and Parkinson's. "Once scientists understand the mechanism, there are hundreds of ways to intervene."[98] That's why, if you insert the term "antioxidant" into the PubMed database, you'll find over ninety thousand references.

There are many antioxidant supplements available. Some are phytochemicals found in grape skins, green tea, lycopene from tomatoes, and the like. Others are vitamins (A, B, C and E). Still others are minerals, such as zinc and selenium.[99] Which ones should we take? Ames will only admit to taking a multivitamin "for insurance" on top of a healthy diet of fruits and vegetables. But he advises about 100 to 120 milligrams of vitamin C a day; anything more and you're literally peeing away your money because that's the most your body can metabolize. With vitamin E, he said probably 200 milligrams is sufficient.[100] He also told me, "I always thought all those high-level B vitamin [B1, B2, B6 and B12] pills were crazy, but now, with new research, I'm thinking there may be something to it for some people."[101] Regarding the two nutrients he tested, "We're not positive about the doses but something like 250 milligrams of ALCAR and 300 of lipoic acid twice a day," he said. "Anecdotally, some people say they feel better."[102] But nobody knows for sure yet. As far as getting too high a dose, Ames said that most antioxidants are very difficult to consume in toxic levels, unlike, say, iron supplements.[103] Selenium is an exception, as is vitamin A, which the National Academy of Sciences' Institute of Medicine has specifically warned against. With vitamin A, 10,000 international units or 3,000 micrograms is considered dangerous, and high doses are associated with both liver damage and birth defects in children of women who took excess supplements while pregnant.[104]

On the low end of the scale, Ames' ALCAR-only study gave the old rats 0.5 grams per kilogram of body weight per day, which comes out to 32 grams of the stuff daily for a 140-pound person. When I asked about this, Ames replied, "We used the maximum amount we could get into the animals in the first study," but "now we have cut the dose by 90 percent and still see the beneficial effects."

Defying Decrepitude

One of the best things about the aforementioned treatments is that they don't threaten to bring down upon us the curse of Jonathan Swift's Struldbrugs, whom Gulliver encountered in his travels. The Struldbrugs lived forever, yet their minds and bodies also forever decayed.[105] Even without biotech, advances in medical care, lifestyles and nutrition are helping us to live healthier lives in old age. A study of the well-being of persons over 65 found that while 6.2 percent of America's elderly were in nursing homes in 1982, this figure had fallen to only 3.4 percent by 1999.[106] Yet these improvements will be utterly dwarfed by what science in general and biotech particularly will deliver. Biotech breakthroughs promise to keep us far healthier, youthful-feeling and capable in our golden years—even if those years are well north of the century mark.

Consider tests on mice, reported in 1998, employing a genetically modified hormone piggybacked onto a virus that had been rendered harmless. When injected into the rodent's right hind limb, the virus spread the hormone throughout the muscle group. The left hind leg was left untouched to be a control. Young mice showed an increase of 15 percent in both strength and muscle mass in the injected muscles. Old mice improved the most, however: they had a 19 percent increase in muscle mass and an amazing 27 percent more power in the treated leg.[107] "Though I expected improvement in the old animals, I never dreamed we would basically preserve them at young-adult levels," said research team leader H. Lee Sweeney of the University of Pennsylvania School of Medicine.[108]

Muscle loss in most of us begins in our twenties and increases by the decade. But if the hormone-injection technique employed on the mice also works in humans—and rodent physiology in this regard is close enough to ours that there's no reason to think it won't—we could routinely make people as strong in their retirement years as they were in their college years. If grandpa was a 98-pound weakling in college, we could make him *stronger* than he was. Scientists not normally given to hype found themselves practically gushing over this development. "These dramatic findings signal a fountain-of-youth opportunity," said neurologist Dr. Leon Charash of the Weill Medical College of Cornell University in New York City.[109] They hold the promise "of helping older people lead healthier and better lives, with less need for medical care."[110] Charlotte Peterson of the University of Arkansas for Medical Sciences in Little Rock said, "It's the first case of a molecule affecting muscle atrophy due to aging."[111]

Such a hormone would not provide a cure for degenerative muscular diseases, but a 20 percent increase in strength could allow sufferers several more years of normality and of life in general—perhaps long enough for a cure to be found. That's why the research was funded by the Muscular Dystrophy Association, to which Charash is medical advisor. Finally, muscle loss related to age leads to a reduced metabolism, which is the main reason people generally fatten as they age. More muscle means less obesity, a condition that has become epidemic in the United States and elsewhere.[112]

The Assault on Alzheimer's

All this is great for muscles, but who wants to be strong and senile?

The major cause of senility in this country is Alzheimer's, and the number of Americans with the disease is expected to more than triple by the year 2050. Currently, one in 10 people over the age of 65, and as many as 50 percent of those over the age of 85, have Alzheimer's disease.[113]

A few medicines are available for treating Alzheimer's, all of which increase levels of acetylcholine, a brain chemical that helps transmit nerve impulses. But they can have uncomfortable gastrointestinal side effects, such as nausea and diarrhea.[114] Further, they do nothing more than slow deterioration somewhat. They "help make the best use of what you have left," but nothing more, said Rudy Tanzi, a Harvard neurogeneticist.[115]

Biotech drugs are going to change all that. Researchers at the University of California at San Diego (UCSD), in 1999, published a study showing that the transfer of a nerve growth factor gene into a skin cell, which was then injected into the brains of old monkeys, did not simply slow the organs' aging process.[116] Instead, "to our surprise, this technique nearly completely reversed" the effects of aging on key brain cells in elderly Rhesus monkeys, exclaimed Dr. Mark Tuszynski, senior author of the study. "We've all heard the dogma that we lose 10,000 neurons a day after the age of 20," said Tuszynski. "Well, that is false. That doesn't happen."[117] He explained that in the cortex, a key part of the brain used in thinking, very few cells are lost with age. Instead, the researchers found it was the control neurons in another part of the brain, the basal forebrain, that were most seriously affected by aging. In the monkey, this part had shrunk and had stopped making regulatory chemicals that affect the cortex. "These cells are like the air traffic controllers of the brain," said Tuszynski. "They control the flow of information in the cortex." The researchers found that about 40 percent of the basal forebrain cells could not be detected in old monkeys, and the other 60 percent had shrunk by one-tenth. But the cells were not dead, Tuszynski explained. Through insertion of genes for nerve growth factor, or NGF, into the brain, the cells were revived and restored to nearly full vigor. "We restored the number of cells we could detect to about 92 percent of normal for a young monkey and the size of the cells was restored to within 3 percent," he concluded.[118]

Just by demolishing the notion that brains lose neurons as they age, the experiment was a tremendous breakthrough.[119] But the researchers knew that restoring neurons wasn't enough. Nerve fibers called axons are vital for transmitting messages between neurons, and they too shrivel up and seem to disappear as we get older. So the UCSD researchers used another set of geriatric monkeys, added some more NGF, and tried to restore the axons of these animals. It worked.[120] "Our research shows that as a result of normal aging, monkeys lose 25 percent of their axons, the threadlike projections that neurons send out into the cortex, the outer layer of the brain where short term memory is retrieved and intellectual processing occurs," said the study's lead author, UCSD assistant professor James Conner. "Following gene therapy, the axons were restored to levels seen in young monkeys, and sometimes exceeded those levels."[121]

Bone Loss and Biotech

If we may soon be able to reverse brain and muscle loss, what else could be in the works? It could well be a medicine that reverses the loss of bone that leads to the ever-increasing risk of fractures. Few aspects of old age are crueler than osteoporosis and none are more inevitable. According to the American Academy of Orthopedic Surgeons, this disease affects 28 million Americans and contributes to an estimated 1.5 million bone fractures each year. One in two women and one in five men over age 65 will suffer bone fractures due to osteoporosis, many of which will be breaks in the hip, spine, wrist, arm and leg that often occur from a fall.[122] Vertebral fractures are less debilitating and costly than those of the hip, but far more common, accounting for about 700,000 osteoporosis-related fractures each year, generally occurring at earlier ages than hip fractures, and actually contributing to the risk of hip fractures.[123] The estimated national direct expenditures (hospitals and nursing homes) for osteoporotic and associated fractures was $17 billion in 2001 ($47 million each day) and the cost has been rising, according to the National Osteoporosis Foundation.[124]

The term "osteoporosis" literally means "porous bone." The disease often develops unnoticed over many years, with no symptoms or discomfort, until a fracture occurs. It often causes old people to "shrink" or gives them a "dowager's hump," a severely rounded upper back.[125] "I've had patients who have lost nine inches," said Dr. Hunter Heath III, the medical director of U.S. Endocrinology for Eli Lilly. In this case, what's happening is that the vertebrae themselves are collapsing from the inside. This same process of bone loss turns many seniors into "glass" people, living in constant fear of a simple fall. Some of them snap their spines just by bending over. "I had a patient in her fifties who broke down and cried in my office, saying she couldn't even get a hug from her husband or son anymore because both of them had broken her ribs while doing so," Heath told me.[126] A life without hugs is a terrible future to contemplate. It may also be a future that soon few will have to endure.

There have already been significant drug advances against osteoporosis, with more pharmaceuticals on the way. All of the nonbiotech drugs approved to fight osteoporosis are known as antiresorptives. "They slow down the loss of bone and make the bone somewhat stronger, but they don't restore it to normal," said Heath. "They do not stimulate new bone formation." This includes Lilly's own drug, Evista.[127]

But what about actually turning back the skeleton's clock? That's the promise of Forteo, given FDA approval late in 2003.[128]

Forteo is a recombinant version of a portion of a naturally occurring human parathyroid hormone that's grown in *E. coli* bacteria. It doesn't just prevent further deterioration, but actually *regrows* bone lost to osteoporosis. It's produced by Lilly and Dr. Heath played a role in bringing it to fruition. "There's been a lot of research to turn on the production of more bone," he told me, "and the compound we're studying is the first in its class."[129] Bess Dawson-Hughes, coordinator of the Osteoporosis and Related Bone Disease National Resource Center, calls Forteo "the most promising agent that [stimulates] bone formation."[130] "I don't think you'd have trouble convincing people with bone fractures to take the drug," she added.[131]

Although Lilly researchers measured the density of the patients' bones using a device called a dual-energy X-ray absorptiometer, it had earlier been found that there's no strict correlation between treatment-induced increases in density and the possibility of breakage.[132] So measuring bone density might not be the best way to determine if the medicine works. Thus, instead of just looking at a "disease marker," the researchers measured what we're really concerned with, the fracture of both vertebral and nonvertebral bones. The results were striking.

The study comprised 1,637 women with an average age of 69. All had advanced osteoporosis. They were divided into three groups: low-dose, high-dose, and those receiving a placebo. The incidence of new vertebral fractures decreased by about two-thirds in women receiving the drug, as compared with those on the placebo. The incidence of nontraumatic nonvertebral fractures was cut by more than half. The researchers also found that patients on Forteo lost less height and had less back pain, which often accompanies osteoporosis.[133] Side effects were mild, said Heath. "We used two doses and at the highest level some had nausea and headaches. About 2 percent had leg cramps."[134] But since the higher-dose women had only slightly better benefits than the lower-dose women, it should be a simple matter for a doctor to lower the dose for a person suffering side effects. Another very pleasing aspect of Forteo is that it seems to improve bone strength much more rapidly than the antiresorptives. Finally, it seems to work best on those who need it most. Women over the age of 65 (who by and large suffer the most from osteoporosis) appear to experience greater increases in vertebral bone density than younger women.[135] Unlike so many of today's drugs, it doesn't fit into the category of "works best if you catch it before . . ."

Some researchers foresee patients first taking Forteo to grow bone and then an antiresorptive drug to maintain bone strength.[136] That's certainly possible. But the average duration of dosing in the Forteo trials was one to two years, which should have you wondering what happened to patients beyond that time limit. Put another way, can users just keep on strengthening bone year after year? "The longest exposure now is three years and there's no evidence that the increase in bone density has stopped," Heath told me, adding that this increase in density doesn't necessarily correlate to fewer fractures. But we are definitely on the cusp of a day when we have octogenarians who stop shrinking and getting humps, and instead have the skeletal structure of a young person.

If osteoporosis is a gradual decay in the skeletal structure, osteoarthritis occurs as a result of injuries and improper constant motions that cause the cartilage to wear away until there's nothing left but bone rubbing on bone. One of the classic aging diseases, it's already a huge problem and it's going to become a lot more so as the population ages. In 1997, the disease ranked behind only chronic heart disease as a cause for adults receiving Social Security disability payments. Fully one-third of all Americans, about 90 million people, suffer some form of arthritis.[137] For many patients, medicines provide little relief; they're never a cure; and usually the condition steadily worsens. But biotech scientists are hard at work developing myriad methods of tissue generation.

One of the earliest breakthroughs came back in 1994, when two brothers, Charles and Joseph Vacanti, met Sean McCormack, a twelve-year-old born with a protruding sternum and no cartilage or bone under the skin on the left side of his torso. His heart could literally be seen beating beneath a wall of mere skin. The Vacantis, now widely regarded as leaders in the field of tissue generation but then seen a bit more as crackpots, harvested cartilage cells from Sean's sternum and seeded them onto a polymer scaffold about the size of a compact disc. With a steady diet of nutrients, the cells quickly multiplied and penetrated the polymer. Within just weeks it was implanted into the child's chest. As he grew, his body incorporated the shield. Today Sean is a star bicycle racer.[138]

Tissue regeneration could also help nature in healing bone breaks. Johnny Huard and other researchers at the University of Pittsburgh have used stem cells to heal the broken skulls of mice almost completely, while researchers at the University of California at Los Angeles School of Medicine healed broken mouse leg bones with a similar technique.[139] Usually our bones grow back on their own, but these techniques could help people whose joints and bones are too badly damaged for this. About one of every twenty fractures won't heal by normal processes.[140] Further, ligaments and tendons often grow back slowly; genetic engineering could speed this up.

Stryker Biotech of Hopkinton, Massachusetts, has developed recombinant human osteogenic protein-1 (rhOP-1), a version of a naturally occurring human protein that has shown a remarkable ability to repair and regenerate bone and other hard tissues.[141] Based on research on tibias (shinbones) in Europe, Stryker has already received marketing authorization in Australia and Europe and humanitarian use approval from the FDA.[142] The company is also developing its products for a number of bone problems including fresh fractures, spinal fusion, periodontal disease and joint repair.[143]

Dr. Jeffrey Bonadio of Selective Genetics in San Diego and his colleagues have used a combination of tissue engineering and gene therapy to repair beagle bones they had sawed open.[144] The gene they used causes emission of the parathyroid hormone, which regulates the amount of calcium in the blood by sending signals to the kidneys, bones and intestines. Too much of the hormone can be harmful, but in the right amount it stimulates bone growth and repair. Bonadio and his team took the plasmid part of the cell, which carries genetic instructions for the cell to produce parathyroid hormone, and implanted it into a spongy material made of cow collagen. These sponges were inserted into fractures in the crippled dogs. The cells then accepted the hormone gene and "became growth factories," said Bonadio, "turning repair cells into local bioreactors."[145] Within two to six weeks after implantation there was an increase in bone formation at the fracture site at a rate much faster than normal.[146]

Bonadio's process overcomes some of the problems that have been commonly associated with gene therapy for bone repair. For example, because the plasmid can't leak out of the sponges, the genes stay exactly where they are supposed to. The principle aim, said Bonadio, is to help the millions of elderly Americans who incur fractures each year because they suffer from osteoporosis.[147] But the technique could be used with anyone who suffers high-impact fractures.

TWO

Starfish for a Day

There are lots of things humans normally regenerate, such as small amounts of bone, most muscle, sinew and skin, but also many parts we wish we regenerated but don't. That's why regenerative medicine is one of the most promising and exciting areas of biotechnology. "This year there are more than 50 laboratories in the United States alone racing to create people-made parts," the popular science magazine *Discover* recently noted.[1]

Some researchers are already experimenting on rodents to try to give them—and later us—the ability to regenerate whole limbs. It's not just starfish that can already do this, but numerous other creatures such as flatworms (planarian), salamanders, newts and zebrafish. Scientists are studying these creatures to identify the genes and proteins that give them these abilities. And they're finding that humans have more in common with them than we might think. "We have the genes planarians use to regenerate their brain, muscle, their entire head," said Alejandro Sanchez Alvarado of the University of Utah in Salt Lake City.[2] Newts and salamanders are vertebrates, as we are. Yet they can regrow many body parts, including tails, limbs and even large sections of their hearts.

Nobody is finding any intrinsic reason why some day humans won't be able to do the same. Indeed, fairly soon, individual body organs that cannot presently regenerate will be able to be directed to do so.[3] Alternatively, organs will be made out of human tissue and used as replacements. "In two years, people will routinely be reconstituting liver, regenerating hearts, routinely building pancreatic islets, routinely putting cells into [the] brain that get incorporated into the normal circuitry," Dr. Ronald McKay, a neural stem cell expert at the National

Institutes of Health, said in late 2000. "They will routinely be rebuilding all tissues."[4] His timeline was obviously off, but we certainly are progressing.

While Dr. Jeffrey Bonadio's research employs gene therapy, studies done at the University of Pittsburgh and UCLA involve cellular-level cloning to reverse joint damage. In 2000, two sets of researchers using somewhat different techniques cloned human corneas, the transparent tissue on the front of the eye, and then attached them to patients whose own corneas had been badly damaged.[5] The challenge to the scientists was to find the right material, something that could grow to the right shape and size but wouldn't be rejected by the recipient. They found their answer in amniotic membrane, which envelopes the fetus in the womb. (It was taken from new mothers.) In cases in which the recipient still had a healthy eye, cells were extracted from that eye and coaxed with various chemicals into becoming corneal tissue. If both eyes were worthless, cells from a closely related donor were used. One of the research teams grew these cells directly onto a sterile membrane, while the other team first grew tissue in a Petri dish and then planted it on the membrane. In both cases, within a mere two or three weeks they had grown a tissue layer five to ten cells thick and an inch across. First one layer was stitched onto subjects' eyes, then a month later another layer. "We are quite a few years out from general application," said one of the researchers.[6] But as a *Business Week* reporter nicely put it, "Bioengineered corneas are a critical advance in tissue engineering, putting us one step closer to some day being able to order up a spare body part as easily as a new fender."[7]

Eventually, whole complex organs such as kidneys, livers, lungs and hearts will be made this way.[8] People can have "banks" of their own vital organs. Or they could use the cheaper alternative of having an organ cloned when the first signs of its failure become apparent. Techniques will be developed to grow organs much faster than they do naturally, such that an adult may have a fully functional heart made in just a few years. Predicts Dr. Annemarie Moseley, formerly with Osiris Therapeutics in Baltimore: "Patients in middle age would come in, before the degenerative process starts, and would get a cell or tissue implant, and degeneration of the structural elements would no longer be part of the aging process."[9]

Bladders, heart valves, breast nipples, penises and teeth have all been created using a scaffolding method.[10] Dr. Anthony Atala, a urologist and director of tissue engineering at Children's Hospital Boston, worked with a team of researchers to make bladders for six dogs.[11] They started with a one-centimeter-square biopsy from each dog's original bladder. Then they isolated the urothelial lining cells and the muscle cells and cultured each type separately, using growth factor to promote speedy cell division. Within about four weeks, the team had about 300 million cells of each type, enough to build their bladders. The cells were then fixed onto a polymer ball the size and shape of a beagle's bladder—the muscle cells on the outside, the urothelial cells on the inside. The cell growth slowed when the scaffolding filled, then the polymer was absorbed into the natural tissue. The new organs worked terrifically. Blood vessels from the surrounding tissue grew into the engineered bladder to keep it healthy. The capacity of the new bladders was about the same as the old.[12] Dr. Atala, along with Curis in Cambridge, Massachusetts, should begin human testing in 2003.[13]

Bladders are among the easiest organs to clone since they're basically just a balloon with a valve. But in December 1999 a Harvard Medical School researcher reported that he had used the beagle-bladder technique to make heart valves that were then implanted in sheep. Human trials of the valves are expected around 2004.[14] Atala, working with H. David Humes of the University of Michigan, has also already built small kidney models that can produce urine.[15] Unlike bladders, kidneys are highly complex structures with as many as twenty different cell types.[16]

Biotech researchers are also making speedy progress in building nipples for women who have lost them in cancer surgery or want larger ones for cosmetic reasons. As of 1998, Curis had already succeeded in growing nipples and associated tissue using human cartilage cells. The underlying breast tissue is grown from a small amount of fat and blood vessel cells taken from the buttocks or thighs. As with the beagle bladders, this is then seeded onto a preformed scaffolding of polymer plastic. Curis hopes to be building human nipples by 2004. While the number of radical mastectomies has dropped considerably, this could make the disfiguring operation a bit less frightening.[17]

Penis loss is far rarer than breast loss, but it does happen due to injury, penile cancer and, yes, the occasional hot-tempered housewife with a carving knife. But more often boys are born without them or with ones that don't function properly to allow them to have intercourse. Dr. Anthony Atala is working on this as well. In late 2002 he announced he'd been able to reconstruct organs for rabbits. The penis comprises three main cylinders wrapped in an outer layer of connective tissue, skin, nerves and blood vessels. The two biggest cylinders, which are made of the spongy material that swells and hardens during an erection, are called the corpora cavernosa. The third tube holds the urethra. The cavernosa are the most complicated part of the penis, so that was Atala's target. He removed them from eighteen rabbits, then used some of the cells to grow new ones, again using the scaffold method. Once they had recovered from the surgery, the rabbits were put in cages with females and, being rabbits, immediately attempted copulation.[18] "They were able to copulate, penetrate and produce sperm," Atala told *New Scientist.* But the greatest importance of this development, he said, is that "the penis is more complex than any of the organs we have engineered so far."[19]

Multifaceted organs like hearts are much further down the road, but work has begun. Dr. Joseph Vacanti, now at Massachusetts General Hospital in Boston, and other researchers took a sheep heart, pumped its vessels full of liquid plastic, then bathed it in flesh-eating enzymes.[20] When Vacanti first saw the cast, he thought he'd blown it—the thing looked like a Styrofoam ball. "Then we looked at it under a microscope and saw: These are all capillaries. It showed us where we're headed," Vacanti said. "In organs like the heart, circulation is structure," he told *Discover* magazine.[21] MIT engineers have used specifications from the cast to design a full heart scaffold. Eventually, it will be seeded and made into a real organ.[22]

Other researchers are trying to build livers from the ground up, using nothing more than a small tissue sample. One such possible liver as described by *Technology Review* would entail etching networks of capillary-like grooves onto palm-sized silicon plates. Two casts lifted from molds of these plates would be sandwiched together, creating a three-dimensional template for interconnected vessels. These would then be filled with thousands of stacked layers of capillaries, liver cells and bile-collecting vessels.[23]

Another method of building organs, believe it or not, might employ an inkjet printer that's just a tad more advanced than the one sitting next to your computer. Therics of Princeton, New Jersey, starts with designs displayed on a computer screen as three-dimensional models, with the ability to include data from laser, X-ray, CT or other scans.[24] The final design is translated into compiled, computerized machine instructions. Then the TheriForm device goes to work, laying down the construction material in a 3-D manner, instead of the 2-D pictures you get out of an inkjet printer. The "TheriForm process can fabricate products composed of virtually any type of material and can combine a variety of different materials in ways that cannot be replicated with alternative technologies," the company claims. The current unit employs multiple nozzles, each printing 800 microdrops per second, which then deposit binder into a powder bed of biocompatible or resorbable materials.

In the short term, Therics is concentrating on relatively simple structures such as joint replacements that will fit recipients perfectly, replacing the primitive joint-and-screw combinations currently used.[25] But just as inkjet printers have progressed from producing nothing more than black type to spraying beautiful color photographs, so too will Therics' technology progress. The only real question is how many organs you'll be able to create from a single cartridge before you have to go to the office supply store to buy a new one.

THREE

Stupendous Stem Cells

Another method of manufacturing biological organs is with stem cells, those that have yet to take on specialized functions.[1] "Stem cells are like babies who, when they grow up, can enter a variety of professions," said Dr. Marc Hedrick of the UCLA School of Medicine.[2] "The baby might become a doctor, a plumber or a fireman depending on the influences in their life, or environment. These stem cells can become many tissues by making certain changes in their environment."[3]

Australian scientists, working with mice, have announced that they have generated a complete and functional thymus. This is a small lymphoid organ in the neck that is critical in generating cells vital to the immune system. But its ability to generate these cells is dramatically reduced by aging, viruses, chemotherapy or genetic abnormalities. The thymus, about the size of a pea, was grown in mice by simply injecting the epithelial stem cells under the skin. Once the organ had grown there, it could be transplanted.[4] "We are very confident that this work will be able to progress to humans within the next three to five years," Jason Gill from Monash University Medical School in Melbourne recently told Reuters.[5]

Yet creating solid organs is only the beginning. Beyond that is replicating the function of cells that constitute blood, skin, heart, nerves, blood vessels, spinal cords or brain tissue. Theoretically, though, stem cells can be used to repair all sorts of tissue, including bone, cartilage, tendons, blood-forming cells and other aged or injured parts, just as a starfish can regenerate an appendage. There are essentially two ways to make stem cells do this. One is to expose them to chemicals that will direct them down the right path. The other is simply to inject them at or near a site that needs repair. To use an oversimplified metaphor, just as water poured into an odd-shaped vessel will nevertheless fill all the voids in that vessel, stem cells seem to have the ability to become whatever they're needed to become. In an injured heart they become heart muscles; in an injured spinal column or brain they become neurons. As Dr. Ronald McKay puts it, the body's tissues are "self-assembling" once their source or stem cells are given the right cues. "I don't know how to make a heart," he told the *New York Times.* "But once you know how to take stem cells and turn them into heart muscle, it's easy."[6]

Given the advanced stage of research on "adult" stem cells (ASCs), the lack of knowledge regularly expressed by prestigious figures during the fierce 2001 debate over federal funding for research on embryonic stem cells (ESCs) was stunning. Former senator Connie Mack, for example, expressed doubt in a *Wall Street Journal* commentary that ASCs even exist.[7] *New York Times* columnist William Safire wrote that "scientists *may* find, in time, that stem cells can be developed from adult cells rather than blastocysts [ESCs]" (emphasis added).[8] In fact, stem cells from marrow have been used therapeutically for many years; the Bone Marrow Donor Program began in 1986.[9] When you hear about marrow donations for leukemia patients, it's these marrow stem cells that they're seeking.

Thus far in North America, all of the approved medical applications for stem cells involve injecting those from marrow or umbilical cords, which then take on the form of the cells that need replacing. It's still fairly low-tech. The next level is removing stem cells, then transforming them outside the body and putting them back somewhere. Europeans can already obtain such a product, a novel skin graft named EpiDex™ from Swiss-based Modex Therapeutics, to treat chronic skin ulcers suffered by over 13 million patients worldwide.[10] With the standard autograft, the top of a sheet of skin is removed from a patient, leaving both the donor and recipient sites with permanently thin skin and scars. EpiDex™ grafts, however, are grown from hair follicle stem cells. Doctors pluck a few hairs and send them off; a few weeks later they receive back a set of skin disks.[11] Clinical trials have shown EpiDex to be far superior in almost every way to the autografts—in terms of area healed, cost and discomfort to the patient.

The retina of the eye will be another organ to benefit from such stem cell development. This little organ is not only complex, it's hard to get to. Rebuilding a cornea is nothing in comparison. The retina is in the back of the eye and works somewhat like the film in a camera. Although it's as thin as onion skin, it has the complicated task of collecting the light that enters the front of the eye and converting it to electrical signals that are then transmitted as information to the brain. Researchers have managed to improve the degraded retinas of rats by injecting them with stem cells from another area in the animals' eyes. The stem cells traveled to the site of damage in the animals' retinas and showed signs of making connections with the optic nerve, which carries the signals to the brain.[12] While the stem cells in the rats' eyes did not turn into full-fledged light-sensing cells, they did show the initial signs of "doing what they're supposed to do," said lead researcher Dr. Michael J. Young of the Schepens Eye Research Institute in Boston.[13] "These cells somehow sense that they are needed, and begin to differentiate into cells that could take on the job of retinal neurons," he continued. "This is very encouraging since there are many blinding diseases of the retina for which there are no cures. We are optimistic that this technique will one day restore vision to those who have been blinded by disease or injury."[14]

Stem cells have also shown promise in saving limbs by promoting blood vessel growth. Scientists at three Japanese universities worked with forty-five people who suffered severe blood circulation problems in their legs. Two-thirds had diabetes, about half had already had a bypass operation in their legs and almost half also had gangrene. Many had sores that wouldn't heal and severe leg pain. In the study, all the subjects had stem cells extracted from their marrow and inserted into only one leg. The patients served as their own controls, in that one group (of 25) received mere saline solution in one leg while another (of 20) received only blood in the control leg. Strikingly, all 45 recipients had improvements in the leg receiving the marrow. Most reported that their leg pain while sitting had simply disappeared, while everybody capable of using the treadmill was able to use it longer. The benefit lasted the entire 24-week length of the study. Toe amputations were avoided in three-fourths of those who were slated to get them, while unhealed wounds improved in 6 out of the 10 patients who had injections. Out of the total of 45 injected people, X-rays showed that 27 clearly grew more blood vessels in the stem-cell-injected leg. In one patient who died of an unrelated heart attack mid-study, leg specimens showed a remarkable increase in the amount of blood vessels in the leg that received stem cells.[15] Dr. William Li of the Angiogenesis Foundation called the results "striking," adding, "This is truly a landmark paper because of its use of stem cells to induce angiogenesis."[16]

Mending Broken Hearts

Heart attacks are the main cause of death in the Western world, with about 1.1 million episodes each year in the United States alone, and over 450,000 deaths. Coronary heart disease in general causes about one in five of all U.S. deaths.[17] A heart attack occurs when a blockage in coronary arteries reduces blood supply to the heart muscle and suffocates cells. There's been tremendous progress in developing medicines and surgeries to prevent these attacks, although the obesity epidemic to a great extent is countering these advances.[18]

Unfortunately, all too often the first symptom of heart attack is death. Nevertheless, most victims do survive their first heart attack. The problem is that once sudden cardiac arrest strikes, it causes irreversible damage. In a process called "remodeling," muscle cells grow abnormally large, extra blood vessels branch out and scar tissue forms. This only further weakens the heart. Contrary to earlier beliefs, it now appears that the heart does repair itself to some extent.[19] But the damage of a heart attack is permanent.[20] Nevertheless, recent studies using stem cells indicate that this may not be the case too much longer.

In side-by-side studies in *The Lancet* in January 2003, two teams of doctors from separate institutions reported promising results in stimulating angiogenesis. This could eventually offer hope to patients with serious coronary heart disease and those unable to undergo bypass surgery. One group from the University of Rostock in Germany injected stem cells into six patients' hearts and found that five had strikingly improved blood flow, indicating the cells had *probably* generated blood vessel growth where the damage was.[21] The other scientists, from the University of Hong Kong, treated eight patients with stem cells from their own bone marrow and also observed an improvement in blood flow to the heart.[22] Previous studies on animals in which cells were given a fluorescent marker showed that the stem cells did turn into heart muscle cells, filling in almost 70 percent of the damaged areas in their hearts.[23] Yet no one knew for a fact that injected stem cells were truly rebuilding human heart tissue until a

French man who had participated in such a study died and could be autopsied. When his heart was examined, the results were clear. Stem cells transplanted from his thigh eighteen months earlier had clearly metamorphosed into heart muscle, which accounted for why his heart showed marked improvement after the procedure.[24] "This is the first demonstration of the concept in humans, and confirms animal findings," said the research leader, Albert Hagège at the Hôpital Européen Georges Pompidou in Paris in February 2003.[25] It was immediately followed up in a Mayo Clinic study that found similar results in four autopsied American women.[26] The issue is settled: Stem cells rebuild hearts.

Stemming Neural Disease

Yet the promise of stem cells goes far beyond regenerative medicine. Sufferers of many autoimmune illnesses are likely beneficiaries, including those with diabetes, lupus, multiple sclerosis, Evans syndrome and rheumatic disease.[27]

Stem cell transplants offer hope for victims of amyotrophic lateral sclerosis, or ALS, an always-fatal neurological disorder that claims about five thousand Americans each year. It belongs to a class known as motor neuron diseases. While it has a genetic component, most instances of ALS have no known cause. The disease occurs when specific nerve cells in the brain and spinal cord that control voluntary movement gradually degenerate. The loss of these motor neurons causes the muscles under their control to weaken and waste away, leading to paralysis. The disease is usually fatal within five years after diagnosis, with physicist Stephen Hawking being the most notable exception to this rule of a relatively speedy death.[28] So far only one treatment has been found that can slow down ALS even slightly, and that's just by an average of three months.[29]

But researchers at the Johns Hopkins School of Medicine introduced neural stem cells into the spinal fluid of mice and rats paralyzed by the Sindbis virus, a pathogen that specifically attacks motor neurons.[30] Normally, animals infected with Sindbis permanently lose the ability to move their limbs, as neurons leading from the spinal cord to the muscles deteriorate.[31] But half of the stem-cell-treated rodents recovered the ability to place the soles of one or both of their hind feet on the ground.[32]

"This research may lead most immediately to improved treatments for patients with paralyzing motor neuron diseases, such as amyotrophic lateral sclerosis," said Dr. Jeffrey Rothstein. The research is being funded partly by Project ALS, an organization founded by Jenifer Estes, a former theater producer from New York who has the disease, and partly by the Muscular Dystrophy Association.[33] "This is the kind of breakthrough we've been working for," said Estes, who has been fighting the disease for three years. "The scientists are our partners in this fight."[34]

"The study is significant because it's one of the first examples where stem cells may restore function over a broad region of the central nervous system," said neurologist Dr. Douglas Kerr, who led the research team. "Most use of neural stem cells so far has been for focused problems such as stroke damage or Parkinson's disease, which affect a small, specific area," he explained. In the present study, however, injected stem cells migrated to broadly damaged areas of the spinal cord. "Something about cell death is apparently a potent stimulus for stem cell migration," said Kerr. "Add these cells to a normal rat or mouse, and nothing migrates to the spinal cord." Somehow the cells know there's no injury to heal. Kerr continued, "After 8 weeks, we saw a definite functional improvement in half of the mice and rats."[35]

Stem cells have also been used against multiple sclerosis, not just slowing its progress but even stopping it altogether.[36] The cells were cultivated from the patient's own blood allowing them to "grow," if you will, a new immune system. "I can write with a pen again," said one happy patient, a college student in the Chicago area. "The hand tremors are much better. I can butter my own bread and eat soup, which I couldn't do before. I can climb stairs again. Before, I couldn't walk a hundred feet. Now, I can walk for miles." Said another, "I have stability in my life. I'm not in limbo-land any longer."[37] Apparently about 85 percent of MS patients who receive stem cell therapy find that it essentially stops further deterioration.[38]

Far more common has been the use of stem cells to repair corneas in people who are far beyond being legally blind. These surgeries, which are being performed in the United States, Italy, Japan, India, Iran and elsewhere, graft stem cells from a donor or a patient's good eye to the injured eye.[39] The cells come from the limbus, a rim around the cornea. They resheath the cornea's surface, or epithelium. Even a tiny graft can expand to fully regrow clear epithelium, restoring sight.[40] It's estimated that three hundred such procedures a year are performed in the United States alone, and the number is increasing because this is the only alternative to plastic corneas for industrial accidents, terrible burns, damage from contact lenses, and even a few rare diseases that cause blindness.

The success rate for surgeries on one eye is about 90 to 100 percent.[41] In one eye, the surgery is "basically a slam dunk," according to its originator, Dr. Kenneth Kenyon, a Boston opthalmologist.[42] "I have some patients 20 years out with good vision," he told the *New York Times*.[43] "I believe these last a lifetime," said Dr. R. Doyle Stulting, editor of the journal *Cornea*.[44] "They are clearly successful and they are permanent."[45] If both eyes need restoration, however, the stem cell transplants stay effective in only half the cases after five years, primarily because of rejection.[46]

The *Times* reported the case of Shawn Smith, a jewelry designer who was cleaning a hundred rough emeralds in acid when the beaker exploded, splashing both acid and emeralds into his eyes. He was so blinded that he could only sense light. But doctors peeled back the thick scar tissue and sutured in limbus donated by Smith's half-brother. In just one week, the stem cells had grown a new transparent surface. After just one more such surgery he now has 20/50 vision.[47]

Researchers in several countries are working on growing sheets of limbus so that corneas can simply be stamped out of them like stamping out a contact lens.[48] A team in Taiwan found that within less than three weeks of harvesting, they could grow two square millimeters of limbal cells into a sheet 10 to 15 times larger. Since the initial study they have used the material on more than 95 patients, with complete re-epithelialization within four days of the transplant. Within a month, the corneas were clear. They say the technique is simple and can be done by any good tissue culture lab.[49]

If stem cells can be used to repair spinal cords and corneas, why not brains? Two studies released almost back to back in late 2000 showed they could do just that in rodents. In one, marrow with fluorescent markers was injected into the tails of mice whose own marrow had been destroyed with radiation. When the mice were later examined, the researchers, led by Helen Blau of the Stanford University Brain Research Institute, were delighted to find that the mice had brain cells that glowed.[50] The other team, led by Dr. Éva Mezey, a neuroscientist at the National Institute of Neurological Disorders and Stroke in Bethesda, Maryland, injected female mice with male stem cells and then located the Y chromosome on new cells in parts of the rats' brains.[51] Since only males carry the Y chromosome, this proved the new cells had come from the donated marrow.[52]

Three years later, these two teams again published studies in the same month showing the same process happening in humans. Both involved autopsied brains of women who had been treated with bone marrow transplants from men. This meant that any cells they found containing the male Y chromosome must have come from the donated bone marrow. The research teams not only found such cells, but found them clumped together, thereby indicating that the stem cells multiplied after reaching the brain.[53] "There's something that recruits these cells," said Mezey. "There's some factor that says: 'Come here; we need you.' Then, they receive further orders as to what type of cell to become."[54] Blau's team concentrated on looking for a specific type of neuron, Purkinje cells, which are located in the cerebellum. This is a part of the brain involved in coordinating and learning muscle movement patterns, and Purkinjes are exquisite in that they receive more connections than other types of neurons and they fire fifty times per second even while you're asleep.[55] "I think the stem cells may act as a repair squad," Blau explained. "They travel through the bloodstream, respond to stress and repair damaged tissues such as brain, muscle and other tissue throughout the body." If we can understand this mechanism, perhaps we can enhance it, she said. You can come up with drugs to mobilize them from marrow and target them to where they're needed. And you can help them to specialize according to the area and disease you're trying to treat.[56]

Other scientists have been showing that it may be possible to get cells already present in the brain to repair damaged areas and restore function. James Fallon, a professor of anatomy and neurobiology at the University of California, Irvine, and his fellow researchers have found that injecting a protein called transforming growth factor-alpha, or TGF-(alpha), into damaged areas of rats' brains stimulated stem cells to grow and differentiate into a massive number of normal, fully developed nerves.[57] The cells were then able to repair damage and restore the rats' movement ability. Injections of TGF-(alpha) into normal rat brain tissue caused no such stimulation.[58] "This study finally shows that stem cells can be induced naturally in large enough numbers and drawn to specific sites of damage, restoring function and replacing damaged cells in the brain," said Fallon. The researchers believe that a small amount of stem cell stimulation may occur naturally to replace damaged brain tissue. However, in cases of a large brain injury, stroke or degenerative disease, the brain's natural repair mechanism may not be able to keep up with so much damage. The addition of more natural growth factors to these areas may give the brain's repair apparatus a boost. "The stem cells already are in the brain and other organs in small numbers," said Fallon. "They can be stimulated in the brain to develop by a growth factor without the need for transplanting stem cells, embryonic tissue or altered cells from outside; instead, we've just stimulated cells that are already there."[59]

A more recent human trial, announced in April 2002, must be viewed with caution since it involved only one person. San Clemente engineer Dennis Taylor had rapidly progressing Parkinsonism until researchers at Cedars-Sinai Medical Center in Los Angeles and Celmed BioSciences of Montreal extracted 50 to 100 of his brain cells, cultured them, and injected about six million back.[60] Only a small percentage of the extracted cells were actually dopamine-producing neurons. Yet using a brain-scanning technology, the scientists found that after a year there was an increase in dopamine levels of almost 60 percent. Surprisingly, while the dopamine later returned to its presurgery level, Taylor maintained an 83 percent reduction in symptoms such as tremors.[61] "Two years ago I couldn't put my contact lenses in without a big problem," said Taylor, who is in his late fifties. "Now it's no problem. And I don't have to take any anti-rejection medication because the cells are myself."[62]

It's impossible to appreciate how truly amazing all this is unless you realize that until recently it was dogma that you didn't grow new neurons. Apparently what happens with severe head trauma and motor diseases like ALS or Alzheimer's is that natural regrowth can't keep up with the degeneration. Maybe soon, with the help of researchers like Fallon and the others whose work I've discussed, this too will change.*

*Incidentally, you may have heard about a horrific experiment reported in 2001 that some people think involved stem cells but did not. Researchers at Columbia University and the University of Colorado drilled holes into the skulls of people suffering from Parkinson's disease and implanted brain cells from aborted fetuses. The implanted cells apparently did survive and differentiate into the right kind of brain cells in some of the younger patients, whereupon, as the *New York Times* put it, they began

A report in the *Journal of Neuroscience Research* in 2000 showed, in fact, there may now be hope for sufferers of "virtually any disorder that destroys neurons, from stroke to brain trauma to degenerative diseases such as Parkinson's and Alzheimer's disease to spinal cord injury," according to lead scientist Dr. Ira Black.[63] Most bone marrow cells grow into red and white blood cells, but those that Black and his fellow investigators used are known as stromal cells. These are stem cells that normally form into the connective tissues of muscle, cartilage, bones and tendons.[64] It had been thought that this was as far as their transformative abilities could go. But Black, chairman of the Department of Neuroscience and Cell Biology at the Robert Wood Johnson Medical School in Piscataway, New Jersey, pushed the envelope.[65] He and his fellow researchers first removed cells from rodent marrow and grew them in Petri dishes, then added an oxidant and growth factors. "Over the course literally of minutes, the cells converted from rather pedestrian, flat, undistinguished stem cells into absolutely typical neuron-looking cells," said Black. "We were disbelieving."[66] Indeed, they were able to convert over 80 percent of the stromal cells. Then they repeated the process with the same type of cells from human marrow.[67]

The researchers said that when the newly made neurons were transferred to the brains and spinal cords of the rats, they appeared to settle in and get friendly with their neuron neighbors.[68] The scientists are now transplanting the altered cells into various areas of the brain and spinal cord in rats. "We know that, at the very least, they can survive within the spinal cord for well over a month," Black said. "That tells us that they are user-friendly in the live animal."[69] Susan Howley, director of research for the Christopher Reeve Paralysis Foundation, which co-sponsored the research with the National Institutes of Health, said, "The possibilities for spinal and brain repair suggested by this study are quite profound."[70]

Meanwhile, Italian researchers essentially did the experiment in reverse, coaxing brain cells from both mice and humans to grow muscle.

---

suffering "nightmarish side effects" including writhing, twisting and head jerking; one man could no longer eat without a feeding tube. But these were brain cells extracted directly from fetal brains to adults; they have nothing to do with stem cells. (Gina Kolata, "Parkinson's Research Is Set Back by Failure of Fetal Cell Implants," *New York Times*, 8 March 2001, p. A1.)

They did this in test tubes, then demonstrated it in live mice. The brain stem cells changed paths to become muscle simply through contact with muscle cells.[71] Some unknown "signal" spurred this change, according to one of the researchers.[72]

A Solution to Sickle Cell?

In late 2001, CBS's *60 Minutes II* ran a segment about a young man with sickle cell anemia who appears to have been cured by stem cells from umbilical cord blood. Although sickle cell has become far more treatable, there's a wide range of illness. Keone Penn, a fifteen-year-old living in Atlanta, had had the disease virtually his entire life and had even suffered a stroke at age five. He and his mother, Leslie Penn, were routinely in and out of hospitals to get blood transfusions they hoped would prevent another potentially deadly stroke. But Keone's pain was "usually so intense that even morphine, Demerol, those heavy-duty medicines don't really touch it," Mrs. Penn told the program.

Then Dr. Andrew Yeager of the University of Pittsburgh School of Medicine, who had heard of the new treatment procedure involving cord stem cells, contacted the Penns and suggested they give the treatment a try.[73] "The goal," he said, was that "these stem cells, which are in a relatively high proportion in cord blood—higher than they would be in our own bone marrow and definitely higher than in our own circulating blood—could then be injected and would take hold and ... make more of themselves, and make a whole new blood factory."[74] Another major advantage of umbilical cords over bone marrow is that marrow cells require an exact match, and finding such a match outside of your direct family is like finding the proverbial needle in a haystack. Umbilical cords needn't have such close matches; Keone's came from an umbilical cord bank in New York that's been storing cord blood since 1992.

Over the Christmas holiday in 1998, Keone was injected with the stem cells. The results may be one of those rare occasions when "miraculous" can be used literally. Within two weeks, his blood type changed from type "O" to type "B," the same type as that of the umbilical cord stem cells. Within a year, doctors declared that there were no sickle cells in Keone's bloodstream. "I love stem cells," young Keone told CBS. "I mean, they saved my life. If it weren't for them, I wouldn't ... you never know. I probably wouldn't be here today."[75] Yet while Keone was presented as being probably the first person cured of the disease, French researchers have been using stem cells to cure sickle cell patients since 1988. At a conference in late 2002 they reported that over the previous fourteen years they had performed 69 transplants, with an 85 percent disease-free survival rate. They started getting even better results since revising their mix of antirejection drugs in 1992. All 30 of the latest transplants appear to have been successful.[76]

Once it dawned on the scientific community that virtually any type of tissue can be regenerated, further evidence for stem cell differentiation and tissue building began pouring in like miners to a new gold strike. Researchers at New York University, the Yale University School of Medicine and New York's Memorial Sloan-Kettering Cancer Center showed for the first time that the human liver can regenerate its tissue with stem cells from a person's own marrow.[77] This study again used as its marking tool the Y chromosome unique to men. The researchers looked at liver tissue from two women who had received bone marrow transplants from male donors, as well as four men who received liver transplants from female donors.[78] Between 5 and 20 percent of the women's liver cells (hepatocytes) and bile-duct cells (cholangiocytes) had Y chromosomes, therefore the new liver cells "could only have come from their male marrow donations," explained Neil Theise, an associate professor of pathology at New York University School of Medicine. All of the four livers transplanted from women to men also showed signs of repair from male tissue, "most likely from [the men's] bone marrow ... given our findings with the female patients."[79] In one case, an amazing 40 percent of the damaged tissue was replaced with new cells from stem cells.[80]

Such regeneration would be lifesaving for those suffering cirrhosis of the liver. Usually caused by viruses such as hepatitis B or C or by alcohol abuse, cirrhosis is the eighth leading cause of death in the United States, killing some 25,000 annually.[81] Mostly because of cirrhosis, about 5,300 Americans received liver transplants in 1998.[82] Currently, the easiest way to get stem cells is extracting them from marrow. But contributing bone marrow is painful and requires general anesthesia.[83] As a result, there are only about seven million bone marrow donors in the world, with about four million in the United States.[84] Further, this method provides only a limited number of cells. But that will change when America's greatest untapped energy supply is put to use. Yes, I'm talking about fat.

One set of researchers from Artecel Sciences of Durham, North Carolina, reported in March 2001 that they had found a way to convert stem cells culled from the fat sucked out during liposuction into various forms of other tissues.[85] One was cartilage, which in natural form is capable of little or no self-repair since it has no blood supply. Cartilage lost to arthritis or injury appears to be gone for good. (Genzyme sells a repair kit that pulls out cartilage from the knees, grows it, and inserts it back. But it doesn't recommend it for arthritis deterioration or for joints other than the knee.)[86] To make the concoction, the Artecel team took two ounces of the connective tissue that lies beneath the layer of fat and extracted the stromal cells. Using both steroids and growth factors, they grew these into a billion cartilage-like cells in two weeks.[87]

"We have grown them into disks, something the size either of a dime or a quarter but thicker," said Farshid Guilak, director of orthopedic research at Duke University Medical Center in Durham, North Carolina.[88] "Those are the appropriate sizes to implant back into the body to repair a defect."[89] The original work was test-tube stuff, but a year later they took fat cells that they had grown into collagen, a main ingredient of cartilage, and implanted them under the skin of mice. These continued for three months to produce collagen and other ingredients in a matrix characteristic of cartilage.[90]

The scientists were able not only to turn fat into cartilage, but also to grow mature fat cells that could greatly aid cosmetic surgery for skin defects from acne or trauma. They had also converted the liposuctioned tissue into bone and blood cells, which could reduce or eliminate the need for matching and extracting bone marrow.[91] "One of the beauties of this system is that since the cells are from the same patients there are no worries of adverse immune responses or disease transmission," said Guilak, senior member of the research team.[92] Because so little fat is needed, you wouldn't have to have a typical liposuction candidate in order to get the raw material. A mere half-pound of fat could provide 50 to 100 million undifferentiated stem cells, enough to replace the cartilage layer of a damaged joint. But extractions of bone marrow only produce a few teaspoons of cells, requiring a lengthy process of cultivation in the lab until enough cells are ready for use.[93]

If you *did* need to get the fat from another patient, there are 600,000 Americans undergoing liposuction every year. It's truly the ultimate renewable resource.[94]

Another set of researchers, headed by the UCLA plastic surgeon Marc Hedrick, has performed similar alchemy with stem cells from liposuctioned fat, growing them into bone, muscle, cartilage and fat.[95] Hedrick is rather pleased that his team is able to do something so useful with something so reviled, comparing it to the transformation of petroleum. "At one time, oil and tar ruined fields and made land unusable," he said. "But then we discovered how to make it the dominant fuel of the industrial revolution. As we go to a revolution with cells and the regeneration of new tissue, you need a source of raw material to drive that. What you fundamentally need is cells, and fat is the only source of a person's own body where you can get lots of cells without causing harm."[96]

We're "just figuring out the best ways to apply this technology," Hedrick said in July of 2001. "But it's already in testing in animals and we expect practical uses to be available within the next five years or so." Further, he added, "We are finding lots of things in fat tissue beyond stem cell therapy that will be really important in the biotech revolution."[97]

Embryonic vs. Adult Stem Cells and Stem Cell Politics

Please note that while all types of stem cells break down into two general categories, embryonic and adult, *all* the above-mentioned stem cell applications use strictly ASCs.

Although supporters of federal funding for embryonic stem cell (ESC) research are often reluctant to admit it, the tremendous progress being made with ASCs could moot the debate. It has long been known that stem cells from marrow could develop into several types of mature blood and marrow cells, but it was also thought that was *all* they could do. That is, you couldn't use a marrow cell to make a brain cell or a heart muscle cell. Moreover, it also wasn't known that, aside from stem cells in marrow, umbilical cords and placentas, stem cells could be harvested from all over the body.[98]

It's still widely argued that ESCs may have greater potential to differentiate or that, in scientific terminology, they are "more plastic."

ESCs, we are widely told, can ultimately be made into any type of mature cell in the body. ASCs, meanwhile, are much more limited in what they can become. Yet not only is this mere theory, the theory is gradually imploding.

First, we don't really know just how flexible ESCs are.[99] A team of ESC researchers in early 2003 published a report in *Science* in which they concluded that previous rodent studies concluding that ESCs were becoming insulin-producing cells were in error.[100] The leader of the research team said he remains optimistic: "We still haven't learned anything that makes me think it will be impossible to turn ESCs into insulin-producing cells."[101] But certainly his phraseology left open that possibility, and to admit that perhaps ESCs cannot become any type of mature cell is to concede that they also may not be able to become other types.

Moreover, there's rapidly growing evidence that ASCs are not just far more plastic than was once thought but may be every bit as plastic as ESCs. The dogma was that only ESCs had the potential to form all three of the embryonic germ layers (as in the germination we associate with plants) from which all mature cells eventually develop: ectoderm, mesoderm and endoderm.[102] But Dr. Darwin Prockop, director of the Gene Therapy Center at Tulane University, declared in a letter to *Science* that "We do not yet know enough about adult stem cells or ESCs to make dogmatic statements of either."[103] He told me, "There's no law of physics or such that I know of that says that [ASCs] are inherently more limited than ES cells."[104]

But there's much more than theory showing that ASCs may be far more plastic than once thought. Recently Catherine Verfaillie and her colleagues at the University of Minnesota's Stem Cell Institute have found stem cells in human marrow that appear to transform into all three of those germ layers.[105] "I think Verfaillie's work is most exciting and translatable into the clinical arena," said Dr. David Hess, a neurologist at the Medical College of Georgia in Augusta.[106] "They seem to give rise to every cell in the body. She seems to have a subpopulation with basically all the benefits of ES cells and none of the drawbacks. I've seen her data and it's extremely impressive to me."[107] Verfaillie calls the cells "multipotent adult progenitor cells," and has isolated them from mice, rats and people. They have already been transformed into cells of blood, the gut, liver, lung, brain and other organs.[108] A small biotechnology company, Atherysys of Cleveland, has obtained the exclusive rights to its commercial applications, but quickly disbursed cells to almost twenty academic scientists and made it clear they are free to use the cells for their research.[109]

I've found no stem cell researcher, regardless of what type of cells he or she works with, who says that Verfaillie's work doesn't have tremendous potential. The worst they'll say is that they're skeptical and that her work needs replication. Yet by the time she announced her findings, there were other studies already providing support to a great extent. Many published reports have shown that various types of stem cells from bone marrow have differentiated in the Petri dish or in the body into cells of tissues of a different germ layer from that which one would normally expect, covering the gamut of all three germ layers.[110] "Taken as whole," Verfaillie noted, "these studies suggest that stem cells with multipotent nature—akin to ES cells—are present in bone marrow and perhaps in other tissues such as brain and muscle."[111]

Since then, two more sets of researchers have found two more types of ASCs that may also be as plastic as ESCs and further support Verfaillie's findings.

In September 2002, researchers at the Robert Wood Johnson Medical School reported the discovery of a cell from bone marrow that, within a mere five hours, they had converted to neurons both in Petri dishes and in rats.[112] Under the old dogma, that was simply impossible; neuronal cells are not supposed to develop from this different germ layer. More importantly, "We found that they express genes typical of all three embryonic germ layers," lead researcher Ira Black told me. "In aggregate, our study and various others do support the idea that one [ASC] can give rise to all types of tissue."[113]

Then, just a little over half a year later, an even more startling discovery was announced, one that may surpass Verfaillie's in some ways. Scientists at the Argonne National Laboratory in Illinois, headed by Eliezer Huberman, found a stem cell in blood that they were able to differentiate into six different cell types from all three germ layers.[114] "Verfaillie's work is excellent," Huberman told me. "There's no question that our work supports hers, but the great advantage with ours is simplicity. We use a completely different type of [stem] cell," one called a monocyte. "Isolation of Verfaillie's cells is a laborious procedure, whereas ours is fairly straightforward. Our purpose was to find a simpler method of isolating stem cells and we did. It should move the field forward very fast."[115]

Even if ASCs do have limits, this would be a limit on each type of cell. But there are so many types of these stem cells already discovered that together they may be able to do everything that embryonic ones can do. You don't need "one-size-fits-all" if you have plenty of sizes. ASCs have already been culled from blood, bone marrow, skin, brains, spinal cords, dental pulp, baby teeth,[116] muscles, blood vessels, corneas, retinas, livers, pancreases, fat and hair follicles.[117] Extremely young stem cells, which don't qualify as "adult" but require no embryo deaths, are in placentas, umbilical cords and amniotic fluid.[118] In fact, different types of cells have been found in two different locations in umbilical cords; one type apparently differentiates into both neurons and another type of brain cell called glias.[119] Unfortunately, umbilical cord blood storage is still somewhat expensive, but scientists in late 2002 announced that by exposing cord blood in the laboratory to a certain protein molecule they were able to increase the number of immature stem cells in a sample of cord blood an amazing hundredfold.[120] It seems that wherever scientists look in the body they eventually find stem cells. Certainly far more types will be found.

Probably many people believe ESCs have an advantage in not evoking an immune response in the recipient. On its face, that might seem reasonable because while fetuses have partial immune systems at best, embryos have no immune *system* at all. But depending on various circumstances, ESCs can produce high levels of a protein that would prompt rejection if they were implanted, and it appears that the more they differentiate, the higher that protein level goes.[121] Researchers hope to engineer embryonic cells that don't evoke the protein, but it's a serious roadblock.[122] Stem cells taken from one's own body, naturally, don't induce any such response. Remarkably though, Osiris has isolated stem cells from bone marrow that don't cause an immune response when transplanted into other animals or even into other species, such as from pigs to rats.[123] This may also hold true of other ASCs. Obviously, in using a person's own stem cells, rejection is never a problem.

Another serious problem with embryonics is their tendency to start "running away," thus causing cancer.[124] "The early-stage stem cells are both difficult and slow to grow, according to lead author Michael Shamblott from the Johns Hopkins School of Medicine and one of the nation's top researchers on embryonic stem cells. "More important," said Shamblott, "there's a risk of tumors."[125] Recent research indicates that the same protein may control the proliferation of both the stem cells and cancer cells.[126] This is by no means an insurmountable problem, but it is one that must be overcome. ASCs have no such drawback.

Unarguably ASCs are much further down the road in practical applications. As of late 2001, "Do No Harm," a D.C.-based coalition that opposes use of ESCs, had listed over thirty different anticancer applications alone from ASCs, all performed on humans and all appearing in peer-reviewed medical literature.[127] Further down the pipeline, Do No Harm lists about one hundred ASC experiments in animals that have shown success against a tremendous variety of diseases.[128]

Therapeutic stem cell use from bone marrow transplants has been common since the early 1990s.[129] The first successful use of umbilical cord stem cell therapy dates back to 1988 and now some seventy different diseases, primarily forms of leukemia, are treated with umbilical cord stem cells.[130] Like bone marrow, cord blood is rich in stem cells, which can invigorate the immune system in the recipient. Studies indicate that cord blood stem cells are even less likely to provoke immune system rejection than bone marrow stem cells because the baby's immune system hasn't developed enough for recognition. It takes about six to nine months for the immune system to develop fully after a baby is born.[131] Further, unlike marrow cells, cord cells are a cinch to collect. Originally, the only stem cells obtainable from cord blood were hematopoietic, meaning they could only grow into blood-forming tissue. But Boston-based ViaCell is pushing cord stem cells into becoming other types. They've already created more hematopoietic stem cells, plus mesenchymal (marrow) and neural cells. "With growth factors we can grow them and with antibodies we can direct them," said ViaCell CEO Marc Beer. Among the diseases its researchers are targeting are stroke, ALS and diabetes. "What ViaCell is doing is creating an ethical source of abundant stem cells with great plasticity," said Beer.[132]

Despite this demonstrated promise, the journal *Nature* in March of 2002 printed two *in vitro* (meaning "outside the body" or "Petri dish or test tube") studies widely interpreted by the popular media as indicating that ASCs are either grossly inferior to what had earlier been believed or even outright worthless.[133] The *Nature* writers basically said their studies showed that quite probably ASCs were not differentiating and multiplying at all; rather, perhaps the cell nuclei were merely fusing and the resulting fusion gave the impression of a new, differentiated cell forming. The media gobbled it up. Agence France-Presse headlined: "'Breakthrough' in Adult Stem Cells Is Hype, Studies Warn."[134] The Australian Associated Press declared: "New Research Tips Debate on Stem Cells," while the *Washington Post*'s subhead flatly announced: "Adult Cells Found Less Useful Than Embryonic Ones."[135] It was terribly damning. And also terribly false.[136]

Stanford's Helen Blau countered with a big "So what?" In a *Nature* commentary, she noted that far from being "mere," "Cell fusion has long been known to achieve effective reprogramming of cells"; in fact, her own laboratory was doing it twenty years ago.[137] Thus, far from showing that adult stem cell research is "hype" or whatever term the particular newspaper or newswire chose to apply, it turns out that cell fusion may both compliment and encourage the differentiation of adult stem cells that's already proven so valuable and promising.

The idea that differentiation wasn't happening at all was simply bizarre in light of myriad studies and therapeutic applications indicating otherwise, including one that appeared in the journal *Blood* shortly thereafter. Showing that bone marrow stem cells can be converted into kidney cells, it pointedly concluded: "The process does not involve cell fusion."[138] "We found no evidence of nuclear material from two cells fusing into one cell," one of the co-authors emphasized to me.[139] Said Tulane's Darwin Prockop, "It may well be that fusion is part of the healing process. But clearly we can take mesenchymal cells and differentiate them into various tissues because it's into bone or fat and it's been done over twenty years."[140] And still the evidence pours in that the fusion phenomenon was grossly overstated. In *The Lancet* in early 2003, researchers looked at cheek cells from five living women who had received bone marrow transplants from their brothers several years earlier. They found cells containing the male Y chromosome, a sign that donor marrow stem cells had differentiated into cheek cells. Moreover, the group found almost no evidence of fusion among the cells in the cheek. Of the 9,700 cells that were examined in the study, only two showed signs of possible fusion.[141]

Just months later it was *Nature*'s prime competitor, *Science*, that again attempted to show that the earth really is flat after all, first with a letter in which the authors from the Baylor College of Medicine claimed they had earnestly tried but failed to find bone marrow cells that had differentiated into brain cells, and later with a paper claiming to show that blood stem cells replenished marrow but appeared worthless for creating other tissues.[142] This second study led to such headlines as United Press International's "Promise of Adult Stem Cells Put in Doubt."[143] Stanford University scientists led by Irving Weissman reported that blood stem cells replenished marrow but seemed unable to generate other kinds of tissue.[144]

That hardly justified a headline implicating *all* ASCs, though it would seem to constitute a real setback. "Blood-forming stem cells from adults make blood," Weissman told UPI. Eschewing the usual cautionary scientific terminology including such phrases as "it appears" or "evidence indicates" or "our particular study has found," Weissman smugly declared: "They [the cells] don't make brain; they don't make heart muscle or any of these things."[145] Any good scientist—and numerous *critics* of Weissman have told me he's an excellent scientist—knows that a single study never proves anything. Weissman certainly knows that, since a study published in *Nature Medicine* in November 2000 showed the opposite to be true. In that case, such cells rebuilt liver tissue when injected into mice. It's not likely he didn't know about that study, considering that he was listed as a lesser co-author.[146]

According to Blau, neither study was worthy of publication in *Science* or anywhere else. The Baylor study, she noted, failed to detect not only neurons but also something far more readily detectable, microglial cells. Yet, "At least 20 reports over the past 15 years have shown that bone marrow transplantation results in readily detectable replacement of a large proportion of microglial cells in the brain."[147] Said Blau, "If they couldn't see those, how could they possibly see neurons?" It would be like failing to detect a tiny virus under your microscope when you also couldn't see the gnat that accidentally got trapped between the slides. Either your microscope is faulty or you simply don't know how to use it.

"As to Weissman's paper, where you look and how you look determines what you see and he doesn't define where he's looking," she said. "Our own experiments have shown there can be a thousand-fold frequency of stem cells depending on where you look." Because he didn't say where he looked, "You cannot replicate his experiments," she said. "You *could* replicate ours but he did not. The other false assumption he made was to look at a fraction of marrow, the hematopoietic part, and he looked in absence of any damage to the body, yet these cells are damage repair cells." In other words, one shouldn't think it remarkable that no ambulance shows up when there's no need for an ambulance.

I asked her if the publication of these two papers was a political act designed to harm the public image of adult stem cells. "That's been a question in many people's minds," she replied. "Why these negative findings should have been published in such a prominent way does suggest a political agenda."[148]

Certainly Weissman is anything but a disinterested party. "Weissman has made it his mission to show adult stem cells don't have any potential," said Georgia's David Hess, who studies the use of bone marrow cells as a way of treating stroke.[149] A year before his *Science* report, Weissman had admitted to the *National Journal* that he was an advocate for federal funding of embryonic research. Six months before the publication he called for such funding in congressional testimony.[150] Weissman is also exceptionally influential. His views on stem cells have been quoted in literally hundreds of news stories over the last couple of years, including the aforementioned William Safire piece.[151] In May 2002, Weissman was named California Scientist of the Year by the California Science Center. Eleven previous recipients of this honor have gone on to win the Nobel Prize.[152] What readers are virtually never told is that Weissman has made millions of dollars through three companies he's founded—Systemix (now incorporated into another company), Celltrans and StemCells—all of which experiment with ESCs.[153] Weissman thinks it's exculpatory that some of the work from these companies has been on ASCs, but the excuse doesn't work because he realizes that his work will do just fine without outside help. It's the ESC experiments and the promise of ESC work that "need" the hype that Weissman so willingly provides. The depth of Weissman's involvement in ESC work became apparent to the public only in December of 2002 when Stanford University announced a new Institute for Cancer/Stem Cell Biology and Medicine, with Weissman as director. The sole purpose of this institution is to use nuclear transfer to develop ESCs for research. ASCs need not apply.[154]

Moreover, it was then that Weissman's willingness to play with facts to serve his purposes became even more obvious. Weissman said his planned research is "not even close" to cloning. But according to the American Association of Medical Colleges, of which Stanford is a member, "Somatic Cell Nuclear Transfer or therapeutic cloning involves removing the nucleus of an unfertilized egg cell, replacing it with the material from the nucleus of a 'somatic cell' (a skin, heart, or nerve cell, for example), and stimulating this cell to begin dividing."[155] That's exactly the type of work the new research institute was established for. Ethicist Ronald Green, a religion professor at Dartmouth University, told the Associated Press that he applauded Stanford's announcement, but said "cloning" is indeed the mostly widely accepted term for what Weissman's team plans to do.[156] Paul Berg, the father of gene splicing, also said he favored the research, but when asked if it was cloning he responded simply: "It is."[157] Yet Weissman maintained his position to the point that the institute claimed on its website that the President's Council on Bioethics supported Weissman's assertion that it was not engaged in cloning and that the council supported the institute's research strategy. This came as something of a surprise to the council, which demanded a correction, saying it does not endorse cloning for biomedical research and that Stanford had tried to "obfuscate" the nature of its potential experiments. Stanford was forced to change its statement, deleting all references to the President's Council on Bioethics. Weissman allegedly was out of the country and unavailable for comment.[158]

Yet Weissman is far from alone in pretending to be—or perhaps more importantly in being portrayed as—a disinterested party when in fact he's a warrior for ESCs. In a piece titled "Mixing Business with Stem Cells," the *National Journal* said that reporters almost never disclose any pecuniary interest that the scientists they quote may have in the subject discussed.[159] One scientist was quoted 124 times, yet only once was identified as being a founder and shareholder of a company working with embryonic stem cells. He insisted that he has no potential conflict of interest because he has "no involvement in the day-to-day running of the company"—which misses the point that his seemingly disinterested statements affect his pocketbook and his career.[160] A scientist who told the Associated Press that Weissman's *Science* study showed "stem cells from the bone marrow will not be a practical source for many cell types needed" to treat disease was identified as the head of Advanced Cell Technology; but readers were never told that his company experiments exclusively with ESCs.[161]

Likewise, when the Australian Associated Press published its piece on the aforementioned two letters to *Nature*, it claimed, " 'New overseas research had boosted the argument for conducting stem cell research on human embryos,' an Australian biotechnology expert said today."[162] It never said that this expert headed a Melbourne company that was just on the verge of "creating cloned human embryos as a source of stem cells" and works exclusively with ESCs.[163] The former U.S. senator Connie Mack, who had displayed ignorance that there was any such thing as a nonembryonic stem cell, was never identified as a hired lobbyist for two different biotech companies.[164] Clearly, many of the allegedly impartial scientists testifying before Congress and trying to persuade the media are merely putting their mouths where their money is. The scientists will claim that their desire for federal funding has no affect on what they say about either the alleged value of embryonic cells or lack of value of nonembryonics, but to believe that is to believe that when somebody puts on a lab coat he ceases to be a human being.

Not surprisingly, all this has tremendously affected popular media coverage of stem cells. When an ESC hiccups, it makes page one; but reports of adult stem cells actually saving human lives are sometimes ignored. For example, in mid-2001 it shook the media world when scientists at the University of Wisconsin School of Medicine announced they had succeeded in turning human embryonic stem cells into blood cells. There are almost five hundred references to it on the Nexis media database, coming from all over the planet. It *was* newsworthy in that it was the first published report that the cells can be turned into other human tissues, but lead researcher Dan Kaufman admitted, "I don't want to give false hope to anybody that we're able to treat cancers or blood disorders now."[165] He added, "I think it's something that we hope will come in the future, but there are a lot of hurdles to overcome in the meantime." Indeed, he said it was too early to tell if the stem cells could actually grow in a person's or an animal's body.[166] Yet for all the media coverage, one would never know that three decades earlier, medical journals were publishing studies on the ability of ASCs to do the same thing and in human bodies, nor that those cells *are* being used to treat cancers and blood disorders now and have been routinely since the 1980s.[167] It's possible to read lengthy, detailed articles on the promises of stem cells that keep the reader ignorant that ASCs even exist.

Some scientists have become quite dismayed at all this, and haven't waited around for writers such as myself to call them up to express their opinions. In a commentary in the *Journal of Cell Science* in February 2003,

British researchers asked in the very title, "Plastic Adult Stem Cells: Will They Graduate from the School of Hard Knocks?" In a good-humored, indeed sometimes humorous, piece the angst nonetheless came through. "Despite such irrefutable evidence of what is possible, a veritable chorus of detractors of adult stem cell plasticity has emerged, some doubting its very existence, motivated perhaps by more than a little self-interest." While certain issues still need resolving, the authors said, "slamming" the "whole field because not everything is crystal clear is not good science."[168]

Even scientists who strongly favor ESC funding readily admit that the issue is highly politicized, with ASCs getting the short end of the stick from research publications, the popular media and the scientific community. Blau and Prockop are among them, as is Patricia Zuk, who works with Marc Hedrick at UCLA in differentiating stem cells from liposuctioned fat. Zuk admits that the "stem cell wars" have the potential to hurt her own research, but she blames the Bush administration for essentially forcing the media and researchers into being dishonest by restricting research on ESCs. (Prockop also puts some blame on "the Christian Right.")[169] "Certainly it's politicized," said Zuk. "I think a lot of embryonic stem cell people are right in trying to protect their jobs. As an adult stem cell researcher I'm not going to tell these people to give up. Further, I feel adult stem cell research would not be where it is as quickly without embryonic cell research. So I can see why they're being critical of the work being done by adult stem cells, though at the same time I think, 'Come on, I'm not trying to take away your job.' "

Zuk puts something of a happy face on the conflict when she speaks of the constant efforts to downplay ASC research and to exaggerate alleged breakthroughs in ESC research. "It makes adult stem cell researchers work faster and produce better science. When somebody says 'I believe everything you're doing is a crock,' I work harder." But isn't she concerned that the disinformation could hurt her funding? "I'm very concerned that my funding could be hurt. It's a very worrisome situation. Why does it have to be them versus us in competing for grants? Science should be a collaborative experience."[170]

The problem is that the research pie is a limited one. After years of tremendous growth, the budget of the largest federal grant maker, the NIH, has practically leveled off. "That means the NIH may be hard-pressed to maintain the funding of existing grants, which typically run for four years each, and would have to cut back sharply on its support for new and innovative proposals," reported the *Washington Post* in February 2003.[171]

"NIH has a lot of discretion into what they can fund," said David Hess. It's not that the people at NIH don't know what's going on and can't tell real science from hype. To the extent the public and the people on Capitol Hill holding the purse strings become convinced that adult stem cell research is essentially worthless while embryonic stem cells have all the promise, money is more likely to flow to ESC research. "Everybody is fighting over the same pie and there's a lot of pressure on NIH to give these scientists [those working with embryonic cells] what they want. And what they want is literally billions of dollars," Hess said. "Certainly one of my motivations is I don't want money from adult stem cell research being pushed into embryonic, though it's already starting to happen."[172]

Ironically, much of the pressure for government funds for ESCs is precisely because ASCs keep appearing more and more promising, especially in the short term. "Adult cells are far closer to commercial application, which is crucial to venture investors," wrote physician Scott Gottlieb, now with the FDA, in a piece titled "Adult Cells Do It Better" in the *American Spectator.* As he explained it, "Given the long lead times necessary to gain approval for new medical technologies, if a company can't get significant results in four to six years, it's generally beyond the scope of a venture capitalist's interest." In addition to speed, "venture capitalists look for a 'technology platform' broad enough to support multiple indications for the same product. That way, if a company fails with one application, there's plenty of room to develop other therapies that can take advantage of the previous research. While the embryonic cells are rumored to have broad potential, so far only adult stem cells have demonstrated wide uses."[173] Some have tried to blame the lack of capital going to embryonics on President Bush's decision to limit the number of cell lines available, but Gottlieb's piece appeared several months before the Bush announcement and suggests that investors know a sure thing when they see it.[174]

Basically, we're looking at a political see-saw. If the value of ASCs appears to rise, that of ESCs falls in comparison. Thus those anxious to see embryonics funded downplay the value of ASCs even to the absurd length of denying their existence. With every breakthrough in nonembryonic research comes the need to turn up the volume of the disinformation. Private money isn't fooled by any of this, but the government can make grants based on political fashion. Numerous companies and individual researchers have become desperate for those grants.

Nevertheless, the list of stem cell researchers who believe that adult stem cells may give us everything we need is long and growing. Dr. Adam J. Katz, a member of one of the research teams that spun fat into body tissues, has declared, "This discovery potentially could obviate the need for using fetal tissue." Added Katz, who works at the Division of Plastic and Reconstructive Surgery at the University of Pittsburgh School of Medicine, "We don't yet know the limits for stem cells found in fat. So far, we have seen promising results with all of the tissue types we have examined."[175] An outside researcher commenting on Katz's work, Eric Olson of the Department of Molecular Biology at the University of Texas Southwestern Medical Center in Dallas, said it's heartening that almost "every other week there's another interesting finding of adult cells turning into neurons or blood cells or heart muscle cells. Apparently our traditional views [that fetuses or embryos are necessary to provide these cells] need to be reevaluated."[176] Ira Black said his discovery that all sorts of cells could be teased into becoming neuronal ones "essentially circumvents all the ethical concerns with the use of fetal tissues," while Markus C. Grompe, a professor of molecular medical genetics at Oregon Health and Science University in Portland, said the same thing after the results of a previously mentioned liver study came out.[177] "This would suggest that maybe you don't need any type of fetal stem cell at all—that our adult bodies continue to have stem cells that can do this stuff."[178]

Yes, many people with terrible illnesses and injuries have testified before Congress that fetal and embryonic stem cell research must be government-funded.[179] This should not be a surprise since most of the public is ignorant of the incredible progress made in using the body's own stem cells. As we've seen, even scientists making huge strides with ASCs, such as Blau, Zuk, Theise, Verfaillie and Prockop, support government money for embryonic research. But it's typical of scientists to want to see more of everything scientific funded, for the same reason you never met a teacher who didn't want more money for education or a general who didn't think more should be spent on national defense. "Most scientists never want a door closed, they want all doors open," remarked Hess. "And anybody who disagrees with that stance is seen as trying to hold up medical progress."[180]

While Hess believes as a scientist that ESCs have no advantage over ASCs, he openly admits he also opposes the technology for ethical reasons. "I and other scientists feel there are ethical constraints on science. If you suggest this, they say you're against progress and for keeping people in wheelchairs, but I think it's a very, very slippery slope and I have a lot of problems with using embryonic stem cells. It's clear to me that life begins at conception. Certainly by blastocyst stage [when the cells start to become useful] it is. I was always taught to first do no harm." He is also bothered by the cloning aspect. Doctors like Hess, or like the more prominent David Prentice (the University of Indiana biologist and stem cell researcher who is the doctor most associated with opposing embryonic stem cell research) show there's a lot more going on here than one scientific method competing with another for attention and funds.[181] Advocates of ESC research are quite correct when they accuse "the Christian right" of opposing ESC work not on scientific but on ethical grounds and using ASC research as their Great White Hope. But it's also clear that prominent abortion proponents such as Anna Quindlen have grossly exaggerated the progress in ESC research as a way of portraying the anti-abortion movement as a bunch of kooks fighting fanatically to prevent cures for a hundred diseases. Quindlen specifically invoked the plight of Michael J. Fox, Christopher Reeve and Ronald Reagan.[182]

So everybody plays the game. Ronald Reagan may be beyond help, but no honest researcher of stem cells of any type will deny that if stem cells are to help relieve the suffering of Michael J. Fox and Christopher Reeve in the next few years, it will be those of the adult variety. To the extent that adult stem cell research loses funding or loses researchers who become demoralized by the full-court press being waged against them in both the scientific and the popular media, it will be the Anna Quindlens of the world, not the George Bushes, who are preventing those cures.

The stem cell controversy is one that technology can render obsolete, but only if politics stands down and lets adult stem cell researchers do their work.

FOUR

Xenotransplantation

Even the hardened Russian police were horrified. It seemed unreal, but there it was on their own videotape. Five-year-old Andrei from a small city south of Moscow is sitting innocently in the back seat of a Mitsubishi SUV. He has no way of knowing he's just been sold for the equivalent of $90,000. He has no idea that the sellers are his own grandmother and an uncle, or that the buyer isn't particularly interested in the whole "package," but rather just the kidneys, eyes, and possibly the heart and lungs as well. Fortunately, the tape ends with the cops moving in to make arrests. Unusual? Yes. But it's not so unusual that there isn't already a specified penalty in Russia's criminal code for selling children's organs.[1]

Ironically, it's precisely because techniques in organ transplantation have tremendously improved that the waiting list for organs has tremendously lengthened. In the United States it now stands at about 80,000 Americans.[2] In 2001, over 6,100 died on that waiting list, including a friend of mine who couldn't get a new liver after his was destroyed by cancer.[3] He was 44.[4]

There is also evidence that organs that are available are not fairly distributed, and much evidence that the shortage has created a grisly market for human parts in many areas of the world.[5]

When there's enough time, making a new organ to replace a failing one could work. Unfortunately, all too often a vital organ is needed immediately. That's where xenotransplantation comes in. Xenotransplantation involves transferring organs such as hearts, kidneys, lungs and livers from pigs or other animals to humans. Pig heart valves have been used in humans routinely since 1964, but they contain no live cells.[6] Cook Biotech of West Layette, Indiana, is already using intestinal lining from pigs for everything from patching up gunshot wounds, to closing up large sores, to regrowing cartilage in knees, to reinforcing the abdominal wall so as to prevent hernias from reoccurring. Damaged tissue is replaced with little or no scarring, mimicking the rapid healing that occurs in an infant's skin. The material is first freeze-dried, then sterilized and cut into sheets. Before application, the sheets are recut to fit the need. The "pig patch" initially seals the area, and then acts as a scaffolding to allow the body to build new tissue. The porcine cells are absorbed. Cook claims there has never been a case of rejection.[7]

The problem arises with the transplantation of whole organs. Whether porcine or primate, rejection is quick and untreatable.[8] The most promising method for overcoming rejection is biotech. It's thought that when human genetic material—preferably the recipient's own—is injected into pigs or other donor animals, the human body will recognize the new organ as human and begin to use it as its own.

Since pig organs are the right size for humans, and since pigs are inexpensive and easy to breed, scientists have long considered them potential donors. But "when you transplant a pig organ into a nonhuman primate, it is rejected immediately," said John Logan, vice president for research and development at Nextran in Princeton, New Jersey (a subsidiary of Baxter Healthcare, based in Deerfield, Illinois).[9] "It just turns black before your eyes, within 30 minutes."[10] The reason is that pig cells have a gene, GGTA1, that helps them make an enzyme that attaches sugar molecules to cells in their bodies. The sugar molecules are instantly recognized as foreign and attacked in primates, and there are no breeds of pig with Nutrasweet or saccharin. In April of 2001, Nextran researchers announced they had found a way to bypass the problem by injecting pig embryos with genes that block the immediate rejection process. Nevertheless, cloning the animals would be more efficient, since scientists could knock out the gene producing the sugar molecule before using the pig cells to create clones, and thus breed sugar-free pigs.

PPL Therapeutics became the first company to clone pigs and then the first to clone transgenic pigs, in April 2000.[11] Over two years passed before researchers at PPL and elsewhere, working independently, were able to clone pigs that lacked a copy of the GGTA1 gene.[12] Just seven months later, PPL announced the cloning of pigs with both copies of the gene knocked out.[13] Shortly thereafter, an American company said it too had developed swine worthy of having pearls thrown before them, with double knockout miniature pigs that theoretically might make better organ donors.[14] But both companies and any that join them will still have to overcome the immune system rejection problem that comes from transplanting organs between humans.[15]

Yet in a vastly different approach, Italian scientists in late 2002 mixed swine sperm with human DNA to create pigs that carry human genes in their hearts, livers and kidneys. The modified sperm was then used to produce litters of pigs carrying the human gene.[16] "What we obtain at high efficiency and low cost is genetically modified pigs expressing the human protein," said Dr. Marialuisa Lavitrano, a researcher at the University of Milan and first author of the study.[17] The organs from the test animals are still not ready for humans, because they retain some genes that would cause immediate rejection. Lavitrano said that five to seven other genes would need to be silenced or replaced by human genes before useful organs could be harvested from the animals. "With our efficiency we think we can add the other genes and breed the animals in about two years."*

One possible problem with xenotransplantation from pigs has some serious researchers worried, namely the transferring of viruses from them to us. We catch viruses from animals all the time; indeed, if you've ever wondered where that new strain of flu comes from each year, the answer apparently is pigs in China. Humans give flu to the pigs, it mutates in the "mixing bowl" of the pig population, and then comes back to us in altered form via birds.[18] SARS may well have followed this route, though there's no proof as yet.[19]

But if you happen to catch a virus, that's one thing; receiving it from an organ transplant is quite another. Moreover, there are a lot of viruses more worrisome than influenza. One study showed that porcine endogenous retroviruses (PERVs), which are integrated into pig genes, can replicate in immunocompromised mice transplanted with pig cells.[20] "We worry about recombination events that could occur in the context

---

*Nextran researchers have already used transgenic pig livers to save the lives of persons with acute liver failure, but the liver was kept outside of the body to do its filtering since its only purpose was to keep the person alive long enough for a human liver to become available. ("Researchers to Study Using Pig Livers as Xenotransfusion Bridge to Transplantation," *Transplant News* 7:20 (31 October 1997).)

of the xenotransplanted cells in a human taking immunosuppressives," said Thomas Okarma, CEO of Geron.[21] Fortunately, PERVs have not been shown to cause disease; instead, they seem to simply set up house and leave the host animal otherwise alone. "Swine fever, hepatitis E, and the Nipah virus have been linked to pigs; however all are exogenous viruses [caught from other animals] and can be excluded from a herd which is bred in a high biosecurity unit," said Khazal Paradis, the director of clinical research at Imutran, a British subsidiary of Novartis in the U.K.[22] "There are no known viruses in the pig genome that have caused disease in man."[23]

But the potential for such viral spread hasn't been ruled out. Happily, for every technological problem there's usually a solution, and BioTransplant of Charlestown, Massachusetts, has announced that it has bred miniature pigs that don't produce PERVs capable of replicating in human cells, while a related company, Immerge BioTherapeutics in the same town, has demonstrated that inbred miniature swine do not transmit PERV viruses to human cells.[24]

Naysaying Longevity

Because there's been so much talk of biotech anti-aging breakthroughs, it was probably inevitable that naysayers would pop up to say it's all bunkum. Nobody in our generation, or even our children's generation, will live significantly longer than they do now, these people insist. "A life expectancy at birth of 100 years, if it ever occurs, is unlikely to arise until well past the time when everyone alive today has already died," said S. Jay Olshansky of the University of Illinois at Chicago at a March 2001 meeting of the American Association for the Advancement of Science. France and Japan will not reach a life expectancy at birth of a hundred years until the twenty-second century, he said, while Americans won't hit the century mark until the twenty-sixth century. "The rise in life expectancy in the future will be measured in days, weeks and months—not in decades, as some proponents of extreme longevity predict."[25]

I must admit I've seen some pretty far-fetched predictions by people who call themselves "futurists." But making predictions about the condition of the human race in five hundred years readily pulls you out of the "scientist" class and plunges you smack into the realm of

Nostradamus. What Olshansky and his fellow nay-centenarians are doing is using circular reasoning along the lines of: "I've never been in a fatal accident, therefore I never will be in a fatal accident." More specifically, they're simply looking at past trends.

Imagine a similar projection of electronic technology from forty years ago. Virtually nobody would have conceived of desktop computers that double in speed every eighteen months, nor bandwidth that doubles far faster, nor could they have imagined the World Wide Web, which didn't come along until 1991.[26] Probably not many people would have thought that shiny hand-sized disks could store movies, several record albums or a library's worth of books. Yet Olshansky is willing to make a prediction for half a *millennium* from now.

Similarly, Leonard Hayflick of the University of California at San Francisco, one of the leading figures in gerontology, wrote in *Nature* magazine in November of 2000, "There is no evidence to support the many outrageous claims of extraordinary increase in human life expectancy that might occur in our lifetime or that of our children."[27] His evidence? Extrapolation data from the U.S. Census Bureau, the U.S. Social Security Administration, and a group of the world's industrialized countries. This says nothing more than "since there were no breakthroughs in the last twenty years, there will be none in the *next* twenty years."

Yet shortly after Hayflick's commentary appeared, researchers announced in *Science* magazine that by causing a gene mutation they succeeded in getting fruit flies to live as much as 85 percent longer.[28] "What's exciting about these findings is that they suggest that there is a genetic system common to all animals that regulates aging," lead author David Gems of University College London told *Reuters Health.* "If we could just tap into the mammalian version of that system it might be possible to retard or even reverse human aging."[29]

Claiming that all of the anti-aging work described in this section will definitely pan out would be foolish indeed. But claiming that absolutely *none* will is far more so. Lifespan extension is a goal that to many of us has an exceptionally high value and offers research institutions and companies awesome financial incentives. Oracle software CEO Larry Ellison's philanthropy alone, the Ellison Medical Foundation, is granting awards of about $20 million yearly to promising anti-aging projects.[30] While most of the foundation's money goes to research on infectious disease, Ellison has plenty of billions more to plow into both. His limit is probably not his generosity but rather the size of the scientific community working on aging-related therapies. "Aging [research] used to be a stepchild, but now a lot of good people are finally getting into it," researcher Bruce Ames told me.[31] As the population ages, longevity research will become more and more attractive to scientists. It all creates an expanding "virtuous circle."

Add to this that we're at a time when knowledge of human genetics is exploding at the same time as raw computing power is. "Biology—including longevity research—is going like a rocket now with help from the genomics revolution and computers," says Ames.[32] "People tend to underestimate how fast the aging field is moving," adds MIT's Leonard Guarente. "We're uncovering the molecular basis of aging. No, we're not at a point where we can intervene in humans yet. But we have every reason to be hopeful that day will come."[33] Or to quote from a man in a different industry who knows a lot about change, Bill Gates: "People always overestimate how much will change in the next three years," even as "they underestimate how much will change over the next 10 years."[34] That's probably true of much of biotech and it's certainly true regarding its effect on longevity.

PART THREE

More (and Better) Food for a Growing Population

ONE

The Defeat of Hunger

Hunger from natural disasters, wars and grinding poverty consigns 830 million people to chronic malnutrition, according to the United Nations' World Food Program. Of these, 791 million live in developing countries. In sub-Saharan Africa, 180 million people—one-third of the entire population—is undernourished, while 200 million children below the age of five are underweight for lack of food.[1] That's why it's so unfortunate that powerful forces in Europe and elsewhere are combating biotech crops with false science, appealing to misguided ideas of nationalism, and often merely tossing around epithets like "Frankenfood." Nobody has a right to inflict malnourishment and outright starvation on another people. Agricultural biotechnology can't prevent wars. It can't stop a typhoon or snuff out an erupting volcano. But it can be a powerful tool to get more nourishment to those who lack it, to supply enough calories and nutrients for everybody now living—indeed, far more than enough for a growing world population. It has astounding potential to provide the world with food that's more nutritious, more delicious, less expensive, and that can be grown using less land and fewer chemicals. The story of golden rice shows what is possible.

In 1999, a research team funded jointly by the European Union and the Rockefeller Foundation announced the development of rice that borrows two genes from the daffodil (giving it a golden color) plus one from a bacterium to give the grain beta-carotene.[2] The body converts beta-carotene to vitamin A.[3] North Americans and Western Europeans rarely suffer vitamin A deficiency, but not so for people for whom rice is a staple, because rice doesn't contain the vitamin.[4] According to the World Health Organization, vitamin A deficiency is a public health problem in 118 countries, especially in Africa and Southeast Asia, contributing to approximately 2 million deaths annually.[5] Between 100 and 140 million children don't get enough vitamin A, and it's the leading cause of preventable blindness in children, with anywhere from 250,000 to 500,000 becoming blind annually, and half of these dying within a year of losing their sight. The deficiency also raises the risk of disease and death from common childhood infections. Improving vitamin A intake could reduce childhood deaths by as much as one-third in the highest-risk developing countries, according to the International Life Sciences Institute, preventing up to a million deaths a year.[6] In pregnant women vitamin A deficiency causes night blindness and may increase the risk of death in childbirth.[7]

But can people in the developing world afford a product that is still in the making after thirteen years of work? Not a problem—or so it first seemed. Those who funded the research and the chief scientist, Ingo Potrykus, emeritus professor at the Federal Institute of Technology in Switzerland, promised to give the rice away.[8] But then things got messy. It turned out that as many as thirty-two companies and other institutions held patents covering seventy different technologies upon which Potrykus relied. It's common practice to allow academic researchers to use these patents without fear of suit, but releasing the product is a different matter. If the rice were meant for sale, the seller would simply negotiate licenses for the patents. Potrykus just wanted to give it away and there was no provision for that.[9]

Fortunately, the patent holders joined Potrykus' crusade. Syngenta Seeds AG of Basel, Switzerland, gave Potrykus its patent rights for use in impoverished nations, so long as the company retained the right to sell it to farmers in wealthier ones.[10] Then the Monsanto Company of St. Louis went a step further and gave away its rights unconditionally.[11] (However many patents it holds, apparently a single one would have been enough to block distribution of the rice.)[12] "I am grateful for this," Potrykus said. "I consider the Monsanto offer important because I can now use this case to tell other companies, 'Look, Monsanto is giving me a free license. Won't you do the same?' "[13] Already by late 2000 India's Department of Biotechnology was planning studies for adapting the plant to use in that country and Potrykus originally hoped to have it in worldwide distribution by 2004, although that now looks overly optimistic.[14]

A Monsanto subsidiary, Calgene of Davis, California, has developed a canola oil enriched with beta-carotene, the patent for which the parent company says it will charge no licensing fee.[15] Getting vitamin A from oil means the vitamins will be in a fat base that aids their absorption. (Certain vitamins, including vitamin E, require a certain amount of fat for proper metabolism.) Further, just a teaspoonful of this oil in the diet could provide the recommended daily intake of beta-carotene for an adult.[16]

This is also true of yet another plant under development, golden mustard. In India, especially in the northern and eastern parts, mustard oil is commonly used for cooking and pickling. Working with Michigan State University and the Monsanto Company, the not-for-profit Tata Energy Research Institute in New Delhi is engineering genes that express beta-carotene into the mustard plant, a crop species closely related to canola, as well as taking the next step of developing strategies to introduce the product into the diets of affected children.[17]

Has this all been good PR for Monsanto? Yes it has, as headlines have made clear.[18] But a good deed is a good deed, and it further "clearly demonstrates that biotechnology can help not only countries in the West, but in the developing world as well," as Monsanto's former president and chief executive Hendrik A. Verfaillie put it.[19]

Stacking the Genes

Biotech gene splicing is relatively new, going back only thirty years in bacteria and twenty years in plants. But genetic engineering is very old. For thousands of years humans have intentionally been making new plants and animals by crossing them with others, hence combining whole sets of genes. There's no such thing as "wild corn," for example. Corn was bred from various grasses about five thousand years ago.[20] It's believed that all types of dogs were bred from the wolf.[21] When you compare a yapping little Chihuahua with a noble Doberman, you know old-fashioned breeding can produce strange results. Gene splicing to engineer plants and animals means that you're no longer throwing together two animal or two plant genomes over and over like rolling the dice until the number you want finally comes up. With that method, even when (or if) you got the trait you wanted, you usually had to "backbreed" to get rid of traits that accidentally rode along.

Certainly there's still trial and error with gene splicing. For example, the science of putting the gene you've selected into exactly the right place on the chromosome is improving but still iffy. You may have to produce hundreds of plants to get one right. But you can grow them all at one time. Thus transgenics is not only already a vast improvement over the old techniques, but it also allows much more opportunity for improvement. While old-fashioned hybridization continues to play a role, it is transgenics that's zipping along at an astonishing rate.

Another major difference between classic breeding and that using selected genes is that usually crossbreeding could only be done within species or between species that are closely related. Successful attempts to cross animals that are of related species but not very closely related result in such creatures as the mule, a mix between a horse and a donkey.

This is also true of plants that have been successfully crossbred, such as modern corn hybrids. The very technique that makes them produce a higher yield will also make them produce seeds far inferior to the parent plant; hence farmers must buy new seeds every year instead of saving and replanting part of their harvest.[22] Even greater differences in species make crossbreeding impossible. In traditional breeding techniques, putting a trait from corn into wheat was about as possible as impregnating a pig by an elephant. Biotech breaks down these barriers to put together genes from organisms that have less in common than Tony Bennett and the Rolling Stones.

One of the beauties of biotech is something called "gene stacking" or "gene pyramiding." This has nothing to do with Dolly Parton's DNA nor the great pharaoh's tombs. Rather, these terms refer to combining genes to add several new traits instead of just one. Not content simply to add genes that coded for beta-carotene to golden rice, Ingo Potrykus also wanted to insert other genes for other traits to help solve another major nutritional deficiency, that of iron. A 1999 United Nations report found that most of the world's people get too little iron.[23] Almost one-third of the world's population is believed to be outright anemic, and about one-fifth of all malnutrition deaths appear to be caused by a lack of iron.[24]

A lot of this is because these people's primary food is rice. And rice has little or no iron, right? Well, that's what I originally thought, too, but in fact rice actually contains a fair amount of the mineral but it also contains a molecule called phytate that locks up about 95 percent of the iron in the plant.[25] This prevents humans from absorbing it. So Potrykus and a graduate student named Paolo Lucca obtained a gene from a fungus that codes for the enzyme phytase, which breaks down the phytate, eliminating the molecular "dam" that blocks iron absorption. And while most enzymes are knocked out by the heat of cooking, this one carries a mutation allowing it to withstand such high temperatures. Yet another gene that Lucca added, which comes from the French bean, doubles the amount of iron in the rice grains. Finally, a third gene he added, from basmati rice, makes a protein that aids iron absorption in the human digestive system. Thus the plant not only has more iron; far more importantly it now has iron in a form the human body can use.[26]

Still other researchers have added genes from wild rice relatives to the best Chinese rice hybrids to get 20 to 40 percent higher yields. One way in which this is done is to make the stalks shorter, thereby diverting growth to the grain itself and making the rice easier to harvest.[27] Another breakthrough has come from transferring corn genes to rice to improve the rice plant's photosynthesis, increasing yields by as much as 35 percent, while University of Tokyo researchers found that by putting two barley genes into rice they could improve yields *fourfold*.[28]

One obvious drawback to growing rice that's apparent to anybody who has so much as seen a rice paddy is that the crop requires enormous amounts of water. Every pound of rice produced requires the consumption of about 600 to 1,300 gallons. Insufficient water supply is a major constraint to productivity, whether in Asia or in California.[29] But Avesthagen, an Indian company based in Bangalore, is developing a variety of transgenic rice that it expects will require only about half as much.[30] The company hopes to begin marketing it by 2006.[31] Yet all this marvelous engineering would be for naught if pests consumed the rice before people could. Here, too, biotech has had dramatic impact. A new transgenic modification of cultivated African lowland rice makes the plant resistant to yellow-mottle virus, a disease "which causes losses of 100 percent of the crop in many parts of the continent," according to Florence Wambugu of Kenya, founder and CEO of A Harvest for Biotech Foundation International.[32]

When the genes discussed are ultimately stacked and restacked, you'll get the vitamin A and iron from Potrykus' plants, the higher-yield genes from the new Chinese rice, and the virus resistance from the new transgenic African lowland rice. Each of these can then be modified to work best in the particular region where it's grown, be it Africa, Laos, China or India.

This rice will not save the world. But it will save innumerable lives, prevent a massive amount of disease, and allow scarce financial resources in the Third World to be spent elsewhere. It will enable countless youths from the underdeveloped world to concentrate on something other than sheer survival so they can grow up to be the doctors, scientists, engineers, architects and other professionals that their countries need to bring them into the First World.

Malnourishment is not nearly as simple as it appears. For example, you can be so well fed that your size 20 dress or size 42 "relaxed fit" jeans are ripping their seams, yet still be malnourished because while you're getting plenty of calories, you're lacking essential vitamins or minerals. We've already discussed the evidence showing that high levels of antioxidants can retard the aging process. But research continues regarding the ill effects of *low* levels of antioxidants. In a significant breakthrough in November of 2000, scientists found that a major cause of malnutrition, and specifically a disease called kwashiorkor, is a lack of antioxidants. As an article about the study put it, "Common sense might suggest that giving children more food would stop malnutrition, but with kwashiorkor, it's more complicated than that. From this study, researchers have determined that it's not the amount of food but the amount of antioxidants in the food that makes a difference."[33]

Kwashiorkor particularly devastates children, afflicting tens of millions worldwide.[34] "It's an extremely serious form of malnutrition," said the study's lead author, Mark Manary, an associate professor of pediatrics at Washington University in St. Louis.[35] "Normally when you're starving, you just get thinner, which is normal because your body is conserving its energy for essential functions," he explained. But "When children get kwashiorkor, their mind is affected and they get large sores on their bodies."[36] Added co-author Christiaan Leeuwenburgh, an assistant professor at the University of Florida in Gainesville, "Now that we've established that oxidative stress may be involved, simple oxidative therapies, like giving the children doses of various antioxidants, can help." He said, "It's very exciting" because it "means that this severe form of malnutrition could be prevented if you could increase the amount of antioxidants in people's diets. For people in Africa and Asia, we're going to try indigenous methods."[37] And guess what? Beta-carotene, the stuff that is being added to rice, canola and mustard, is a major antioxidant. (Canola is a form of rapeseed developed in Canada, the name being an amalgamation of "Canada" and "oil.")

"Indigenous methods," incidentally, doesn't mean supplying people with vitamin and mineral pills from GNC or the Vitamin Shoppe, as suggested by a vociferous opponent of biotech at an Organization for Economic Cooperation and Development meeting in Edinburgh, Scotland.[38] In some geographic areas, distributing pills can work; in others it cannot. In any case, nutritionists in *developed* countries are proclaiming that we're relying *too much* on vitamin and mineral supplements and putting too little emphasis on an overall wholesome diet.[39]

But it's not just malnourished people in poor countries who can benefit from genetically engineering food. About 90 percent of Asians, 75 percent of all blacks and many whites are intolerant of cow's milk.[40] That's because their bodies don't produce enough of the enzyme lactase, which is needed to digest the milk protein lactose. In all, this includes about 50 million Americans, along with a few billion people in Asia and Africa—give or take.[41] A project under way at INSERM, the French medical research agency, is working to eliminate the problem by giving cows a gene that will cause them to manufacture their own lactase, which will be present in their milk.[42] Essentially, the lactase will predigest the lactose, making the protein, calcium and other minerals available to everybody.

Many other people cannot enjoy wheat, oat, rye or barley products—from bread to beer—because gluten, a subcomponent of a substance in those grains, wreaks havoc on their intestines. (As with lactase deficiency, this is "intolerance" rather than allergy because it doesn't provoke the immune system.) British researchers are working on a process to leave most of the gluten intact while removing only the part that causes illness, the alpha-gliadin. This would leave the part that is important in baking and for other uses.[43]

Biotech Attitudes and Acceptance

There are many reasons for the reluctance of Europeans and others to accept transgenic crops. One is that, while many of their fears of its downsides may be exaggerated or fictitious, they havn't yet seen much of an upside. That's because virtually all such crops developed to date are known as "inputs." That is, they are *put into* the crop itself to help it survive and thrive, but don't otherwise *put out* value such as more flavor, freshness or nutrients. Those that do are "output" traits. Ironically, the very first biotech food sold was an output, the Flavr Savr tomato. While delivering on its promise to taste as fresh as garden-picked tomatoes, it suffered in the marketplace not so much from consumer reluctance as from the fact that the harvesting machines then in use had been made to pick hard tomatoes, not soft, already ripened fruit.[44]

As of March 2001, 73 transgenic variations of 16 different crops had received regulatory approval in one or more countries. In addition to some of those already discussed are such products as melons, squash, sugar beets, wheat and carnations.[45] Most of the variations have been related to crop protection; however, within five years about half of the products on the market will be related to food or oil enhancement.[46]

Just as understanding of the human genome increases at an explosive rate, so too does that of plants. DNA sequence data for the tomato alone is over 70 times higher than the entire sequence database for *all* life forms was in 1992.[47] As we'll see, input traits can be a real boon to farmers, by lowering costs and reducing the time they must spend in the field. They can help the environment in many ways, such as by reducing pesticide runoff and topsoil erosion, and by helping to ensure that only the right bugs are killed while helpful ones go unmolested. Already they are helping malnourished populations, in the form of food donations from excess crops.

Some consumers, especially in Europe, have also unfortunately been misled into believing biotech crops may be less healthy for us. Not only is this untrue, but in some ways they're already healthier. That's because insect-damaged plants are often attacked by fungi called mycotoxins. "Lower mycotoxin concentrations in [corn with built-in insect resistance from *Bt*] clearly represent a benefit to consumers," said Gary Munkvold, a plant pathologist at Iowa State University.[48] "Studies show [the plants engineered to control] European corn borer damage to kernels usually have very little Fusarium ear rot, and consequently, lower *fumonisin* concentrations."[49] The International Agency for Research on Cancer and the U.S. National Toxicological Program classify *fumonisin* as a possible human carcinogen, so less insect-chomping could mean less cancer for us.[50]

What consumers in the West have seen so far is perhaps a slight price decline that makes the world's cheapest food a bit cheaper, and we don't even know for certain that this is the result of biotechnology. It seems sure that the more benefits consumers perceive from biotech foods, the less hostile some will become and the more they will see the implacably hostile ones as rebels without a cause. Indeed, in a mid-2000 survey, one thousand Americans responded thus to "hypothetical" scenarios:

- 75 percent support using biotechnology to develop foods that stay fresh longer.
- 80 percent support using biotechnology to make foods more nutritious.
- 81 percent support using biotechnology to develop crops requiring fewer chemicals.
- 81 percent support using biotechnology to develop crops that need less land and water.
- 82 percent support using biotechnology to develop trees that grow faster.[51]

If the respondents had known that the first three were already in practice and the other two were well along in development, the approval rates would probably have been even higher.

Actually, the quickest way to reassure a consumer about biotech food is simply to inform him that he's already been eating it. A survey of Americans sponsored by the Pew Initiative on Food and Biotechnology in 2001 found that only 19 percent said they had eaten genetically engineered foods, while 62 percent said they had not. It also found that a mere 29 percent considered such foods "basically safe," while almost as many considered them "basically unsafe" and about half didn't have an opinion. Nevertheless, once these people were informed that more than 70 percent of the food in grocery stores contains genetically engineered ingredients, almost half said they considered the food "basically safe," with those remaining in the "basically unsafe" category dropping to a mere 25 percent of those surveyed.[52]

Certainly one of the best measures of consumer acceptance of biotech food is that every year, more and more such crops are planted. Between 1996 and 2001, the area planted commercially with transgenic crops increased about 35-fold, from 4.19 million acres to about 145 million acres, close to three times the total land area of the U.K.[53]

The increase in area of transgenic crops between 2001 and 2002 was 12 percent, or about 15 million acres. More than one-fifth of the global crop area of soybeans, corn, cotton and canola is now biotech. Nearly 6 million farmers in 16 countries chose to plant biotech crops in 2002, up from 5 million farmers in 13 countries in 2001.[54]

By far, the majority of these crops were grown in the United States, covering about 72 million acres or 68 percent of the world total. Following are Argentina, Canada and China.[55] Almost all of Argentina's soybeans are now transgenic.[56] Brazil might be a major leader in soybeans except for a four-year government-imposed moratorium, temporarily lifted in March 2003.[57] Not that this had been stopping the nation's farmers from trying to cash in on a good thing. Reports from Brazil repeatedly indicate that farmers there are smuggling in herbicide-resistant transgenic soybeans from Argentina.[58]

For a while, things looked rather gloomy for farmers growing biotech crops in the United States, not so much because Americans were rejecting this innovation but rather because it seemed the rest of the world was, especially Europe and Japan. Headlines in 2000 like "Farmers Plan to Cut Back on Biotech Plantings" featured such gloating as Margaret Mellon of the antibiotech Union of Concerned Scientists declaring that the year "probably represents a turning point for the technology."[59] A scholar at the Worldwatch Institute in Washington, D.C., gleefully declared in an *International Herald-Tribune* op-ed (entitled "After Four Seasons of High Growth, Transgenic Crops Are Now Wilting") that "After four years of supercharged growth, American farmers are expected to reduce their planting of genetically engineered seeds by as much as 25 percent in 2000 as spreading public resistance staggers the once high-flying biotech industry."[60]

But the predictions were based on surveys that, for whatever reason, weren't reflecting farmers' actual seed-buying practice. Rather than being driven off the beachhead, biotech crops in the United States were in fact consolidating their hold. Biotech plantings held about even in 2000 and then burst ahead. Transgenic soy went from 54 percent of total soy acreage in 2000 to 80 percent in 2003. Biotech corn grew from 25 percent to 38 percent, while transgenic cotton blossomed from 61 percent to 70 percent in 2003.[61]

Globally as well as in the United States, the most common biotech crop is soybeans, followed by corn, cotton and canola. The most common trait in these plants, representing about three-fourths of biotech crops planted in 2001, is herbicide tolerance. This means putting a gene into a crop so that a specific herbicide you're using kills the weeds but not your crop. Insect-resistant crops accounted for about 15 percent of the transgenic crops planted, while those pyramided with both herbicide tolerance and pest resistance were 8 percent of total acreage.[62] The day will come when all crops have multiple added traits, but for now it makes sense to add in one gene at a time, see how the crop does for a couple of years, add in another, see how that does, and so on.

The global market for transgenic crop products has also been growing. Global sales from transgenic crops totaled roughly $75 million in 1995, but tripled within a year and then again the next year. By 1999 they were estimated at around $2.2 billion. This is about a 30-fold increase from 1995 to 1999. The market is projected to reach approximately $25 billion by 2010.[63]

The Neocolonialism Myth

India was also a curious holdout in this otherwise bullish market, since even while resisting the *planting* of transgenic crops, it has been a leader in *developing* them. I suspect the two phenomena are related. Many Indians still feel a burning resentment of colonialism and are open to arguments that involving Western corporations to such a great extent in so important an area as agriculture is a form of neocolonialism.[64] Indeed, some of the most bitter opponents of agricultural biotechnology are upper-crust Indians who have made it clear they would literally choose starvation for many of their countrymen over allowing them to purchase transgenic seed from other countries. In at least one instance, they even tried to keep famine-threatened countrymen from receiving *donated* grain from the United States because part of it was transgenic corn and soybeans.

The underdeveloped world's most prominent biotech critic is Vandana Shiva, who in her position as director of the Research Foundation for Science, Technology and Ecology in New Delhi claimed that the United States and the "giant multi-nationals" were using Indians as "guinea pigs."[65] This was in 1999, when American families had already been eating the same food as what was introduced into India for four years.[66] In what sounds just like Marie Antoinette's legendary sneer of "Let them eat cake!" Shiva also has blasted golden rice by saying that better alternatives are "liver, egg yolk, chicken, meat, milk and butter."[67] (The main difference is that the French queen never said any such thing,* but Shiva did.) The assertion is all the more insensitive considering that the predominant religion in India is Hinduism and many Hindus refuse to eat meat. People like Shiva readily take advantage of reporters' ignorance—and the fact that it's easier to pick up the phone and interview her than a poor Indian farmer scratching at his few acres to barely feed his family. Shiva told one Associated Press journalist, who dutifully relayed it to readers, that bioengineered crops may work in the short term, but because they are costly and deprive the soil of moisture and nutrients, ultimately their benefits cannot be sustained. Actually, at this point no biotech crop takes up more moisture and nutrients than what it replaces—which is also true of many of the Green Revolution crops from the 1960s still being used today. Moreover, as we've seen, many biotech crops are under development that will require far less of both water and nutrients.

The AP reporter also repeated Shiva's admonition that "farmers have been driven into steep debt buying the expensive seeds, which must be bought anew every year since they cannot be replanted as organic seeds are." But there's no such thing as an "organic seed." The term "organic" refers only to how plants are grown. Most of the corn grown in the United States is hybrid, and hybridization *does* require purchasing fresh seed every year in order to maintain yields; but the transgenic qualities of a crop are maintained generation after generation.[68] Shiva told the reporter that "farmers also must invest in costly irrigation systems and pesticides to accommodate the new system which, she said, often wipes out their savings."[69] Again, this isn't so. No transgenic crop benefits any more from irrigation than its nontransgenic counterpart. Finally, for all the miracles that transgenic crops will be able to perform in the future, at this point their main advantage is precisely that they require much less pesticide usage.

---

*Not only is there no evidence she ever said it, but the expression was floating around the vernacular even before she was born. ("Did Marie Antoinette Really Say 'Let Them Eat Cake'?" *The Straight Dope,* at http://www.straightdope.com/classics/a2_334.html.)

Such rabid xenophobia and insensitivity to the needs of the poor in one's own country helps no one. Fortunately, Shiva is very much in the minority in her casual attitude toward saving and improving the lives of her countrymen. Despite her efforts, India is becoming a powerhouse of agricultural biotechnology research. Read carefully this assertion of Prime Minister Atal Bihari Vajpayee: "Biotechnology is a frontier area of science with a high promise for the welfare of humanity. A new generation of biotechnology, *developed as a result of intensive work in India,* has opened up research of *national relevance* [emphasis added]. I am confident that the fruits of biotechnology would be harnessed for the benefit of millions of poor people as we move into the next millennium."[70] But whatever country's lab it began in, these developments are ones in which the whole world can share.

Chimeraplasty—From Rats to Mennonites to Plants?

Some foods in the pipeline will be changed when the genes they already possess are modified by being switched on or off so as to express or not express certain traits. This has been dubbed "transgenomics" or "chimeraplasty." Making it a bit more understandable, C. S. Prakash, professor in plant molecular genetics and the director of the Center for Plant Biotechnology Research at Tuskegee University, calls it "native gene surgery."[71]

It's not every day that a medical technology moves from humans to plants. But such was the case with chimeraplasty. It was pioneered in rodents and then humans, especially the Amish and Mennonites, to fix genetic diseases.[72] These religious sects have an extraordinary rate of hereditary illness because the estimated 75,000 sect members are descended from just 47 families.[73] One extremely rare disease from which they suffer is Crigler-Najjar, a bizarre liver ailment. Crigler children suffer severe jaundice and must sleep under special rehabilitating lights for up to sixteen hours a day to prevent deadly brain damage.[74] The gene has been successfully fixed in rats, and now doctors are trying to alter it in humans so the children can have normal lives.[75]

Taking its cue from chimeraplasty work on rodents and people, Kimeragen of Newtown, Pennsylvania (now part of ValiGen) is developing it for use on plants.[76] The idea is to get plant genes to express proteins that they're already capable of coding for but they don't code for because those genes are "switched off." To switch them on, researchers send a molecule made up of both DNA and RNA, called a chimera, into the cell. In reports released together in 1999, scientists announced the first successful application of chimeraplasty in two plants, corn and tobacco. Because it can make only subtle changes to a plant's genome, chimeraplasty is less versatile than conventional methods of genetic engineering. And it won't quiet antibiotech groups who claim their only objection to biotech food is that it sometimes means combining two completely alien species.[77] But some important things should be possible with chimeraplasty. "Imagine you want to cultivate a rice plant that has wide leaves instead of skinny leaves," said Richard Jefferson, the director of Cambia, a nonprofit plant biotechnology research center in Canberra, Australia.[78] "In West Africa, where rice is becoming a very important crop, there's a serious problem with weeds," he explained to *New Scientist* magazine. "Because of this, West Africans tend to grow *Oryza glaberrima* rice, because its wide, droopy leaves shade out the weeds. *O. glaberrima* is not, however, their preferred rice. Many people prefer classic Asian rice, a different species named *Oryza sativa*, but they don't grow it since Asian rice has skinny leaves, which allows weeds to proliferate. But what if we could get Asian rice to grow wide leaves?"[79] Jefferson thinks chimeraplasty would fit the bill.

Already, gene silencing is proving an effective adversary to a disease that can devastate crops, called "crown gall." Spread by a bacterium, it attacks numerous perennial fruit and nut plants, including those bearing apples, grapes and walnuts. The "gall" is actually a tumor. Researchers in the Department of Pomology at the University of California, Davis, were able to reduce the number of tumors in tomato and *Arabidopsis thaliani* plants by 90 percent, not by stopping the plants from becoming infected but by silencing the genes in the plants that the bacteria hijack to create the tumors.[80] (*Arabidopsis* is a member of the mustard family and a distant cousin of the cabbage; it's also called thale cress.)[81] "Usually when you try to prevent disease, you look at stopping the bacteria or other disease-causing agents at the 'front door' and preventing them from entering the plant," said the lead researcher, Professor Abhaya Dandekar.[82] He's now using the gene-silencing technique to produce crown gall resistance in walnut trees.[83]

Just as most of Asia is heavily dependent on rice and thus vulnerable to deficiencies of any nutrient that rice doesn't provide, the same is true of corn dependence in many African and Latin American countries. Corn has nearly enough protein, for example, to sustain those who eat little else. Scientists at Rutgers University, however, have devised an approach to pump up the protein content in corn without the currently controversial aspect of splicing in a new gene from another organism. They increased the plant's ability to produce more of its own naturally occurring protein by raising the level of one amino acid, methionine, a common building block of protein, but one that the body cannot manufacture by itself.[84] "In industrialized nations where diets are rich in meat and other plant sources of protein such as soy, this is no problem. However, for the poor kids in Bolivia and elsewhere, steak and chicken are rarely on the table," said one of the Rutgers scientists. "With our discovery, there is now a chance for them to get the protein necessary to ensure their health."[85]

No Sex Please, We're Biotech

A technique that may soon revolutionize crop biotech is apomixis, which means obtaining seeds without sex. Some of these puritanical plants reproduce themselves but don't produce seeds in the process. Others, ranging from the pesky dandelion to the delectable blackberry, do in fact produce seeds without fertilization. They're natural clones. Apomixis occurs naturally in over 35 plant families and more than 300 plant species.[86] While fertility researchers keep trying to improve methods for humans to have *sex without offspring,* agricultural researchers are working to make major cultivated crops have *offspring without sex.*

One advantage of seeds without fertilization is that farmers would get their new seeds every year from their own crops.[87] Unlike hybrids, in which the seed quality becomes poorer each year, these crops would produce identical seeds season after season.[88] "Today's hybrid corn retains its hybrid vigor and desirable genetic traits for only a single generation," explained USDA plant geneticist Bryan Kindiger. "With an apomictic hybrid, these desirable traits could be maintained indefinitely." Apomixis will also prevent the propagation of diseases from crops such as cassava and potato that carry viruses when their cuttings are planted.[89] Further, since most apomictic plants neither accept pollen from other plants nor produce any themselves, this means there would be almost no risk of unintentional gene transfer to weeds or to other crops. Obviously, farmers would love this. But no seed company would ever make plants that require no seeds, right?

Cambia's Richard Jefferson said he originally thought this would be the case, but then discovered that at some seed companies "some of their top people dream of apomixis. It means they would no longer have to make such massive investments in seed improvement technology, and then worry about recovering it. Seed innovation and production is a costly and time-consuming business, whether it's biotechnology-based or conventional."

Then how would these companies make money? It's simple, said Jefferson. "Apomixis would cut the time needed to evaluate new lines, and dramatically reduce the cost of hybrid seed production. You could be releasing hundreds of new varieties a year, each adapted to localized conditions."[90] Basically, seed companies would become more like electronics companies, which get people to abandon perfectly good appliances by introducing new models that do something better than the old ones. Nobody's forced to buy a new DVD player to replace one that already works, and no farmer would be forced to buy the improved corn, wheat or rice seeds. But if the seed keeps improving, enough farmers will buy them so that seed companies can continue to make a profit.

Research to make crops apomictic has been going on for decades using traditional crossbreeding. But with biotech, scientists are finally closing in. Researchers at the International Center for Maize and Wheat Improvement near Mexico City have been working on an apomictic relative of corn named *Tripsacum* and have found a gene that might be responsible for apomixis. "It's called *elongate,* and we're in the process of cloning it," Olivier Leblanc, the head of the research project, told *New Scientist* magazine in 2000.[91] "People say apomixis [will be figured out] in the next five years," said Brian Johnson, an advisor to the British government, in late 2000. "Apomixis is a winner from all perspectives."[92]

While work continues with such techniques as chimeraplasty and apomixis, biotech crops that are currently available commercially use transgenics, which you'll recall is simply transferring a gene or genes from one organism into another. In plants, it involves three steps.[93]

In the first step, the desired DNA is "snipped" out of the donor organism. This DNA is then introduced into a tiny "factory" to make copies. This factory is the *E. coli* bacterium that has proven so useful in making recombinant pharmaceuticals. But in this case, *E. coli* has the added advantage of having not only a chromosome but also a smaller piece of DNA we've discussed earlier called a plasmid. These plasmids have the unique ability to replicate themselves without cell division. When they do so, they produce many copies of the desirable genes.

In step two, these new bacterial plasmids are inserted into the plant's genome. This can be done in one of two ways. The earlier method was to "mate" the *E. coli* with another bacterium. What's neat about this second bacterium, *Agrobacterium,* is that it can transfer its DNA into the plant's genome. *Agrobacterium* was probably the first gene splicer, beating scientists to the technology by a few hundred million years. When it transfers its DNA naturally, it causes crown gall disease. Now scientists have disabled this particular strain of *Agrobacterium,* so we can take whatever trait was encoded on the *E. coli* and let *Agrobacterium* insert it into the plant.[94]

But some crops, such as rice, wheat and corn, were not initially as susceptible to *Agrobacterium.* For these, a new technology was invented, although it can be used even where *Agrobacterium* would work well. With this method, the *E. coli* cells are broken apart to recover the engineered plasmids. The plasmids are then coated onto tiny gold or tungsten particles and fired from something known as the "biolistic particle delivery system," which may sound like something developed as part of President Reagan's "Star Wars" (Strategic Defense Initiative) program. But actually, it's just a gunlike device fired at a distance of about 13 centimeters.[95] The plasmids are muzzle-loaded into a .22-caliber shell, and the gene gun is pointed at a screen covering the plant tissue in a Petri dish. This is not a whole piece of a plant, but rather a culture of embyroid tissue or some other part of the plant that hasn't yet differentiated into leaves, a stalk or anything else. The shell is fired, dispersing the plasmids as they smash into the screen. To the eye it looks like a splash of water.

Now we're at step three, in which the cells are treated first with plant growth factors that induce them to form shoots, then with a different set of growth factors that produce roots. After this, the plantlets are transferred to soil and tested to see if the gene "took" properly.[96]

Currently, the maximum number of genes that can be shot into a plant's DNA is about two or three. Since ultimately we'll want to be stacking a lot of genes in all crops to give them many desirable characteristics, this is not good enough. But there's no physical law determining how many different genes you can transfer, and indeed researchers have shown the ability to transfer as many as thirteen genes at once into a rice plant. A few of the test plants were sterile and sometimes the genes didn't go to the right part of the chromosome, but the mechanics are there and will improve.[97]

TWO

Safe Biopesticides

A biopesticide is any material derived from living organisms used in pest control, such as bacteria, plant cells or animal cells. Biopesticides are usually used for crop protection, though they may also be used to attack pests like mosquitoes that prefer fresh blood to plants. They are toxic only to the targeted pests (or in some cases, close relatives of those pests) and do not harm humans or other mammals, birds or fish. They can be powerful tools against the insects, weeds, bacteria, viruses and fungi that believe they have a greater claim on our crops than we do.

Strangely enough, it seems that biotech pesticides were first used not by us but by ants to protect the fungus they collect, grow and eat. Researchers have found that just as we need to protect our crops from insects, those insects—specifically Central American ants—need to protect their fungus farms from an insidious parasite called *Escovopsis.* They concluded that keeping a fungal ant farm going "involves a continual struggle to suppress the specialized fungal parasite." The way the insects do this is with a potent antibiotic produced by a fungus known as an actinomycete. The ants work in teams. One consists of "leaf cutters" that cut down leaves and bring them back to make composting piles. From these piles grow the colony's food. But sometimes evil fungi come along and gobble up the good fungi in the compost heaps. When this happens to the ants' farms, they don't suffer foreclosure; they starve. To prevent this, there's another ant team, the "farmhands." These insects cover themselves with actinomycete and go about their work, spreading it throughout the nest, which keeps *Escovopsis* at bay.[1]

Researchers think it's no accident that most of our own antibiotics come from various types of actinomycetes.[2] In other words, ants beat us to both antibiotics *and* biotech. Ants have made use of the organisms in the world around them to protect their food supply and their lives, just as we are now doing.

The most comprehensive evaluation of the impact on U.S. agriculture of crops developed through biotechnology, which came out from the National Center for Food and Agricultural Policy (NCFAP) in June 2002, found that six crops genetically engineered to fight insect or weed pests increased yields by 3.8 billion pounds in 2001, saved growers $1.2 billion and reduced pesticide use by over 45 million pounds. Estimating the impact of twenty-one other crops that have been approved but not yet employed or are under late-stage development, it found total annual yields could be increased by over 14 million pounds, costs could be cut by about $2.5 billion and pesticide usage reduced by over 162 million pounds.[3] "The study shows every region in the country stands to benefit from development of the new varieties evaluated in this study," said chief author Leonard Gianessi. "In nearly every case we evaluated, biotechnology provides equal or better control of harmful pests at reduced costs," he added. Moreover, "We are still in the early stages of realizing the impact of biotechnology on food and fiber production in this country."[4]

Yet yield increases in underdeveloped countries may be far greater because they are starting from a much lower baseline; that is, they gain more because they have so much more to gain. A 2003 study published in *Science* looked at four hundred field trials in seven different Indian states and found that use of cotton with the anti-insecticidal *Bacillus thurengienesis* (*Bt*) bacterium spliced into it increased yields by up to 80 percent compared with nontransgenic varieties. That absolutely blows away increases in developed countries, which are more on the order of 10 percent.[5]

The first crop of *Bt* cotton in South Africa's KwaZulu-Natal province boosted the yields of black farmers by 50 to 89 percent compared with conventional varieties. Per amount of seed, the yield was all the greater, with increases up to 129 percent, because fewer seeds are needed for the *Bt* variety. Moreovever, labor and pesticide poisonings were reduced. "This was the first study in sub-Saharan Africa. It's not trial data, it's real farm data," said Stephen Morse of the University of Reading in the U.K. "We were not expecting differences as big as this. The farmers were glowing, they were very happy."[6] Small wonder that practically overnight, seven of every ten South African cotton farmers switched to biotech varieties.[7]

The lead author of the *Science* magazine study of Indian crops, Matin Qaim, also told *New Scientist* that "You could even argue that the results would be more impressive for food crops."[8] That's because cotton is a cash crop, allowing the current growers of the nonbiotech varieties to purchase pesticides, whereas for subsistence farmers who cannot afford any pesticides, the switch to a pest-resistant biotech crop could give even greater yield increases. Moreover, noted Qaim, a professor at the University of Bonn in Germany, insect pests are more numerous in the tropics than in temperate regions.[9]

Bacteria That Hate Bugs

One advantage of many biopesticide products already on the market is that they can reduce the need for pesticide sprays. *Bt* provides a nice example of how. That this bacterium produces various insect-killing toxins has been known for a century, since it was identified in 1901 in the larvae of diseased silkworms.[10] Within three decades it had been incorporated into effective insecticidal sprays against caterpillars, mosquito larvae and beetle larvae.[11] When the era of gene splicing began, genes from *Bt* were natural candidates for inserting into the plant genome itself.[12] That way, from the moment that stalk of corn or cotton plant sprouts and attracts the attention of a predator, it already has protection and keeps it until the day it's harvested and even beyond. Currently there are approved *Bt* forms of corn, cotton and potatoes, with other crops such as *Bt* peanuts waiting in the wings.[13] As of this writing, over one hundred U.S. field trials are under way with *Bt* crops, including tomatoes, grapes, cranberries, alfalfa, walnuts, canola, eggplant, peanuts and even trees like poplar and spruce.[14] One of the most appealing things about *Bt* is the incredible number of insecticidal toxins it contains. More than 120 different insecticidal proteins have been found in *Bt* so far, and there are probably at least as many more yet to be discovered.[15]

Built-in crop protection like *Bt* isn't going to put insecticide sprays out of business any time soon. Spraying has become an impressive science in its own right. Computers, satellite imaging, weather forecasts—all of these are used to apply just enough chemical and no more.

But there are always variables. If a farmer uses a chemical that's supposed to stay on the crop itself and suddenly it rains, he has wasted a lot of time, fuel and chemicals. What if he sprays insecticide and there's a sudden cold snap? A lot of the bugs might have died anyway. Because of variables like this, there is no perfect time to spray, and repeated sprays are often necessary. Too few sprayings and your yield gets clobbered. Too many sprayings and even a bumper crop may not compensate for your losses in chemical costs and tractor fuel.

Bt crops are far from perfect. For example, *Bt* only kills certain insects, not necessarily all those preying on the crop. It doesn't kill bacterial or viral or fungal predators, it can't stop a weed, and rodents couldn't care less about it. Further, insect infestations vary each year. What if the farmer pays extra for *Bt* seed and the insects essentially don't come calling that year? Look at a line graph of European corn borer infestations in Illinois and you'll see it looks like a heartbeat monitor attached to somebody watching the shower scene in *Psycho,* sometimes going from an average four bugs per stalk one year to less than half a bug the next.[16] If the farmer plants expensive insect-resistant crops and it's a low-bug year, he loses. He has already paid extra for the seed, so it doesn't do any good that he has to spray less. But if it's a high-bug year, he wins. What the farmer *always* gets, though, as one *Bt* specialist explained to me, is peace of mind. It's like insurance. Few of us will pay a year's premium for life insurance and then curse our fortune because at the end of that year we're still alive.

Further, as with anything, there's a learning curve and it can be safely assumed that each year farmers will get better at using their *Bt* tool and have less need for spraying. Yet even in the earliest years of *Bt* crops, they were already paying off. The June 2002 NCFAP study found that increased use of *Bt* corn eliminated about a million acres' worth of crop spraying in the United States—following on the heels of an NCFAP study from a year earlier, funded by the Rockefeller Foundation, that came to the same conclusion.[17] *Bt* cotton eliminated the use of 2.7 million pounds of insecticide in the U.S. in 1999.[18] In Australia, overall pesticide use has dropped by one-half, according to that country's Commonwealth Scientific and Industrial Research Organization (CSIRO), with much of the credit going to biotech cotton alone.[19] "The technology is helping us overcome our greatest issue of pest control, with remarkable reductions in chemical use," said Jim

Peacock, head of plant sciences at CSIRO, in November 2000.[20] The aforementioned Qaim study found that use of *Bt* cotton cut pesticide spraying by two-thirds in India.[21] According to an estimate by the International Service for the Acquisition of Agri-biotech Applications, *Bt* cotton alone now eliminates the need for 33,000 tons of insecticide worldwide, or 40 percent of the current global use.[22]

By the year 2009, biotech row crops* in the United States will be responsible for a yearly reduction of 13 million pounds in insecticide use and 45 million pounds in herbicides, according to projections by the Kline and Company consulting firm.[23]

But it's predicted that the greatest saving will come with corn. "After resistance to corn rootworm is incorporated into seeds that already resist European corn borers, Kline estimates that the market [for pesticides] will drop by 70 percent," the company stated in its October 2000 report.[24] Less insecticide used also means less getting into water supplies. From 1996 through 1999, the USDA's Agricultural Research Service analyzed runoff samples for insecticides from both *Bt* cotton and non-*Bt* cotton fields. Researchers looked especially for those called pyrethroids and organophosphates because these are widely used in the area. While runoff from non-*Bt* cotton sites contained slight amounts of the insecticides, runoff from *Bt* cotton sites had essentially none.[25]

Even these data may well underestimate the usefulness of integrating *Bt* into a crop's genome. Each year, stored-grain insects cause multimillion-dollar losses for U.S. producers of stored commodities such as corn, wheat, rice and grain sorghum. Yet the most effective pesticide for stored crops, methyl bromide, is scheduled to be phased out in 2005.[26] What to do? Work by entomologist John Sedlacek at Kentucky State University indicates that two different insects that feed on stored grain throughout the world, the Indian meal moth and the Angoumois grain moth, seem to have a strong dislike for *Bt* corn.[27] So transgenic *Bt* appears to protect crops from the very beginning to the very end. No one chemical could ever do that.

---

*Row crops are those planted in continuous lines spaced evenly so they can be cultivated with a tractor and harvested with a combine or other mechanical equipment that can operate on two or more rows at a time. Usually when we think of crops, we think of row crops.

And much greater savings can be had. According to the USDA, corn rootworms account for more pesticide use on U.S. row crops than any other insect and cost the nation's farmers up to $1 billion a year in control expenses and yield losses, hence the nickname "billion-dollar bug."[28] Feeding studies and field trials of corn with two different strains of *Bt* incorporated into it have found that the transgenic corn isn't just a substitute for chemical spraying but is actually far superior and becoming more so.[29] That's because the rootworm has recently begun to adapt to crop rotation pest control by sneakily laying eggs that skip a year before hatching. Further, in some areas it has started becoming resistant to spraying.[30] According to the 2002 NCFAP study, this latest-generation *Bt* corn could not only put an end to these clever strategies but eliminate the need for spraying insecticide at time of planting on 23 million acres in eighteen states.[31] Use of *Bt* for rootworm control would also help birds, because rootworm insecticides are normally soil-applied granules that birds sometimes mistake for seed—a mistake they often don't get a chance to repeat.[32]

Monsanto began selling its YieldGard Rootworm seed in 2003, after receiving EPA approval in which an official of the agency declared, "What this decision means is that the environment will have literally millions of pounds of very toxic pesticides not being used." (Dow AgroSciences of Indianapolis, Indiana, and Pioneer Hi-Bred International of Des Moines, Iowa, are also collaborating on built-in rootworm protection, but are further back in research.)[33]

Biotech seed companies are sometimes criticized even by those friendly to them for putting their first emphasis on developing transgenic crops that give added value to the farmer rather than to the consumer. The reason they have done so actually makes a lot of sense: that's where they had the greatest expertise. It's comparable to developing the electronic calculator before you invent the electronic computer. That said, consumers apparently are interested in more than greater freshness or more nutrients in their foods. In an August 2002 poll commissioned by the International Food Information Council (Washington, D.C.), 71 percent of respondents said that all things being equal, they would be "very likely" or "somewhat likely" to buy crops genetically engineered "to be protected from insect damage and require fewer pesticide applications."[34] Apparently the "farmer-friendly" versus "consumer-friendly" dichotomy is false. Consumers like crops that make farmers' jobs easier and reduce chemical use.

In the United States, *Bt* cotton means the farmer will save time, money and fuel by not having to spray so often, and he'll save on worry. In China, *Bt* cotton can actually save lives. "Until Monsanto introduced a genetically altered cotton plant here in the buckle on the Chinese Cotton Belt three years ago," began a *New York Times* article in late 1999, "farmers like Ma Yuzhuo sprayed their fields with tons of organophosphate pesticides to kill bollworms, grubs that feed on immature cotton bolls. So toxic is the compound, which is similar to the basic ingredient of nerve gas, that many people die from exposure to it each year, though the Chinese government will not disclose the number of fatalities."[35] A report in the journal *Science* put the figure as high as ten thousand farmers annually.[36]

Explaining that Monsanto's cotton, named Bollgard, carries the *Bt* gene, the *Times* continued, "the plant produces a protein toxic to bollworms, reducing the need for spraying pesticides, raising yields and allowing Mr. Ma to survive." Before Bollgard came along, he and his family "sprayed their fields 20 to 30 times from May to September because the worm had grown so resistant to the pesticide." With such an incredible number of sprayings and without the protective equipment of American farmers, he and his sons repeatedly fell violently ill. Ma, concluded the article, "said his yield had increased as much as 50 percent since he started planting the modified cotton. And, he said, by not buying pesticides he saves $80 an acre—and his life."[37] Indeed, the *Science* report found that because of the use of *Bt* cotton, the incidence of pesticide-related poisonings plummeted 79 percent.

More Grain on Less Ground

In addition to reducing pesticide use, it's been shown that biotech crops can increase yields. Without the *Bt* corn planted in 1999, there would have been an extra half-million acres of crops lost to worms that year, according to the 2001 NCFAP report.[38] Increased crop yields mean more land can be turned back to nature. This was already happening in the United States and Western Europe long before biotech crops came on the scene. Contrary to popular assumptions, forests in these areas have been expanding for decades.[39] But the process will greatly accelerate. That explains why Sierra Club executive director Carl Pope declared in 1998, "I strongly endorse [the] call for renewed commitment to governmental and philanthropic funding of agricultural research, including research into conventionally bred or bio-engineered new varieties of crops." Pope explained that "A massive increase in such research is . . . absolutely critical. Only then can the promise of high-tech breeding be combined with the social and environmental needs of the world."[40]

As nice as it would be to accelerate the speed at which land in industrial countries is being returned to nature, it would seem far more important to stop and reverse the trend in impoverished nations of more and more land being cultivated as farmers hack and burn away the rainforests, deplete the soil, and then because of this depletion are forced to hack and burn away more. Ironically, Greenpeace admits that the refusal of the Brazilian government to allow legal planting of soybeans is destroying the world's largest rainforest; it's also admitted it doesn't care. Reuters quoted William Laurance of the Panama-based Smithsonian Tropical Research Institute as saying, "Soya farming really is emerging as the critical driver of Amazonian deforestation," and that "Historically, the Amazon has been nibbled away at the edges, but now what's been happening is like somebody going right in and chopping it right up." Greenpeace had no disagreement with any of that, but its quoted representative said the greatest threat to Brazil is that "Contamination [from biotech beans] is coming in from Argentina, one of the main GM soya producers. A lot of the soya in the south of Brazil is already contaminated." Apparently, tough choices have to be made, and when the choice is between Brazilians growing biotech plants or losing their rainforest, well, "Rainforest, we're going to miss you."[41]

In October 2000, Chinese and Filipino researchers announced that they had been able to make two common strains of rice resistant to insects by inserting a *Bt* gene. These strains were hybrids that had been introduced into a famine-threatened China in 1976. As such they were among the more spectacular successes of the Green Revolution,[42] becoming quickly adapted throughout Asia. The hybrids produced about 20 percent more food than the rice strains they replaced.[43] In addition to preventing famine, it's estimated that the new hybrids made unnecessary the development of an additional 37 million acres of land.[44] But this came with a price. The new strains were more vulnerable to disease and pests, especially the stem borer. With corn, the stem borer can be controlled by insecticides, but with rice it's impossible because the larvae remain for only a short time on the outer surface of the plant before they penetrate the stem.[45] For such a job, inserting the *Bt* gene into the plant was made to order. Now the insects would be killed while they were doing the boring, and the vital rice would be saved—if the experiment worked.

There was every reason to think *Bt* rice *would* work, since *Bt* was already a commercial success in corn, soybeans and cotton. But you can't say for sure until you try it in the field. Thus, the new type of rice was planted in Wuhan province in central China in order to see how it would stand up to both the yellow stem borer and another nasty little cropmuncher, the leaffolder. The verdict? The transgenic plants "exhibited excellent protection against extremely high, repeated infestations," reported *Nature Biotechnology*.[46] Specifically, the yield was 29 percent higher for the *Bt* version than for the non-*Bt*. The result is that a lot more people get fed with the same amount of land, or a lot less land is cultivated to feed the same number of people.[47]

Beneficial Bugs

Bt crops, or any others with insecticide built in, are also far better for beneficial bugs, such as lady beetles and spiders. Why?

Insecticides have truly come a long way from the one-size-kills-all types we formerly used in the industrialized world such as DDT or, going back even further, various forms of arsenic. DDT quickly swept aside previously-used pesticides in great part because it was safer. But more than that, the chemical's ability to zap pretty much anything that crept and crawled was considered an advantage, precisely *because* it killed such a wide variety of insects, and *because* it persisted in the soil. Then we realized we didn't want to kill all the insects, just those devouring our crops. And we wanted chemicals that broke down quickly after doing their job rather than persisting and moving up the food chain from bugs to fish and birds. So as our chemistry improved, we were able to invent insecticides that quickly break down in the soil and that kill specific types of bugs while leaving others alone.[48]

But not all of the problems have been solved. For example, the European corn borer that wreaks havoc on corn in the United States and many other countries is so close a relative to moths and butterflies that if you spray your crop to get the borer, you're going to kill the other insects as well. Nobody cares much for moths, especially those of us who like wearing wool. But butterflies, these we like. The ultimate specific insecticide is one that kills only predators even while leaving beneficial insects alone. That's what built-in *Bt* does. By definition, it can only kill a bug that munches on the crop, which excludes insects such as butterfly larvae. You can't possibly design a spray-on insecticide that's better at sorting out the good bugs from the bad.

More nonpredatory bugs left in a field create better conditions for wildlife, especially birds. As the *Telegraph* of London reported, "Farmers in America are reporting increasing numbers of birds of prey and other wildlife in their crops of genetically modified cotton, soya and maize. After three years of practical experience with GM [genetically modified] crops, they say they have seen an upsurge in hawks, owls and other birds returning to their land since they switched much of their production to GM varieties." Explained the newspaper, "The recovery has been linked to increasing insect life on farms which cut back on pesticides sprayed previously in repeated heavy doses to protect conventional crops." The article told of an Illinois farmer who had noticed that owls had returned to a nest on the farm for the first time in years.[49]

Then you have the beneficial bugs, such as the lady beetles and spiders. Spiders not only leave our crops alone, they eat the insects that don't. A recent large trial in India compared the impact on beneficial bugs from the country's most widely used cotton crop with Monsanto's *Bt* cotton, Bollgard. The title of an article on the outcome was lengthy but pretty much said it all: "Study Finds Biotech Plant Kills Bollworms, Spares Other Pests: Yields Better Health, Economic Benefits." While both sets of crops required some spraying, the Bollgard crop needed little enough that greater numbers of spiders and beetles were found on those plants.[50] By leaving beneficial bugs in peace, you create a "virtuous circle." The less you have to spray, the less you have to spray.

Better *Bt*

Bacillus thurengienesis actually refers to a number of types of bacteria within a family.[51] The similarities between types of *Bt* can make it easier for crop technicians to work with, but the *differences* determine what type of toxins they produce and why one may be more or less effective against certain insect species. Most produce one or two toxins, meaning each insect predator must be dealt with one at a time. It would be great if there were a strain of *Bt* that could knock off several types of insects that chew into the plant. As it happens, scientists have found one *Bt* strain with an amazing eighteen different toxins, allowing it to kill moths, beetles, mosquitoes and even nematode worms. The multiple toxins will also make it quite difficult, or even impossible, for insects to become resistant. This super *Bt* was discovered in Egypt, where researchers found it in dead larvae from the pink bollworm, which rapaciously devours cotton from deep inside the boll where it's difficult to reach with chemical sprays.[52]

"We were amazed and very happy when we stumbled on it," Yehia Osman, head of the team that isolated the bacterium at the Agricultural Genetic Engineering Research Institute in Giza, Egypt, told *New Scientist* magazine. "This is the most potent *Bt* strain yet found, and has the most diversified host range," he said. "It's ideal to have a single isolate with this much activity."[53] Not only is this a source of different *Bt* toxins for genetically engineered crops, but because organic farmers still use *Bt* in spray form, it would be a real boon to them and their customers as well.

There's also no reason to stick with only *Bt* if other insect killers can be found to work when inserted into a plant's genome. One alternative is avidin, a protein originally found in egg whites. As discussed in an article in the USDA journal *Agricultural Research,* entitled "Avidin: An Egg-Citing Insecticidal Protein in Corn" by a punning reporter, the protein avidin kills bugs by preventing them from absorbing a vitamin they need called biotin.[54] Avidin may have a similar role in chicken egg whites, protecting the embryos from disease-causing organisms that require biotin.[55] "As a biopesticide, avidin is better than *Bacillus thurengienesis* in corn because it has a knockout punch that hits a broader range of insects," said chemist Karl Kramer of the USDA Agricultural Research Service (ARS).[56] While that is true in the lab, it remains to be seen whether it's also the case in the field. Moreover, some people are allergic to eggs, and careful testing would be needed to make sure that avidin proteins don't cause allergies. To avoid any problems that could result from avidin being expressed in corn pollen, Kramer said that molecular techniques will use specific promoters that place it not in the pollen but rather in the seed, roots, stems or leaves.[57] If it works in real-life conditions, it quite possibly would be combined with the *Bt* gene to give plants an added layer of protection. Further, when incorporated into the plant's genome, avidin would have the same advantage of *Bt* in killing only insects that try to feed on it, thus leaving most beneficial ones alone. And avidin, like *Bt,* protects groundwater because it's biodegradable and can't be washed away by rain.

We've been discussing ways of altering plants to make them resistant to insects. But why not consider altering the insects? Actually, it's been done for years. Fruit flies are irradiated by the millions to sterilize them before they mate with unsuspecting fertile flies. But the practice has limited use because the treatment leaves the flies in a slightly weakened state. In a newly infested area, they work fine in "soaking up" the available females, but in places where the insects are already entrenched, they can't compete with the stronger, nonirradiated fertile fruit flies.

A better technology is on the way, though. It uses something named piggyBac, a piece of DNA that hitches a ride on a baculovirus, which is a virus that targets insects. According to ARS insect physiologist Paul Shirk, piggyBac can be used to introduce genes to mark a population (perhaps through changing its eye color or making it glow under ultraviolet light) so scientists can track and learn about it. This will allow them to spread certain genes into an insect population either to alter their behavior in some beneficial way or to create sterile insects for use in sterile-release pest control programs. ARS geneticist Alfred Handler is investigating the use of piggyBac to transfer genes as an improvement on the fruit fly sterility programs.[58]

Probably the most notorious fruit fly around today is the Mediterranean fruit fly, or Medfly. One of the most destructive insects in the world, it feeds on a multitude of fruits and vegetables, and costs places like California and Florida a fortune in monitoring and spraying. Although spraying has been thoroughly studied by federal agencies and given a clean bill of health, advocacy groups often successfully spook people about the dangers of its use near their homes.[59] An alternative to spraying would relieve everyone. Since piggyBac was first discovered in moths, other obvious targets include the moths whose larvae feed on corn and trees (corn borers, rootworms and gypsy moths). Boll weevils and mosquitoes are also considered potential victims.[60]

The voracious pink bollworm is the target of Dr. Robert Staten and others at the USDA Animal and Plant Health Inspection Service laboratory in Phoenix. They've used piggyBac to create males that can pass a fatal flaw on to any egg it fertilizes. The idea is to fly over cotton fields and drop millions of modified males, enough to crowd out wild male suitors. Although Staten already releases radiation-sterilized moths, like the irradiated fruit flies they can't compete in places where the pest is already entrenched. To take territory back from the bollworm, indeed to eradicate it (it's not a U.S. native, anyway), the researchers have created moths with a gene that would be dormant as long as the moths were in the lab and on a special diet, but become active once they were released. Activated, the gene produces a substance that disrupts cell specialization. While eggs might be produced, there would be no offspring. The outdoor trial run began in mid-2002.[61]

Herbicide Tolerance

Another way of getting higher yields of one type of plant is to ensure lower yields of another, namely weeds. At one time, ploughing up weeds, or tilling, was state of the art. But then along came the "Dust Bowl" of the mid-1930s, and the ploughed-up, nutrient-rich topsoil was whisked away to all corners of the globe. Many farmers simply abandoned their land.[62] Aside from such outright catastrophes, tillage promotes a constant loss of topsoil to forces of wind and water, and topsoil can take literally a thousand years to form again.[63]

As of 1990, American farmers were still losing almost 1.7 billion tons of soil a year to erosion.[64] But in what the *New York Times* labeled a "Quiet Revolution on the Farm," farmers have made a slow but dramatic shift toward a reduction in tilling.[65] A no-till system can reduce soil erosion by 90 percent or more, offering tremendous advantages.[66] It allows a buildup of natural debris that composts into fertilizer, reduces fuel costs because tilling is very energy-intensive, and allows rows to be planted closer together, thus yielding more crop from the same acreage.[67] Sedimentation is also the most prevalent pollutant in streams identified as environmentally impaired, and when soil runs off into water, it carries both pesticides and fertilizer with it.[68]

The weeds still need to be kept at bay, and the main weapon is herbicides. But as with spraying for insects, spraying to kill weeds is a lot more involved than it might seem. A single crop can have numerous weed enemies, and an herbicide that devastates one might leave another feeling virile and hungry.[69] In any case, for all the advantages of reduced tilling, it led to a significant increase in the use of herbicides.[70]

The typical way farmers fight weeds is with soil-applied "pre-emergent" herbicides, which keep weeds from sprouting. Generally they need one active ingredient for grassy weeds and one or more for broadleaf weeds. Monsanto helped make things easier with Roundup, the trade name for the chemical glyphosate. Now made by several companies in addition to Monsanto, glyphosate is "postemergent," meaning it's sprayed on weeds that have already sprouted. It controls a remarkable range of grasses and broadleaf weeds—over 125 common weeds, according to Monsanto. Roundup also has good environmental characteristics, such as having extremely low toxicity to people and animals. It binds well to the soil until it completely deteriorates, so very little can run off into water supplies.[71] Another good reason to use Roundup is that it can replace the use of multiple herbicides because it kills so many varieties of weeds. Roundup would have been truly ideal except for one thing: it doesn't have the ability to single out weeds from desirable plants. Spraying it completely over a field would be like soldiers calling in an air strike on their own position when they're in danger of being overrun. But what if you could drop bombs on your position while you were hunkered down in a bunker with reinforced concrete walls ten feet thick?

That's what Monsanto accomplished when it inserted a Roundup Ready gene into plants, allowing them to produce an enzyme making them tolerant to the effects of glyphosate. The crops would be safe inside their bunker, even as the weeds all around them sputtered and died. By refusing to be an "equal opportunity destroyer," Roundup—employed in combination with these transgenic crops—eliminated the need for multiple active ingredients and reduced herbicide spraying trips over the field.[72] Earlier efforts to breed glyphosate-tolerant crops conventionally had failed.[73] But now Monsanto has produced and commercialized soybeans, corn, canola, cotton and sugar beets that all carry the "concrete bunker" gene, with wheat and other crops on the way.[74] During testing and use of Roundup Ready there have been no problems with harm to the plants, toxicity to animals, allergenicity or nutritional inferiority.[75]

Getting a little closer to home, the Scotts Company is developing Roundup Ready grass. Whatever size lawn you have, there is an advantage for both the user and the environment in spraying a single herbicide over the top. Obviously, it will be a tremendous boon to golf courses as well.[76]

In the United States, 80 percent of all soybeans are now herbicide-tolerant.[77] In the nation with the second-largest acreage of transgenic crops, Argentina, an amazing 95 percent of the soybeans grown have this trait.[78] It certainly would seem that herbicide-resistant crops would lead to more no-till farming, but do we know for a fact that they do? Yes. A 2002 report from the Conservation Tillage Information Center in West Lafayette, Indiana, shows conclusively that "There is a clear association between sustainable tillage practices and biotech crops."[79] Conversely, "Farmers who don't use herbicide-tolerant seeds are not as likely to engage in conservation tillage."[80]

No-till farming, and hence herbicide-resistant crops, are even more important in the developing world, where soil depletion and erosion have become mammoth problems and there aren't masses of fallow cropland lying around as there are in the U.S. Excluding the United States, worldwide soil losses from erosion were 24 billion tons in 1990. India was losing 5,188 million tons a year, while China was losing 3,991 billion tons.[81] And there's no reason to think the situation is improving.

At least one researcher who consults for antibiotech groups has produced a study he says shows that use of Roundup Ready crops actually increases herbicide use.[82] But other studies indicate the opposite.[83] A 2001 report by the National Center for Food and Agricultural Policy found that U.S. growers of herbicide-tolerant transgenic soybeans in 1999 saved an estimated $216 million over conventional herbicide treatments. It credited five factors for farmers' use of the biotech seeds: less complicated weed control, broader-spectrum control, less crop injury, more flexibility in timing herbicide treatments, and less concern about herbicide carryover to rotational crops.[84]

But rather than subject ourselves to a battle of the dueling studies, let's use a bit of common sense. Roundup Ready seed costs significantly more than nontransgenic seed, because Monsanto charges a per-acre "technology fee."[85] Further, farmers who normally save some of their soybean harvest to use as seed for next season have to agree not to save Roundup Ready soybeans and to purchase new seed every year. Yields are measurably higher with Roundup Ready, but not dramatically.[86] None of this is much of an incentive to buy Roundup Ready. Yet farmers have been falling all over each other to buy the herbicide-resistant seed. The fact that the seed is so popular despite being more expensive has led agbiotech opponents to suggest that farmers have fallen for Monsanto's devious propaganda. I've interviewed and otherwise interacted with farmers for years in my work, and I have yet to meet one who merited the epithet of "hick" or "hayseed." I've found that they rank up there with the smartest businessmen around. And the reason is that they *are* businessmen, and they're in an unforgiving industry in which lesser competitors have been forced out for two centuries. In 1900, about 40 percent of Americans were farmers; now it's 2 percent.[87] What does that say of the fitness of the survivors?

"Groupthink" is not a characteristic of these businessmen. They don't care what works for their neighbor; they care about what works for *them*. Farmers might gamble on a product one year, but if they find it's inferior, they'll drop it. To think they can be outsmarted year after year by seed companies using slick marketing managers and colorful brochures is a prejudice that permeates the thinking of city-bound activists, as well as many journalists. But Old McDonald not only knows his own job, he's probably a lot smarter in general than those who think they know how to run his farm better than he does.

While Monsanto was a pioneer in herbicide-resistant transgenic crops, it's not without competition. Pioneer Hi-Bred and its partner, DuPont, are selling LibertyLink corn, which has been genetically engineered to ignore a somewhat different type of herbicide than Roundup.[88] A small amount of American transgenic cotton called BXN from Bayer CropScience carries protections from the herbicide Buctril.[89] Other companies sell the over-the-top herbicides that work with Roundup Ready crops but are a different formulation from what Monsanto uses, such as Syngenta's "Expert."[90] So farmers do have a choice, and those choices will continue to expand. One of those options will always be to use non-transgenic crops and old-fashioned weed spraying, if they so desire.

As the table below (taken from a preliminary study of thirty different crops) shows, biotech crops may still be in their infancy, but they are already vastly superior in saving farmers and consumers money, increasing yields and decreasing pesticide use.

POTENTIAL FOR BIOTECH TO IMPROVE CROP PEST MANAGEMENT IN THE UNITED STATES

Crop	Change in Production	Change in Pesticide Use	Production Increase
Virus-Resistant Papaya in Hawaii	prevented loss of $17 million industry	no pesticide available	not applicable
Virus-Resistant Tomatoes in Florida	none	88,000 lbs./yr. reduction (88%)	$10.7 million/yr.
Herbicide-Resistant Tomatoes in California	none	4.2 million lbs./yr.	$30 million/yr.
Virus-Resistant Citrus in Texas	could prevent loss of entire $48 million industry	no pesticide available	not applicable
Insect-Resistant Sweet Corn in Florida	22 million lbs./yr. increase	112,000 lbs./yr. reduction (79%)	$1.3 million/yr.
Virus-Resistant Raspberry in Oregon/ Washington	could prevent loss of 10 million lbs./yr.	371,000 lbs./yr. reduction (50%)	$2.5 million/yr.
Insect-Resistant Cotton in U.S.*	260 million lbs./yr. increase	2.7 million lbs./yr. reduction	$99 million/yr.

Source: Leonard P. Gianessi and Cressida S. Silvers, "The Potential for Biotechnology to Improve Crop Pest Management in the U.S. (30 Crop Study)," National Center for Food and Agriculture Policy, Washington, D.C., June 2001.

*All figures for cotton are estimated.

"Superweed" Fears Confirmed—and Confronted

The worry over weeds developing resistance by being pollinated with herbicide-resistant crops, such as the Roundup Ready varieties, is yet another instance of the double standards applied to biotech. Research has indicated that plants with superior genes may be able to pass them on to surrounding weeds through pollination if the weeds are of a closely related species.[91] This finding has received tremendous media attention, with headlines like "Superweed Fears 'Confirmed' " and "Fears Grow as 'Superweed' Resists Killers."[92] In his article "Apocalypse When?" in *New Scientist* magazine, Jeremy Rifkin seized upon this as an example of a term he has diligently tried to promote, "genetic pollution." He used as his first example the "build-up of herbicide resistance in weeds."[93] But as plant biology professor Paula Jameson at the Massey University Institute of Molecular Biosciences in New Zealand[94] pointed out, "We've been breeding resistance to all sorts of things for years, using classical techniques. This is not a problem unique to a genetically-modified plant."[95] Yet few people ever hear this common sense. They believe this problem is intrinsic to biotechnology, often because that's exactly what they're told.

In any event, biotech in the case of "superweeds" has already unveiled *newer* technology to help address the problems of older technology. There are at least eight techniques under development or already in use with some plants that will prevent biotech crops from spreading their genes.[96] At least a couple of studies, one using the grain rape (an oil-producing crop with bright yellow flowers that's common in Canada and Europe) and another using tomatoes, have involved inserting the newly introduced gene (the one you're worried about spreading) not into the nucleus, as is standard practice, but rather into its chloroplasts.[97] These are the parts of the plants important to photosynthesis. By inserting the gene here, researchers greatly reduce or eliminate the risk of its being expressed in the pollen and from there making its way into weeds. There are also some low-tech ways of dealing with this in the meantime, such as having extremely good weed control. Or you can do what the Calgary firm of SemBioSys Genetics has done.[98] Cognizant that there are lots of weeds closely related to transgenic canola, it switched its research over to another excellent oil producer, safflower, which has no weed relatives, at least in North America.[99]

Another way of preventing "gene jumping" would be to sterilize the bioengineered plants. That's a technology that Monsanto and other companies were working on until antibiotech activists dubbed it "The Terminator" and "suicide seeds," and made tremendous political gains by claiming it would force poor farmers to buy seed every year instead of being able to replant some from their own crop. For people who have never gotten closer to a farm than watching *Green Acres,* this is somehow "nightmarish," as syndicated columnist Molly Ivins put it, while Britain's Christian Aid claimed it would destroy whole societies in the developing nations.[100] Those who know a bit about crops (and both Ivins and Christian Aid weren't even aware that the technology was theoretical and had not actually been developed) realize that American hybrid corn essentially puts farmers in the same position. These hybrids were first sold in the 1930s.[101] Then, as now, more costly seed that couldn't justify its expense would have been quickly terminated by the farmer. Nobody is at the mercy of seed suppliers; seed suppliers are at the mercy of the market.

"Terminators" would also be an efficient way to protect the developers' property rights and huge investments. An agricultural company that may expend tens of millions of dollars in developing and field-testing a single new crop faces a quandary: How do they sell that seed at a price the farmer can afford, but then keep him from reusing those seeds—that is, using seed from the last crop to plant the next, on and on? One method now used is to have farmers sign written agreements not to do so.[102] The "terminator" technology would accomplish the same thing without the contracts, the detectives, the snitches and the litigation. "All it is, is a way for seed producers to protect [their] products, the same way makers of music CDs and computer software programs do by copyrighting them," explained Harold Collins, vice president of technology at Delta & Pine Land, the Mississippi company that co-owns the patent on terminator technology with the USDA.[103] The comparison is apt. Illegal freeloading downloaders aside, nobody considers it wrong to pay $16 for music encoded onto a silvery disk that costs less than $2 to stamp and package. Lots of people buy software packages consisting of a CD or two and a paperback manual for perhaps $1,000, without complaining that the physical package itself probably cost less than $10. Software developers and film and music studios make it clear to purchasers that they own the plastic and the book, but that's all. Attempts to make copies are dubbed "piracy" and dealt with seriously when caught.

Nonetheless, Monsanto gave in to tremendous political pressure in 1999 and said "*Hasta la vista,* baby!" to its effort to develop the so-called "terminators."[104] Hopefully, though, the technology will be resurrected.

In any event, a ten-year study has shown that, far from creating "superweeds" through unintentional pollen transfer, there's nothing "super" about transgenic crops themselves when left on their own and not tended by a farmer. A team of British scientists found that transgenic beets, corn, potatoes and a type of canola were no better at spreading and persisting in the wild than their nontransgenic counterparts. "It puts the last nail in the coffin of the idea that all genetically engineered plants are terrible weeds," said Norman Ellstrand, a biologist at the University of California at Riverside, who had no stake in the study.[105] In the experiment, researchers sowed thousands of biotech and conventional seeds (or for potatoes, tubers) in twelve different natural habitats and under a variety of conditions in each habitat. After a decade they found that whether genetically engineered or not, the crops did not stand much chance of survival against England's native fauna in any of the twelve habitats. In fact, nearly all of the crops of both kinds disappeared within the first few years of the experiment. "The bottom line is with these genes in these crops, there's no difference," said Mick Crawley, lead author of the study.[106]

Why weren't they super survivors? Because they weren't intended to be. The beets, corn and canola were given genes that provide herbicide tolerance. In the wild, where no herbicide is sprayed, this is useless. The potato was given insect resistance from *Bt.* This might seem that it would provide a boost in survivability, but insect pests are much more damaging to cultivated crops than to plants in the wild, where insect infestation is less of a problem.[107]

The Papaya Story

One of the great success stories of agricultural biotechnology, although it received far less notice than environmentalists' scare stories, is how it saved the papaya crops of Hawaii and other areas from an especially destructive virus.[108] Papaya ring spot virus (PRSV) is a devastating disease. It's extremely hard to control, and once a plant is infected, it's lost. Avoidance has been the only method to prevent crop destruction. That means not planting papaya in areas known to have the virus, and trying to prevent viral spread to uninfected regions.[109] About three decades ago, PRSV wiped out the papaya on the island of Oahu. The industry had to move into new geographic areas, with much of it concentrated in the Puna region of the Big Island of Hawaii. The area was kept PRSV-free for a long time, but in 1992 the inevitable happened. PRSV struck the Puna region and spread like a lava flow. In just four years, fresh papaya production dropped by over one-third. The industry was dying.[110]

But researchers at the University of Hawaii, Pharmacia (now part of Pfizer), Upjohn and the USDA were already at work on the solution, a commercial variety of papaya engineered to be preinfected with the viral coat of a mild strain of the virus.[111] The researchers began field trials of the transgenic papaya in 1992, the same year the virus struck the Puna area. The results were spectacular: the plants were immune. Unfortunately, paperwork problems were more daunting than the scientific ones. The scientists needed to get the new papayas registered, licensed and distributed to farmers before they were forced to say "Aloha!" to their farms and livelihoods. Happily, since the USDA was already participating in the research, it was able to give speedy approval. The EPA and FDA gave the fruit clearance in late 1997. The researchers also had to get licenses from patent holders, but companies were sympathetic to the growers' plight and it helped that the licensing would be restricted to Hawaii. By the planting season of 1998, farmers were receiving the new seeds. The new papaya, named Rainbow, continues to remain PRSV-free, has similar or higher yields than industry standards, and tastes and looks like a real papaya. Statewide papaya production, having fallen 45 percent from 1992 to 1998, increased by 35 percent from 1998 to 2000.[112]

The Best Genetic Defense

The story of the strange little bacterium *Xylella fastidia* illustrates the cross-benefits of biopesticide engineering, how countries can combine resources to fight a common enemy, and how stopping something that ravages one crop can help other crops. It even shows how understanding a bacterium that attacks plants can help us understand those that attack humans.

Xylella travels inside an insect with the curious name of "sharpshooter leafhopper."[113] *Xylella* lives in the bug's gut and throat, and when the insect punctures leaves to feed on the sap, some of the passengers disembark and take up residence. In orange trees, this causes a disease called citrus variegated chlorosis (CVC) and the infected orchards bear smaller, tougher-skinned fruit that's unsuitable for processing. It has now infected over 60 million orange trees in Brazil, one-third of the country's total. Brazil provides about a third of the world's fresh oranges and half its juice.[114] A single Brazilian state, Sao Paulo, produces fruit for 1.15 million tons of concentrate a year and provides jobs for 400,000 people, according to Joao Carlos Setubal, coordinator of the Bioinformatics Laboratory at the University of Campinas in Brazil.[115] "It has been estimated that five million trees must be destroyed yearly in Sao Paulo state because of CVC," he said.[116] But it's not just orange trees that suffer. Related strains of the bacterium cause Pierce's disease of grapevine, alfalfa dwarf, phony peach disease, periwinkle wilt and leaf scorch of plum. Other strains are associated with diseases in mulberry, pear, almond, elm, sycamore, oak, maple, pecan and coffee.[117]

Nor is only Brazil affected by the disease. Argentina's oranges have been infected and agricultural specialists are calling the bacteria the greatest threat in the history of California "viticulture," agriculture having to do with grapes. First reported in the state in 1989, the bacteria have spread throughout southern California and gained a foothold in the Central Valley. They are now slowly pushing their invasion northward toward the vineyards of Napa and Sonoma counties, threatening the state's $12 billion grape and wine industry.[118]

That's why the USDA, the American Vineyard Foundation and the California Department of Food and Agriculture, together with Brazil's Citrus Plant Protection Fund (Fundecitrus), put up half a million dollars to have *Xylella*'s genes mapped.[119] After two years of work by 200 mostly Brazilian scientists in 34 different laboratories, the bacterium's entire genome was mapped. It was a first for an organism that causes plant disease.

Within just three months after these results were published in the July 2000 *Nature* magazine, other scientists had used the work of Setubal's team to devise a brilliant strategy for defeating *Xylella*. Specifically, they figured out how to alter its DNA so that it could infect the host plant harmlessly. "We have simply developed a technique to mutate the bacteria from a pathogenic to a nonpathogenic strain—one that would grow on the citrus tree but not stimulate the disease symptoms," said Patricia Brant Monteiro, an engineer with Fundecitrus.[120] That's one system that will be tried. Alternatives would be to engineer antibacterial protection into the plants or make plants with nutrients that slow the progress of the disease.

But this is hardly the end of the benefits from the Brazilian gene-sequencing project. Using what's been learned, these and other researchers are attacking another plant buster, *Xanthomonas citri*. This bug, which causes citrus canker, has already cost the Florida citrus industry half a billion dollars just to destroy infected trees since it was spread by a tornado in 1996.[121] One company, Integrated Plant Genetics in Alachua, Florida, is using the genomic information to try to create a genetic defense, by altering the genes of the trees so that the bacteria can't spread.[122] Lauding the Brazilians, Arizona State University's Charles J. Arntzen said, "This sort of information is going to open up crop protection strategies the way genome sequencing is opening up new pharmaceutical strategies to control infectious diseases" in people. That's because it illustrates how bacteria that cause human diseases develop resistance to antibiotics. "It really is fascinating to see how many similarities there are between plant and human pathogens," Arntzen continued. "That is what is coming out of a lot of genome sequencing. We can study one organism and learn something about another one."[123]

As we've previously seen, bioengineering doesn't necessarily entail taking a gene from one organism and putting it into another. Sometimes researchers simply work with the genes already present to make plants better for us or harder on other organisms.

Fresh alfalfa sprouts, crunchy bean sprouts or any of the half-dozen other raw sprouts sold today are no longer just for the tofu-eating set. They're widely used to add texture and taste to soups, salads and sandwiches and provide small amounts of protein, fiber and vitamin C. But if not grown and processed properly, they may be high in something else, too, namely bacteria such as *Salmonella* or the form of *E. coli* that causes disease. A single outbreak of radish-sprout-related *Salmonella* sickened 10,000 Japanese in 1996.[124] A CDC survey of just a two-year period turned up six sprout-related outbreaks, with almost 23,000 suspected illnesses. This is quite a potential problem in the United States because these sprouts are "produced primarily by small operations," said ARS microbiologist Amy Charkowski. Many of these mom-and-pop "sprouters," as they are known in the industry, can't afford expensive equipment to kill the germs. Further, "many sprouters run organic operations."[125] This means they won't use irradiation or a chemical called calcium hypochlorite, which the FDA recommends for sanitizing sprouts. Since organic farmers use manure on their fields instead of conventional fertilizer and there's no place *E. coli* and *Salmonella* would rather live, organic crops are the ones that need disinfecting the most. The FDA currently advises all consumers to cook sprouts before eating them, and further says that the young, the old and those with compromised immune systems shouldn't eat them at all.[126]

Charkowski, who works for the Food Safety and Health Research Unit of the ARS in Albany, California, is doing something about this. She's analyzing *Salmonella's* genome to discover what genes the bacterium activates when it contaminates the sprouts. In some experiments, she has genetically engineered strains of *Salmonella* by knocking out genes one at a time. This will tell her which gene or genes reduce the germ's ability to infect the plants. She has also inserted a gene from jellyfish to make the germ glow bright green.[127] She's using the glow to track the genes that *Salmonella* affects as it grows. Charkowski has already found a strain of *Salmonella* that's only about "one-tenth as effective in colonizing sprouts as conventional *Salmonella*," she said. "Once we know which *Salmonella* genes are critical in an invasion, we may be able to develop a strategy to activate and amplify the natural protective response by the sprouts."[128]

"The genes that *Salmonella* activates when it invades sprouts are likely the same as those it uses when it colonizes other fresh produce and perhaps meats and poultry," said Charkowski. "That means the food safety strategies developed from our genetic studies may help protect these other foods from *Salmonella*, as well."[129]

THREE

Survival of the Transgenic

For the next few decades, the biggest problem with food plants will be growing enough of them. Until now, the tremendous growth in yields has come primarily from getting more crops from land that was already pretty good to begin with, although even going back to ancient times we've improved it with fertilizer. Now there is an increasing need to get better yields under bad conditions and, further, to get yields in conditions that heretofore were completely inhospitable.

Sometimes these conditions are natural; sometimes they're manmade. A scary preliminary study that appeared in May 2000, using satellites and other systems, found that nearly 40 percent of the world's agricultural land is "seriously degraded." The Consultative Group on International Agricultural Research, a global agricultural research network, found that soil degradation has reduced food production on about 16 percent of the world's cropland.[1] Almost three-fourths of farmland in Central America is seriously degraded, the researchers said.[2] Degradation occurs because of erosion from flooding, chemical effects such as nutrient depletion, and damage from the waterlogging or compaction of soil. Yet even as more and more land degrades, world population climbs. According to U.N. projections, there will probably be about three billion more people before world population levels off in the middle of the twenty-first century, about a 30 percent increase over current figures.[3] Without high productivity, "You have to bring more land into agriculture, and we don't have that land," said Phil Padey, one of the Consultative Group's researchers. "We've got to learn to manage this resource carefully."[4]

One way of getting more food out of land is to make the crops resistant to cold. That way, early freezes won't destroy the crop, or the plants can be grown in climates too cold to support them currently, or you may be able to squeeze in an extra growing season by being able to plant earlier and harvest later. Many techniques are being evaluated to do just this. The Chinese claim to have inserted cold-tolerance genes from fish into vitamin-rich beets, while British researchers have discovered an "antifreeze protein" for carrots that they're experimentally incorporating into other crops.[5] A 1999 study found that simply increasing a plant protein that regulates gene expression, known as a transcription factor, made various plants more resistant to heat, salty soil and cold; 96 percent of the genetically engineered plants survived freezing, while fewer than 10 percent of the nonbiotech variety did.[6] Mendel Biotechnology of Hayward, California, is also experimenting with gene expression to develop cold-tolerant vegetables, corn, soy and canola plants.[7]

Researchers at Queen's University in Kingston, Canada, have isolated a powerful gene from larvae of the yellow mealworm beetle that keeps the worms from freezing to death during the winter.[8] They believe it is a hundred times more powerful than the "antifreeze genes" found in flounder, a fish particularly good at protecting itself in cold waters. One application, they said, could be to use it in prepared food to prevent "freezer burn."[9] Research is also under way to insert the gene into plants so as to keep them from freezing. To find the substance, the researchers first needed to screen over one hundred proteins. The proteins were then cloned and spliced into bacteria. "We are producing 10,000 times more protein than we ever could from natural sources," said Queen's chemist and paper co-author Peter Davies.[10]

Agricultural Research Service scientists are also focusing on cold-related chlorophyll breakdown in canola seeds. If an early frost hits the plants while the seeds are still green, they produce less oil. This can cost North American canola growers up to $150 million annually. But "many plants break down chlorophyll even after a freeze," said C. John Whitmarsh, a plant physiologist at the University of Illinois in Champaign-Urbana. "That's what happens in the fall when chlorophyll disappears and other leaf pigments become visible, creating spectacular fall foliage."[11] So Whitmarsh and others are looking for mutations in a close cousin of canola that will cause the seeds to keep turning brown, even if they've been zapped by one of those "Siberian Express" cold waves. Basically, they're freezing the plants to see which have seeds that turn brown and which don't. Eventually, genes from the brown-seeded plants can be inserted into canola.[12]

If cold weather makes the growing season too short, another way of dealing with the problem is to make plants that grow faster. Professor James Murray of Cambridge University's Institute of Biotechnology in England and his colleagues put a gene from a flowering weed into tobacco plants, making the tobacco grow much more quickly.[13] The gene produces a protein that causes the tobacco plant cells to divide far faster at the tips of roots and shoots. It is a technology that can be used in other, more crucial plants.[14]

With a technique showing that even in the biotech age discoveries can come about accidentally, researchers at Hebrew University in Jerusalem announced in 1999 that they had more or less stumbled upon a way to accelerate the growth of cotton, potatoes, tomatoes and corn, and even make several types of trees grow as much as 50 percent faster than normal.[15] In this case, Oded Shoseyov of the Hebrew University's Agriculture Faculty was experimenting with ways to eliminate the seeds in tangerines. He and other researchers inserted something called a cellular binding domain (CBD) gene into the fruit plants because, according to theory, this would inhibit the growth of the plants, inhibit the growth of seeds, and perhaps provide fruit that required no spitting. Yet just the opposite happened. "In one test tube we inserted the CBD gene into the pollen of the citrus plant, and in another we had the plant without an inserted gene," Shoseyov explained. "The next day, we saw a clear difference between the two tubes. To our surprise it was the citrus pollen with the CBD that was longer. We thought that we'd mixed up the labels."[16] While the Israelis found they could readily repeat the experiment, it was four long years before they finally found out why it worked. Proteins on the surface of plant cells produce strings of cellulose that bind together, with the slowest-growing of these strings holding all the others back, like the proverbial slow ship in the convoy. The CBD gene keeps the strings from binding, thereby permitting each to grow as fast as it wants and allowing the entire plant to grow much faster.[17]

Another way of getting your food faster is persuading a fruit tree to start producing sooner. Spanish researchers have made orange trees that bear fruit in their very first year, even though the nonproducing "juvenile" period for citrus trees is normally anywhere from six to twenty years. The special trees were engineered with two genes taken from *Arabidopsis thaliana.* Other scientists have had some success inducing early flowering in other plants with the same genes. But this was the first full-scale success. "Both types of transgenic citrus produced fertile flowers and fruits as early as the first year. These results open new possibilities for domestication, genetic improvement, and experimental research in citrus and other woody species," said José Martinez-Zapater, of the Spanish National Institute of Agricultural and Food Science and Technology, who led the research.[18]

Other researchers are making plants more heat-tolerant, while a Canadian company is making them grow with less sunlight.[19] When increased resistance to cold and heat and less need for sunlight are combined with faster growth rates, it means that farmers limited to a few types of crops by weather and soil may be able to switch over to others that are more profitable. Regions that are almost entirely dependent on one or two crops may be able to grow a dozen types and to rotate them, diversifying their risk in cases of market downturns even as they keep pests off balance. Even farmers with small parcels may be able to plant numerous crops side by side.

When Rome finally conquered Carthage in 146 B.C., it burned the city to the ground and sowed the earth with salt so it would remain barren. High saline levels in soil are still a special problem for irrigation farmers, because irrigation water inevitably contains some amount of salt. Over the years this builds up, sometimes to the point where the land must be abandoned. Yet an estimated 40 percent of the global harvest comes from irrigated land, and already salinity has rendered some of that unsuitable for crops.[20] Land that has never been irrigated but at one time was under the sea could also be salty enough to reduce yields. Such acreage could be brought into productive use or kept that way if plants were made to withstand high levels of salt. Researchers at the University of Toronto and the University of California at Davis accomplished this with mustard and tomato plants.[21] The "proof of the concept" came with the mustard, in which the scientists cloned a gene that coded for a protein that appeared to make the plants more saline-tolerant. It worked. With the cloned gene, an extra amount of the protein was emitted and the plant was able to withstand levels of salt that would otherwise have killed them.[22]

Now that it had been demonstrated in one plant that this gene was crucial to salt tolerance, two years later it was introduced into tomatoes. The transgenic fruit could grow in brackish water or in soil with as much as fifty times the salt content that would have destroyed non-transgenic tomatoes. Further, the tomato actually extracts the salt (into the leaves, not the fruit), thus desalinating the land as well as producing fruit. There's every reason to think this same mustard gene could allow any number of other plants also to withstand water with high saline contents. Indeed, the tomato team is already working to do the same with canola.[23]

Other researchers have tapped into bacteria with special talents for resisting extreme conditions, logically named extremophiles. While actually isolating a particular bacterium might be quite difficult, it's easy to find where they live. You simply go to some place with extreme natural conditions. If you want bacteria that can withstand heat, you go hunting in hot springs or a volcano vent. If you're looking for those that resist cold, you look in frigid places. Researchers who wanted bacteria that could tolerate extremely salty conditions went to salt flats such as the Great Salt Lake in Utah and to the Dead Sea, and found an extremophile named *Halobacterium* NRC-1. Other researchers then decoded its genetic structure.[24] The *Halobacterium* can thrive in salt levels ten times higher than ocean water; thus the gene can be spliced into a plant that should then theoretically be able to grow in the saltiest of water.[25]

While there's probably no country that couldn't benefit from plants with added salt resistance, for places such as Bangladesh and Israel they would be a godsend. In Israel, freshwater supplies are already overexploited and farmers are forced to use salty water for irrigation. It's not surprising, therefore, that it was scientists at Hebrew University who succeeded in isolating a gene that helps the common European aspen, *Populus tremula,* to grow in salty soil. They believe the protein from the gene may protect cells by attracting water molecules and also by binding to other cell proteins, though they don't really understand the exact mechanism involved. The researchers then increased the tree's salt tolerance by giving it more copies of the gene. Although normal aspens shed their leaves about five days after being exposed to highly salty water, trees with the extra genes went for ten days before dropping them.

While the technology is still in the early stages, the Hebrew University team believes it has tremendous potential. The hope is that it might eventually be applied to protecting other kinds of plants. "We expect that the same group of protective proteins will be relevant in fruit trees and tomatoes, and there is some evidence for their homology [similarity]," said lead researcher Arie Altman, director of Hebrew University's Institute of Plant Sciences and Genetics in Agriculture.[26] But Altman believes many plants already have the gene. If so, it may not be necessary to transfer a new gene into crops. "Once you have a molecular probe for the gene you can use it also in traditional breeding techniques to speed up selection," he said. "By developing molecular tools you can screen out the trees in which the gene is being expressed the most."[27] Altman remarked, "What we call desertification is a worldwide problem and it's already very severe. It's conquering the world and we have to stop it."[28]

Foiling Aluminum

Other soil produces poor crops because of high levels of certain minerals, such as aluminum. The most abundant metal on earth, it's highly toxic to plant roots, especially in acidic soil, which readily releases aluminum ions into the groundwater.[29] For some major crops such as corn, aluminum toxicity is second only to drought as a cause of poor yields, reducing production by up to 80 percent.[30] Acidic soils cover well over half of the world's 8 billion acres of otherwise arable land, including about 86 million acres in the United States, according to ARS plant physiologist Leon Kochian.[31] This is one reason more tropical rainforest is being slashed and burned every year; over 40 percent of the land in tropical regions is acidic.[32] Even without aluminum, too much acid harms plant growth by tying up phosphate, an essential plant nutrient. How do you compensate for a low crop yield if you're in the rainforest? You burn down some trees and expand your fields, though your per-acre yield remains low because soil in the rainforest is fine for trees but lousy for crops and grazing land. Then you exhaust that land, burn down some more trees, and on it goes.

Biotech may have the answer. Scientists have long known that the roots of some plants excrete citric acid to make the phosphate soluble, though most crops can't turn this trick. Along with his colleagues, Luis

Herrera-Estrella, a botanist at the Center for Research and Advanced Studies in Irapuato, Mexico, took a gene for citric acid production and inserted it into tobacco plants. He found that when just lightly fertilized, the plants grew 20 percent larger than tobacco plants without the new gene. The plants achieved maximum growth using half the fertilizer that normal plants needed. The researchers later inserted the gene into papaya, with similar results. They concluded, "This finding opens the possibility of applying this technology to important crop plants, such as maize [corn], rice, and sorghum, which are often grown in acidic soils in which aluminum toxicity is a major problem."[33] Commercial development could come as early as 2004. If it works anywhere near as well as it seems to, it could be a tremendous boon to cash-poor farmers in impoverished nations because they could get the benefits of fertilizer at far less cost. "Insoluble sources of phosphate fertilizer are cheaper and we could use much less," said Herrera-Estrella. Insoluble phosphates would also be less likely to contaminate the water table. The researchers said that while they have filed for patents on the biotech plants, they will enforce them only against those who can afford to license them. "We want to give it for free to small farmers," Herrera-Estrella emphasized.[34]

Cornell University researchers used a different technique. They intentionally induced plant mutations with a chemical and then screened these mutants for aluminum tolerance, finding two that could thrive in soils containing four times the level of aluminum that stunted the growth of normal plants.[35] Properties from these plants may soon be transferred into food crops for testing.

Two other metals, cadmium and copper, build up in soils contaminated either by heavy fertilizer use or by industry. They harm plants by producing free oxygen radicals, which damage plant cells just as they do ours. In this case, they play musical chairs. When essential metal nutrients such as zinc get up during the music, the cadmium and copper quickly plop into the open seats before the music stops. The zinc is "out." Ultimately it's the plant that loses the game. Some plants deal with these nasty metals by binding them up and sequestering them within their cells. Now at least three different research groups have identified the gene that allows them to do so, in *Arabidopsis thaliana,* wheat and yeast. It's only a matter of time before they're spliced into other food plants that lack them.[36]

More Fertile, Less Fertilizer

Along with making crops more resilient to negative minerals or chemicals, biotech researchers are making them require fewer positive ones.

While the mineral aluminum is devastating to many crop plants, one of the three most important plant nutrients is also a mineral, iron. But unlike the other two, phosphorous and nitrogen, you can't fertilize with iron. That's because the problem isn't one of abundance but rather of solubility. You could force-feed a lot of crops—including corn, wheat and rice—with iron tablets and it wouldn't do any good; if they're grown in high-alkaline soil, they simply can't absorb the metal. Unfortunately, about one-third of all arable land is too alkaline for these cereals. But other grains, such as barley, have no trouble taking up iron.[37] So researchers at the University of Tokyo took two genes from barley and introduced them into rice plants, using *Agrobacterium* to make the transfer. The result was that the transgenic rice plants produced more than four times as much grain as the nontransgenic ones in the same soil. There's no reason to think the same technique won't work with other vital grains.[38]

It has long been known that adding phosphorous and nitrogen to soil can tremendously increase crop yields. But besides costing money, these nutrients may also become pollutants. That's because when fertilizer runoff empties into a basin, whether a lake or a bay or a gulf, it can cause "algae blooms," which can choke off fish and other aquatic life.[39] A federal government task force has recommended that almost $5 billion yearly be spent to reduce nitrate runoff into the Mississippi River system, because some scientists believe the nitrate fertilizer is causing a depleted oxygen area (melodramatically called "The Dead Zone") in the Gulf of Mexico. Aside from the sheer cost, this would involve submerging "merely" 20 million acres of prime cropland and forcing the remaining farms to use one-fifth less fertilizer.[40] University of Florida scientists have developed a protein that could allow Mississippi Delta farmers to keep their land and keep it productive. Ironically, the process they have patented uses algae itself. They found that a certain algae gene, inserted into crop plants, can boost yields by almost one-third because the new strain converts nitrogen fertilizer far more efficiently. In preliminary trials, genetically modified wheat plants with the new trait were more robust, grew larger and produced significantly more grain than conventional wheat plants on the same amount of nitrogen fertilizer. Or, farmers can grow the same amount of wheat with one-third less fertilizer.[41]

There aren't a lot of places that complain about having too much fresh water, but there are degrees of neediness. In the United States, we complain bitterly when temporary droughts force us to let our lawns go brown. In poorer nations, especially in Africa, drought can mean death. Conversely, crops that require less water can mean life. A New Hampshire research team conducted a worldwide analysis of surface and shallow-aquifer water supplies, concluding that almost 1.8 billion people live under what it termed "severe water stress." The team also calculated that within a quarter-century, population growth and other factors will force us to use almost 10 percent less water per person, even though water shortages are already common.[42] Many partial solutions will be needed, including better water management and more plants that convert salt water to fresh water. But a tremendous boon will be plants that require less water.

One method is being co-developed at two different institutions in Japan. To get the gene they needed, researchers went to some hot springs. These, along with geothermal sources like Old Faithful in Yellowstone Park, are favorite hunting grounds for biotech researchers because the extremophile bacteria that can live with such tremendous adversity must have numerous special survival traits. In this case, the Japanese took a gene from a hot springs bacterium that excretes an enzyme allowing it to absorb carbon dioxide more rapidly. The scientists modified the gene, inserted it into *E. coli* and found that this new *E. coli* could now "breathe" twice as quickly as before.

So what? Well, a plant growing in a dry climate finds, much to its dismay, that if it opens a single pore to take in carbon dioxide, it loses about five hundred water molecules through evaporation. Double the carbon dioxide absorption rate and you cut in half its water loss, and thus its water requirement.[43] "This technique alone will not ensure a blooming of the deserts. But the idea is to use it in conjunction with other techniques we are developing, such as (breeding) plants with a more efficient utilization of water," said molecular biologist Hirokazu Kobayashi.[44] "We want to produce a kind of desert super-plant."[45]

Or the Japanese could use the technology in conjunction with techniques others are developing, such as the mustard plants genetically engineered at the National Research Institute on Plant Biology in New Delhi. The Indian researchers inserted a gene from a weed relative of the mustard plant to make the new one be able to thrive on as little as half the amount of water that the old mustard plant needed. When they know enough about how the process works in mustard, the New Delhi researchers are planning to use a gene from the highly resilient pearl millet to make rice and wheat crops that need far less water. They are also using gene stacking or pyramiding to make plants resistant to high and low temperatures and salinity. Ultimately, they plan to distribute the seeds to farmers at the same price they were paying for the inferior, nontransgenic varieties.[46]

Cornell University researchers and Korean biologists have fused together and then inserted two different *E. coli* genes that tremendously improve the ability of plants to resist many types of stresses that would seem to be unrelated. This includes resistance to drought, cold and salt, and improved uptake of micronutrients such as zinc and iron. The type of rice they used was of the Indica varieties, which represent 80 percent of rice grown worldwide. But they've repeated the process in other plants, and believe that ultimately they can do so with a huge range of crops including corn, wheat, millet, soybeans and sugarcane. That's because the genes promote production of something plants already have, trehalose, though most have little of it except so-called "resurrection plants" that can survive prolonged droughts in the desert. Nor is stress required to make use of the increased trehalose. Even in crops grown in optimal conditions, increased yields of 20 percent were observed. None of the change in the plant occurs in the edible part. Moreover, in a gesture that should silence critics, the researchers are patenting the various processes but only so that they can make the information available free of charge to anybody who wants it.[47] Most companies can do something like this only occasionally, as they depend on profits to provide their "seed money," if you will, for other projects. But when scientists can get enough support through research grants, they can give the gift of nutrition to literally billions of people.

Lighting the Way

Of course, there's already a way to tell if crops aren't getting enough water. They begin to wilt. Then they die. No farmer would allow the situation to get that far if he could help it. The problem is that with a lot of crops, such as potatoes, by the time they show their very first sign of not getting enough H_2O, it's already too late. Yields can be cut by as much as two-thirds because the plant has gone into drought mode and is compensating by slowing its growth rate. An alternative is to *overwater* crops, wasting water and again causing trauma that lowers yields. But what if the plant could tell the farmer it needed water before damage was done?

Scottish researchers have fashioned a potato plant that glows green when it's thirsty. They inserted a jellyfish gene that makes the leaves appear fluorescent under the light of a hand-held monitor. The gene makes the leaves glow when it is triggered by a protein that forms as the plant becomes dehydrated. A small number of these potatoes would be planted near the crop to act as sentinels, telling the farmer whether watering is necessary and thereby saving the cost of needless irrigation as well as maximizing crop yield. "The best-placed organism that can tell you what is happening in terms of environmental insults like dehydration and mineral depletion is the plant itself," explained research head Antony Trewavas of the University of Edinburgh.[48] Potatoes were their test crop, but the same could most likely be done with other tubers or root vegetables such as carrots and turnips.

Then there's the other side of the coin: plants that suffer from flooding. Obviously, too much water isn't good for any crop. The biggest difficulty, however, comes with those such as rice that require a certain amount of submersion but not too much. Research on this problem is well under way. One important breakthrough came when California scientists found a new way to look for flood-tolerant plants at the International Rice Research Institute (IRRI) in Manila, the Philippines.[49] The old-fashioned—which is to say, current—method of finding such a plant is about what you would expect. You try to drown it, and if you don't succeed, the plant is considered flood-tolerant. The problem with submerging plants is that it can take as much as two months per generation to get a result. Then, if you cross this plant with another to see if that improves the situation, you've got to wait another couple of months, and so on.

But that will no longer have to be the case. In analyzing the IRRI's plants, University of California at Davis scientists found something special not in the genes themselves but on the uncoded DNA on chromosomes near the genes. They came to realize that when they saw specific patterns in the DNA (using a chemical test), a gene inevitably would be nearby indicating the plant could withstand long periods of submersion without succumbing. Normally rice can survive about a week underwater, but the flood-tolerant types the Davis researchers developed in conjunction with the IRRI and Chinese scientists survive at least twice as long. One early success was giving flood tolerance to high-yield types of Indica rice widely grown in Southeast Asia. Formerly, farmers in oft-flooded areas have dealt with varying water levels by growing varieties of rice with taller stalks. But as we have already seen, the taller the stalk, the less nutrients there are for the rice. The new flood-tolerant types also have an advantage in that high water levels will kill the surrounding weeds.[50]

Demonstrating that you never know what the next breakthrough in agricultural biotech will be or where it will come from, scientists at the University of North Carolina at Durham, working with colleagues at other universities and with what is now Bayer CropScience, have found a way to adjust the size of a plant's cells even while it's growing.[51] Using tobacco for their practice run, they altered a key gene that creates bigger leaf cells without changing the overall plant size. Although the work is preliminary, biology professor Alan M. Jones, author of the study published in *Science,* told the *Durham Herald-Sun* that among its promising applications are:[52]

- It could toughen the grain crop stalks so they could not be broken and made unharvestable by a single strong wind.
- It could help a grower confronting changing market prices because a tomato grown for paste and ketchup usually contains small cells, while a slicing tomato has large, water-filled cells. The farmer could change these to meet demand for either product.[53]
- It could similarly make a tree's wood either better for furniture or better for paper.

"Before long, we should be able to control the size of wood cells in trees," said Jones. "Since soft wood is better for making paper, for example, we could grow softer woods by increasing the size of cells in a tree trunk, or we could produce harder woods for furniture by making their cells smaller."[54]

To do this, Jones and his colleagues identified a protein they believed to be the receptor to auxin, a major plant hormone that causes cells to grow, expand, and change form. Then, Jones explained, "We over-expressed—or excessively turned on—a special gene we created that makes [the auxin receptor] in tobacco. Expression of this special gene, we call a transgene, can be turned on or off by feeding plants the drug tetracycline. When we added tetracycline to the transgenic plants, what we found in both cultured cells and whole plants was remarkable."[55] The larger cell walls don't weaken the plant because they produce more cellulose to compensate for having less structure. In addition to the traits Jones described, he said it could also lead to food crops that have greater resistance to overly wet or dry conditions.[56]

Freshness from a Flower

It's great to be able to raise more crops on less land, but this does nothing about one of the most serious problems of agriculture: losses between the farm and the consumer. "In many Third World countries where they grow enough food to feed themselves ... bad transport arrangements make it difficult or even impossible to move perishable material long distances," said Peter Meyer of the Institute for Plant Biotechnology and Agriculture at England's Leeds University.[57] "Theoretically the world grows enough food to feed itself completely, the problem is getting it to the people who need it," he added. "This ultimately could change all that."[58]

That's why Meyer, Elena Zubko and their colleagues are working to engineer vegetables to stay fresh far longer than they do now. They spent four years searching for such a "freshness" gene that could be transferred from one plant into many others, finally finding it in the petunia. They inserted it into a tobacco plant that stayed alive in a mere glass of water for six months and even began sprouting new shoots.[59] Tobacco is one of the last plants you'd want to keep fresh because the leaves must be dried to be of any use. But remember, tobacco to scientists is just a green guinea pig. Any number of vegetables could have the same results, staying fresh for perhaps months without special storage. Moreover, while this would be of greatest benefit to the underdeveloped nations, as would any new crop developments, it certainly would come in handy anywhere.

FOUR

Foods That Taste Too Good to Be Healthy

When he was president, George H. W. Bush angered nutritionists everywhere by announcing, "I'm president of the United States and I'm not going to eat any more broccoli!"[1] His comment, which brought an immediate reproach from his wife, Barbara, touched on an important issue. There seems to be a rather unsettling correlation between what a lot of us think tastes good and what's good for us. A recent survey in Scotland, for example, found that people considered healthy eating to be "regimented, boring, and tasteless." If that's how people feel, they're not going to eat wholesome food. Indeed, they don't. Another Scottish survey found that half knew they should eat five or more portions of fruit or vegetables a day but only 23 percent did so.[2] One result is that Scotland has one of the highest rates of heart disease in the world. Toying with flavors could make all of us both happier and healthier by making foods high in nutrition more palatable. Biotech scientists are working on it, as indeed they are on a host of foods that will provide direct consumer benefits.[3]

The science of improving food taste is as old as the hills, involving the actual flavor that hits the taste buds, as well as smell, texture, creaminess, temperature, pungency and other qualities.[4] Flavor itself is considered to consist of the perception of four qualities: salty, sweet, sour and bitter, all of which have a receptor area on the tongue. (Some researchers argue for a fifth quality, *umami* or the "savory" taste of soy.)[5] Certainly taste is to a great extent subjective, as evidenced by pubs that claim to offer three hundred different types of beer. Taste is very much an acquired thing. Europeans have traditionally liked hazelnuts and loathed peanut butter, while Americans tend to love peanut butter and only in more recent years have seen fit to dignify the hazelnut by dropping the American name for it, "filbert."[6] Australians love a yeast extract spread named Vegemite, which to Americans both looks and tastes like tar.[7] But there are some generalities: People tend to dislike bitter tastes and to like sweet tastes. Fat is universally beloved, both for its taste and for its feel in the mouth.[8]

With a little adjustment here and there, said one researcher, we can "make healthy food taste better or make good-tasting food healthier."[9] As the *Washington Post* put it, "It's part of a new discipline that might be called 'consumer genetics.'" The *Post* reported, "There are plans to create seasonings to make vegetables more palatable to young children by blocking specific tastes that overwhelm their palates. Some companies foresee additives that precisely mimic the taste and feel of rich foods without the fat," while others are "interested in developing artificial sweeteners that can survive cooking, as some of today's popular ones cannot."[10] Among the earliest products envisioned are medicines, diet sodas and coffee that would lack their characteristic aftertaste because they would contain compounds that momentarily block the tongue's perception of bitterness.

Traditionally, good old chemistry has been the way new flavors are found. But this is slow, expensive, and subject to the limitations of human taste tests at every step of development. Now genomic-based companies are implementing faster, more efficient methods. Noted the *Post,* "As they learn the precise structure of proteins in the tongue or nose that detect taste and smell, they can copy those proteins. The copies can be used to build robotic testing systems that can screen tens of thousands of new chemical compounds a day. If a new compound binds tightly to taste or smell proteins, it's a clue that it might elicit a strong sensory perception."[11]

One of the key players in this is a San Diego company, Senomyx, which has licensed patents believed to cover virtually all the human genes that permit detection of bitter tastes. In June of 2001 the company announced that it had identified 347 human olfactory receptor genes, which it believes represent substantially the complete set of functional receptors related to smell.[12] Each of the 347 genes is thought to encode a unique protein that controls the recognition of odorants, of which it's believed there are about 10,000 that humans can identify.[13] "The goal is not to manipulate genes, but to use clever chemistry to fool those genes into doing something slightly different," according to Charles Zuker, a professor of biology at the University of California, San Diego. "Just find a molecule that blocks the receptor for that taste, like a lock and a key. I gum up the lock, and the key won't fit. It lets you mimic the taste of wonderful wine or the smell of a rose."[14]

"You have to have human tasting," said Gene Grabowski, a spokesman for the Grocery Manufacturers Association.[15] "Biotechnology enables us to speed up the pace of experimentation and increase the variety. It doesn't change the fact that we have human beings designing food."[16]

Some of these food products are already available. Many more in the pipeline will be a tremendous boon to consumers in terms of nutrition, taste, freshness, extended shelf lives, or some combination of these. Sweeteners are high on the list.

America is a nation of sweetener-holics. From just 1980 to 1999, U.S. intake of sugar and other caloric sweeteners rose from 123 pounds per person per year to over 158, a development that has contributed to the obesity epidemic.[17] It is surely also contributing to the epidemic of diabetes, cases of which have grown 33 percent from 1990 to 1998, with a 76 percent increase among people in their thirties.[18] Sugar is especially insidious in that unlike fat, protein or complex carbohydrates, it takes up so little space. A 16-ounce bottle of iced tea with 200 calories takes up no more room in your belly than a 16-ounce bottle of water. Even in baked goods, sugar has very little volume and thus lacks the ability to give the eater that certain sense of satiety. But the Dutch may help bail us out of this one, with a beet that produces real sugar but provides only half the usual calories and may also help guard against heart disease and promote digestion.

Andries Koops and others at the Center for Plant Breeding and Reproduction Research in Wageningen, outside of Amsterdam, implanted a gene from the Jerusalem artichoke into a sugar beet.[19] This fools the beet into transforming its high-calorie sucrose sugar into nondigestible fructans. "The sugar is less sweet than sucrose, let's say 25 percent, but you can use it as a sweetener with the added advantage that it can't be digested," said Koops. "Humans have the enzymes to digest sucrose but not fructans."[20] Since more fructan sugar is needed to achieve the same level of sweetness, a product made with the new sugar would end up with about 40 percent to 45 percent of the calories of a sucrose product. True, aspartame, sucralose and saccharin have far fewer calories, but fructan sugar has bulk, so you can bake with it. Moreover, fructan fiber encourages a friendly environment in the gut for beneficial bacteria such as *lactobacillae* and *bifidus.*

Heart-Smart Foods

The only things we love as much as sweet things are fatty things. (Or, best of all, sweet things that are fatty!) If obesity is the only consideration, then you should know that all fat has the same number of calories to the ounce. That said, some naturally occurring fats (such as fish oil) are far better for us in terms of cholesterol and atherosclerosis, while others (such as palm oil) are far worse. One of the healthier oils is canola, another is soybean.[21] Since so much of our canola and soybeans are now transgenic, biotech has already made a real contribution toward a "heart-smart" diet. It has also taken the next step, fashioning oils that will allow us to have our cake and eat it too. Among those already available are:

- High-oleic sunflower seeds that produce oil low in trans fatty acids. Many experts believe trans fats are an important contributor to heart disease, though the issue is still hotly debated.[22] (Trans fatty acids arise from the process of hydrogenation; high-oleic oils do not have to be hydrogenated and therefore have little or no trans fat.)
- A soybean oil with the remarkably uninspired name "LoSatSoy," meaning that it's low in saturated fat. This too will help prevent heart disease.[23]
- High oleic peanuts modified in the same way as the aforementioned sunflower seeds. They also provide longer shelf life for peanuts and products made with them, such as peanut butter and candy.[24]

Calgene is working to reduce canola oil's already low saturated-fat content by more than half.[25] Another Calgene project will make healthier margarine. Currently, plant oils must be hydrogenated, which means adding hydrogen atoms to produce stearates, saturated oil that gives vegetable oils solidity so they can be spread. But this leaves a residue of unsaturated trans fatty acids. A Southeast Asian fruit called mangosteen is known to carry an enzyme that could be used to make stearate, so Calgene researchers spliced the mangosteen gene into a canola plant. It has the desired physical characteristics but, since it needs little if any hydrogenation, it has up to two-thirds less trans fatty acids.[26]

As rice is to much of the underdeveloped world, so are tomatoes to much of the "overdeveloped world." What's the healthiest condiment usually found on hamburgers, French fries and pizza? On hamburgers and fries, it's ketchup. On pizza, it's tomatoes in the various forms of paste, slices or those tangy little sun-dried pieces. Tomatoes, processed and fresh, are ubiquitous in our diet.

Despite the old expression that we only like the taste of things that are bad for us, it so happens that these tasty tomatoes are also packed with vitamins, antioxidant lycopene, and flavonoids. (Flavonoids are chemicals found only in plants that are widely thought to have healthy benefits, but so far nobody's been able to package them effectively in a pill.)[27] Because of all this, breeders have been tinkering with tomatoes for a long time. Now they're tinkering using transgenics. Among the new types are three tomato breeding lines from the USDA that contain about 10 to 25 times more beta-carotene than typical tomatoes.[28] A British study found that such tomatoes can help stave off cancer and heart disease, and do so more effectively than pill supplements.[29] Another variety has improved pectin so that the fruit remains firm longer and retains more pectin during processing into tomato paste.[30] Among tomato augmentations expected within the next few years are improved flavor, color, shelf-life, and more antioxidant vitamins. In one experiment, British and Dutch scientists transferred a gene from the petunia into a tomato. Petunias are pretty and teeming with flavonoids, but aren't the tastiest things around. By moving one petunia gene into the fruit, they created tomatoes with almost *80 times* the flavonoid content.[31] Another trait being added to tomatoes will make them easier to harvest because the fruit ripens more uniformly.[32] Inserting a single yeast gene into tomato plants rewarded researchers with fruit that had more lycopene and a longer vine life, and that made better juice.[33]

Other fruits ripe for improvement by biotech in the next few years are firmer and sweeter peppers, and ripening-controlled bananas and pineapples that will provide tastier fruit that is easier to harvest.[34] Strawberries are being made to stay firm longer with the same technology that makes the plants more resistant to disease.[35] Raspberries are also being engineered to protect them from raspberry dwarf viruses, which make the fruit small and crumbly, suitable only for juice or puree.[36]

Vitamin-Packed Plants

Getting more vitamins into foods we enjoy is a major goal of biotech engineers. As it happens, those who pop vitamin supplements are also those most likely to eat well, while unfortunately nonpoppers tend to eat more junk.[37] Moreover, with at least one important vitamin, E, it's almost impossible to eat enough food containing it to meet the levels many nutritionists now think we should be ingesting. The recommended daily allowance of vitamin E ranges from 10 to 15 international units (IUs). However, studies show that an intake of 150 IUs or more may lower the risk of cardiovascular disease, cataracts and some cancers, and it may slow the progression of some degenerative diseases such as Alzheimer's.[38] "Just to get to 150, we'd have to eat about four pounds of spinach a day or take in 3,000 calories of soybean oil," said Dean DellaPenna, a biochemist at Michigan State University.[39] So DellaPenna, along with doctoral candidate David Shintani, decided to do something about it.

They knew that in the United States, approximately 60 percent of dietary vitamin E intake is from vegetable oil, primarily from soybeans. Americans and many people in other countries eat lots of soybean oil, but the main problem is that vitamin E in soybeans occurs in four different forms that vary in the amount of vitamin E activity: alpha, beta, gamma and delta. In soybean oil, 70 percent of the vitamin E is in the least potent form, gamma, while only 7 percent is in alpha, the most potent form. The other primary oilseed crops—corn, canola, cottonseed and palm oils—are no better.[40] DellaPenna and Shintani took the mustard plant, *Arabidopsis thaliana,* known as the "white rat" of the plant world because it produces new generations quickly, has a fully mapped genome, and doesn't take up much space.[41] They located the gene that makes the single protein needed to convert the gamma of vitamin E to the alpha. They inserted the gene into bacteria and cloned it, then put several copies of the gene back into the plant. This increased the amount of alpha vitamin E almost tenfold. Because the pathway to convert gamma to alpha is the same in soybean, corn and canola oils, it's a good bet that the same can be done with them.[42] "With this latest discovery, we're more than halfway to the 150 IU level, which I believe we can achieve in the near future," said a happy DellaPenna.[43]

There's also nothing to stop researchers from using similar techniques to pump up levels of other vitamins in other plants. For example, those at the University of California at Riverside altered a gene from wheat and inserted it into both corn and tobacco, so that they produced two to four times as much vitamin C as ordinary tobacco and corn. The ramifications could be tremendous for providing more vitamins from crop plants both for humans and for livestock.[44]

"When we debate GM foods we should keep an empty chair at the table for the people who will benefit—or stand to lose out if this work is not done," said DellaPenna. "That empty chair represents 2.5 billion people who are vitamin deficient, and who live on less than $2 a day. They can't afford to buy supplements."[45]

Java junkies will also have something to rejoice about in five years if either of two groups of researchers succeeds. Both are trying to extract the caffeine from coffee beans without steaming, stripping or washing the beans in organic solvents. "The decaffeination processes, particularly with organic solvents, do not just take out most of the caffeine, they also take out some of the aroma and flavor," said Alan Crozier, lead researcher for one of the two teams. "So to an espresso addict like myself, decaf tastes like dishwater."[46]

There's nothing intrinsically wrong with caffeine. It's one of the most thoroughly researched ingredients in our food supply and tends to increase alertness, provide energy (which is why the International Olympic Commission treats it as a controlled substance), and keep semi-trailer truck drivers from plowing into your car at three in the morning because they've fallen asleep at the wheel.[47] It's also an ingredient in several types of medicine such as decongestants. That said, it can also make you nervous, irritable and anxious, and keep you awake not just on the road but in bed.[48] Crozier, a professor of plant products and human nutrition at the University of Glasgow, worked on the decaf process with colleagues in Japan.[49] First, they had to isolate the enzymes involved in the manufacturing process inside the plant. "That was the difficult part," said Crozier. "It took us six years."[50] Then, using antisense technology, they turned off the gene that directs the plant to produce caffeine. "All you're doing is switching off caffeine, not the compounds that affect flavor and aroma," Crozier explained. "It should taste like the real thing.[51]

"With the right commercial funding we can produce the first no-caffeine plants within three years," he said in August 2000. "But it will be five years before anything starts to appear on supermarket shelves."[52]

There's a good chance there will be a competing brand, though. Researchers at the University of Hawaii and Integrated Coffee Technologies had earlier announced they had also used antisense technology to turn off caffeine production in a coffee plant, albeit using a different gene. In collaboration with Monsanto, Integrated Coffee hopes to have a commercial harvest in 2006.[53]

Transgenic Animals

To date, most transgenic engineering has involved crops. In part, that's because it's simply easier to work with crops than animals. But there may also be some hesitancy because of public opinion. "People feel different fundamentally about animal biotechnology than they do about plants," said Tom Hoban, a professor of sociology and food science at North Carolina State University who researches public perceptions of biotechnology and food issues.[54] "They're going to attract a larger set of critics because they're dealing with animals."[55]

But cloning is already being used to perpetuate lines of top-quality domestic animals such as cows. You'll probably never eat a cloned cow or pig, but you'll be eating their offspring.[56] (Cloning is also being used to preserve species on the verge of extinction.)[57]

You may soon be eating transgenic salmon as well. Americans have been developing more of an appetite for fish, but even under controlled conditions it still takes three years to raise a normal fish from egg to a nice pink (and healthy) salmon steak.[58] That's about to change, though. In research that began with trying to use flounder genes that would help salmon survive in frigid waters that would normally kill them, Canadian scientists also found a gene that speeds salmon growth. The re-engineered fish grew four to six times faster during their first fourteen months. The combination also worked on other fish such as rainbow trout, which grew eight to ten times faster than normal in their first year.[59]

The first such fish to hit store shelves will probably be the AquAdvantage, a salmon from A/F Protein in Waltham, Massachusetts. It grows to market size in only eighteen months, half the normal time.[60] The super salmon alone could greatly relieve the problem of declining salmon populations in the Pacific Northwest. It could also convert this healthful fare from a yuppie food to one that readily competes with red meat and chicken in all households.[61] As stocks increase, it could directly alleviate the worldwide problem of overfishing. Finally, ground into fish meal it would provide a cheap source of protein for billions of people who are deficient in this nutrient. It's a winner for the environment, and for the peoples of both the developed and the developing countries. A/F Protein says it's also working on faster-growing varieties of other fish including Arctic char, trout, tilapia, turbot and halibut.[62] Elsewhere, scientists are using gene transplants to create fast-growing trout and oysters that can withstand viruses such as those wreaking havoc in the Chesapeake Bay.[63]

As one might expect, there have been cries of "Frankenfish!" and talk of biotech fish escapes into the wild species that would cause some unprecedented genetic disaster.[64] But neither American nor Canadian regulators have so far raised concerns about the safety of the new fish. Indeed, Canadian government-funded studies of the transgenic salmon have found they appear to be almost identical to ordinary ones except for their fast growth.[65] Nonetheless, the company is taking precautions such as sterilizing all the fish. Further, "Our fish have no environmental advantage," said Elliot Entis, chief executive of A/F Protein. "All work ever done on this kind of modification suggests you give your fish a disadvantage [in the wild] when you give it these traits."[66] Wild fish thrive by staying small in their youth in streams, then growing in a spurt when they leave for the ocean. A fish that grows quickly from the start is "remarkably unsuited to survive in the wild," Entis said. "That means that these problems are self-limiting," because an escaped AquAdvantage fish wouldn't likely live long and breed successfully.[67] Evaluations of both accidental and intentional fish releases—those in which authorities are *trying* to get a new fish population into an area—indicate that, as one review put it, "it is difficult to genetically impact established natural populations of fish."[68]

Meanwhile, Mother Nature has served up all sorts of fishy beasts that can do considerable damage, and we learn to deal with them. Look no further than the monster discovered in a Maryland drainage pond in June 2002. Early reports of a huge creature with a snakelike head and razor-sharp teeth were treated as just another fish tale—until one was caught. It turned out to be a native of Asian waters, a northern snakehead, a pair of which somebody had chucked into the pond. Unfortunately, the pair were male and female and they spawned thousands.[69] The creatures grow to be three feet long and as one Interior Department spokesman put it, "They eat literally anything that's living," including even ducklings.[70] Ultimately there's nothing left but the snakeheads.

The snakehead has no trouble surviving cold winters and it has one especially hideous trick up its fin. By virtue of a primitive lung, it can crawl out of the water and slither across dry land until it finds another home. It—and its offspring—can then eat their way through the new habitat and repeat the process as necessary.[71] The story of the snakehead shows that, yet again, opponents of biotech food judge it by standards that would banish nature itself from the planet.

FIVE

Developing the Underdeveloped Nations

The oft-repeated refrain of "Who needs it?" concerning biotech crops applies only to the industrialized nations, all of whom have well fed if not grossly overfed populations, virtually all of which (the U.S. aside) will soon begin declining.[1] The developed nations also suffer little from global calamities such as tuberculosis, malaria and cholera. For us, biotech food will be very nice and, like air conditioning or luggage with wheels, we will soon wonder how we ever did without it. For those in the underdeveloped nations, it's literally a matter of life and death.

About 790 million people in the developing world are currently undernourished, meaning they can't meet their basic need for energy and protein.[2] On the upside, despite terrifying bestsellers warning of massive famine such as Paul Ehrlich's 1968 book *The Population Bomb*, the food supply has actually been growing faster than the world's population.[3] Because of the Green Revolution—an international effort to increase crop yields by developing superior plant strains and scientifically applying pesticides and nutrients—between 1961 and 1997, food supplies per capita increased by almost a fourth.[4] But we need to maintain this progress even as the world's population grows, and must try to do so without destroying forests or other heretofore unplowed land, but rather relieve some of the pressure to the ecosystem (in terms of pollution, nutrient stripping of soil, and so on). As the architect of this revolution, Norman Borlaug, has said, even without biotech, "Increases in crop management productivity can be made all along the line: in tillage, water use, fertilization, weed and pest control, and harvesting. However, for the genetic improvement of food crops to continue at a pace sufficient to meet the needs of [the growing world population] both conventional technology and biotechnology are needed."[5] Said Borlaug, "It is access to new technology that will be the salvation of the poor, and not, as some would have us believe, maintaining them wedded to outdated, low-yielding, and more costly production technology."[6]

PUBLIC HEALTH PROBLEMS PARTLY CAUSED OR AGGRAVATED BY INSUFFICIENT FOOD OR POOR NUTRITION, AND BIOTECH'S POTENTIAL IMPACT[7]

Problem	Current extent (year)	Likelihood that biotech crops would reduce the problem
Undernourishment	825 million people (1994–1996)	Very high
Malnutrition	6.6 million deaths/yr. for children under 5 (1995)	Very high
Stunted Growth	200 million people (1995)	High
Iron-Deficiency Anemia	2 billion people (1995)	High
Vitamin A Anemia	260 million people (1995)	High

Sources: World Health Organization, *The World Health Report 1999* (Geneva: World Health Organization, 1999); Food and Agricultural Organization, FAO Databases, at: http://apps.fao.org; FAO Databases, "The State of Food Insecurity in the World," at: www.fao.org/FOCUS/E/.

Kenya's Florence Wambugu (founder of A Harvest for Biotech Foundation International) goes further to argue that agricultural biotechnology is the solution to the famine, environmental degradation and poverty besetting her continent. "I'm not saying that transgenics alone will solve all the problems. But it will lead to millions of tons more grain," she said.[8] James O'Chanda, a lecturer in biochemistry at Kenya's University of Nairobi, agrees: "We want to create an enabling environment where African people can participate and benefit from biotechnology in a responsible and sustainable agriculture." O'Chanda has described the growing concerns in Europe over the safety of biotech crops as "Europe's way of propagating fear and scare to prevent us from reaping the benefits of biotechnology."[9]

Such a statement may or may not be a bit paranoid, but certainly it's "Eurocentric" and simply wrong to force Africa to use an analysis of potential risks versus potential benefits based on Western needs. Or as Calestous Juma, a Kenyan who is special advisor to Harvard University's Center for International Development, put it, "For the world's developing countries, one of the greatest risks of genetic engineering is [the risk of] not being able to use this technology at all."[10] Sadly, denying underdeveloped nations access to these improvements is the intention of some agbiotech opponents. Others just don't seem to understand that while Europe's population will soon begin to decline, the developing nations will continue to grow until at least the middle of the century.[11] Famine *can* effectively limit growth; there's no argument there. But is this really the form of population control that biotech critics want?

The White Man's Burden?

The British group Christian Aid, in its strongly titled and widely covered 1999 report *Selling Suicide,* said that transgenic herbicide-tolerant crops should be kept from India because they allow such efficient killing of weeds that they will "remove one of the few cash-earning employment opportunities for women," namely walking down crop rows stooped over to yank weeds.[12] Not that the authors would want *their* wives or sisters or daughters to be locked into such a menial, back-breaking task. Nor does it seem to occur to them that by freeing up time for the farmer and his family, these crops are providing opportunities that heretofore couldn't even be imagined. "We can attend to other things besides staying in the field," a South African cotton farmer told a researcher at the International Service for the Acquisition of Agri-biotech Applications. "Our standard of living is very much improved when we have money to send our children to school."[13] Freedom from labor in the fields would result in more education and better employment, two factors shown to limit family size.

Others have referred to this mindset of telling people of the underdeveloped world what's best for them as a throwback to the idea of the "White Man's Burden," a renewed effort to impose foreign values for "saving souls" and saving the environment. It is "another crusade, reminiscent of those which led to Western imperialism in the past," said Indian-born UCLA professor Deepak Lal at an international conference in India in December 2000. Whatever its professed aim, its practical effect will leave "little hope for the world's poor," he said. Opposing access to agricultural biotech for developing nations is not the only goal of these people and their groups, said Lal, but it's an important one.[14] Florence Wambugu put it more strongly: "Those who protest biotechnology do so with a full belly."[15]

The most horrific example of this (to date, at least) was in the summer of 2002 when famine gripped Africa. The United States rushed to deliver massive amounts of corn to several countries, with about 17,000 tons sent to Zambia alone. And there it rotted while perhaps millions were threatened with starvation. Why? Because about 30 percent of the corn was transgenic. Initially, the Zambian government claimed it was afraid that some of the corn would get released into the wild and mix with nonbiotech varieties, hence destroying opportunities to export to countries (in Europe particularly) that don't allow imports of biotech corn. That Europe actually imports no African corn was of no moment, nor were offers of other countries to mill the corn so it couldn't be planted.[16]

The Zambian government took this position because of influence by environmentalist groups, especially Greenpeace and Friends of the Earth, which declared the food to be "poison."[17] Friends of the Earth insisted—believe it or not—that the food donation was an effort at "recolonization," while Greenpeace asserted: "As long as supplies of non-genetically engineered grain exist, nobody should be forced to eat genetically engineered (GE) grain against their will."[18] Neither group was swayed by the insistence of the director of the U.N. World Food Program that "There's no way we can help them if they don't accept the food," and that "No one is going to step up with donations of non-GM [genetically modified] corn to fill the gap." The director added, "This is food we have complete confidence in," while the World Health Organization also said there was no evidence that such food is dangerous and U.N. officials pleaded with the Zambian government to reconsider its position.[19]

U.S. Aid for International Development administrator Andrew Natsios accused the environmental groups of endangering the lives of ultimately millions of famine-threatened Africans by encouraging Zambia and other governments to reject biotech food aid from the United States. "They can play these games with Europeans, who have full stomachs, but it is revolting and despicable to see them do so when the lives of Africans are at stake," Natsios told the *Washington Times.* "It doesn't make sense," said Tony Hall, U.S. ambassador to the United Nations for food and agriculture agencies. "You don't have this kind of argument [about food] when people are starving to death. And it's only going to get worse." Hall met with agriculture and social welfare officials in Zimbabwe, telling them, "You know, with your action, you're going to kill your people. You don't have any other food in this country."[20] The environmentalists "are using big-time, very well-organized propaganda the likes of which I have never seen before," he said, and this was contributing to the deaths of "millions of poor people in southern Africa through their ideological campaign."[21]

Let's face it—anybody who deals in a major way with the underdeveloped world can be tagged as a meddler and perhaps an opportunist. The biotech companies say they want to help; the antibiotech organizations say only *they* can help. But organizations, like people, can best be judged not by what they profess, but by what they do. And it appears that what the biotech companies are doing in the name of profit with a bit of altruism tossed in is better than what their opponents are doing allegedly in the cause of truth and righteousness.

Consider, for example, the activists' claim that the developing world will not be helped by biotech because it is unprofitable in comparison with selling technology and seed to relatively rich Western farmers.[22] Never mind the obvious counterargument that use of biotech crops can increase a farmer's income and the more he earns, the more he has to invest. The allegation has already proven false. About one-fourth of the world's transgenic crop acreage is now planted in underdeveloped countries.[23] In terms of actual numbers of farmers, more than three-fourths of those in the world who plant biotech crops live in developing countries.[24] Further, the rate at which biotech crops are planted is about five times faster in the underdeveloped nations than in the industrialized ones.[25]

A look at some of the individual biotech crops being developed also shows that many have limited economic value in the West. One such is the aforementioned golden rice enriched with beta-carotene, and another is golden mustard. How many of us in North America, Europe or Australia cook with mustard seed oil?

In early 1999, Monsanto also announced it had a "working draft" of the rice gene sequence that, when completed, would be made available to the public, thereby passing up patents worth probably hundreds of millions of dollars.[26] But others had the same idea and soon Syngenta, along with Beijing Genomics Institute and the University of Washington Genome Center, published drafts of such a sequence in *Science*.[27] The sequences will be used to find genes responsible for desirable traits such as high yield and disease resistance. One of the authors said that not only could it be used to genetically engineer plants, but "it will also accelerate traditional breeding by allowing scientists to throw away all the junk fast."[28] The rice genome sequence will boost efforts to improve other crops because the arrangement of rice genes is quite similar to that of other cereals such as barley, wheat and corn. "This is such an explosion of sequence information, researchers are running around in a frenzy," said Pamela Ronald, a plant pathologist at the University of California, Davis.[29] Combined, these cereals provide about 90 percent of our calories.[30] Further, in December of 2002, DuPont announced it was making its proprietary wheat genome data available to public and private researchers without restriction, more than doubling the amount of wheat genome information available.[31]

For several years Monsanto has also had a program of bringing scientists from the developing countries to its Missouri research institutes strictly to work on developing new crops for their own parts of the world. Monsanto then ensures that these scientists go back home to continue their work.[32] One success story was Kenyan scientist Florence Wambugu, who developed a transgenic sweet potato resistant to the devastating feathery mottle virus, a significant development because sweet potatoes are a staple in Africa. "Monsanto had the technology to attack viruses and were looking for an opportunity to work on an African root or tuber crop. It offered to train and support someone and donate the intellectual property rights to Africa," she told *New Scientist* magazine. Monsanto approached her, and she went to the company's labs with seven sweet potato varieties familiar to Kenyan farmers. "I had to learn everything from the beginning," said Wambugu. "Transferring the gene into the sweet potatoes took me about three years. Then we selected virus isolates from the fields in Kenya and sent them to Monsanto to test them on the sweet-potato varieties in the greenhouse. It's taken me ten years to reach the point where we are about to begin field trials in Kenya."[33] That work came to fruition in the summer of 2000, when Kenyan farmers first planted her transgenic sweet potatoes. It was expected that the new strain would increase yields by as much as 60 percent with no use of pesticides. Former U.N. ambassador Andrew Young was on hand for the celebration, pronouncing that "Africa is on the verge of a tremendous revolution."[34]

I asked Wojciech Kaniewski, who has spent ten years in Monsanto's Scientific Outreach Program, to tell me more about this. "The program started in 1991 when Florence came to Monsanto," he explained. "She was here almost three years and several other Kenyan scientists have followed in her footsteps." Working with Wambugu, he said, "Our first step was to use an American strain of feathery mottle virus just to see if the technique worked. We got a gene from severely infected sweet potatoes, cloned it, and inserted it into another sweet potato plant to induce protection. Essentially we vaccinated it." It worked on the American strain, so Monsanto "imported feathery mottle strains from Africa," Kaniewski said. After that, it was a fairly simple matter to duplicate the work and "immunize" African sweet potato plants. Kaniewski added, "We expect that none of the plants will be infected with the virus now; that there will be complete protection. We can't guarantee it, but that's our goal." In any event, other transgenic strains of sweet potato resistant to feathery mottle are in late stages of development.[35]

The transgenic plants are now essentially the property of the Kenyan people or anybody else in Africa who wants them. "We have an agreement with KARI [the Kenyan Agricultural Research Institute] that we will help them using only technologies that we own [patent rights to] and we will give them away for free," Kaniewski said.[36] Monsanto will not [financially] benefit." But American farmers might. "If what's growing in our test plots in South Carolina works, we might give the technology away to American farmers," he told me.

Give it away? How do you explain that to shareholders? "Sweet potatoes are a crop that Monsanto doesn't think it could make much money out of, and anyway we don't have all the patents," Kaniewski realistically explained.[37] "I have requests from Asian countries as well. But we'll need to import strains from those countries. We have five countries that we're helping, Indonesia, Malaysia, Thailand, Vietnam, and the Philippines, and we're cloning genes from each of their viruses." The program works with other crops as well. "Since 1991 we've had a white potato program for Mexico. We've tested three varieties of potato in Mexico. We've inserted three different genes for protection." Monsanto is also working with the Kenyan Agricultural Research

Institute to develop a virus-resistant cassava root, another African staple, and is helping several countries to develop virus-resistant papaya.[38]

Why would a company sponsor developing world researchers, especially since there's not much PR value in a program that virtually nobody's heard of? "It's not a big expense but we are making a lot of allies by doing it," Kaniewski told me. "In these countries they will be more accepting of our corn and soybeans." In other words, Monsanto is doing well by doing good. Or at least it hopes to. But Kaniewski did suggest another motivation for developing crop protection for underdeveloped countries: "If they don't have it, they will starve to death."[39]

Monsanto is not alone in reaching out to the underdeveloped world. Syngenta's contribution on rice has already been noted and the company has also given away to Vietnam a gene that produces insect resistance in sweet potatoes, one of about half a dozen or so such charitable arrangements the company says it has.[40] And, as we've seen, numerous countries waived license fees to make golden rice a reality. The U.S. Department of Agriculture has also begun a program to encourage American colleges and universities to work with researchers from developing countries on biotech. Several institutions started such programs on their own. Tuskegee University, for example, guided by C. S. Prakash, has established research ties with the African National Agricultural Research Service in Kenya, Ghana and Senegal.[41] A South African researcher working at Tuskegee University, Chantal Daniels, is developing biotech sweet potatoes containing up to five times the normal amount of protein.[42]

Developments like these make agbiotech opponents absolutely livid by devastating their claims that biotechnology is predatory globalism, benefiting only the developed world, and that it has almost no value in underdeveloped countries. In such a worldview, the most significant developments are only cynical public relations. "Golden rice is part of an exercise to make the technology acceptable," Margaret Mellon, director of the agriculture and biotechnology program of the Union of Concerned Scientists, said dismissively.[43] "The timing of this [golden rice announcement] is so clear," remarked Charlie Kronick, head of Greenpeace U.K.'s anti–genetic engineering campaign. "People are talking about the potential benefits of the second generation of GM crops when almost no questions raised by the first have been answered. You don't have to be paranoid to think the tactics are deliberate."[44]

But, in fact, Ingo Potrykus' lauded work with golden rice began in 1990, five years before biotech crops were commercialized and even longer before they became an issue for activists.[45] It popped out when it was ready, not when it would do Greenpeace's antibiotech campaign the most harm. Further, golden rice is just one of many Western-developed biotechnologies that will help poor countries far more than wealthy ones, so it hardly makes sense to spew all one's venom on that single plant. "The reaction that really matters comes from the developing countries," said Potrykus. "They are saying 'we don't have to be told by spoiled white countries what to do,' and rightly so. If you have blind children in the street and a technology that can help, then you have no doubts."[46]

True enough, golden rice and other breakthroughs *do* stifle the message of groups like Greenpeace and the Union of Concerned Scientists. On the other hand, Greenpeace, for instance, has long been known for opposing industrial progress for the very sake of it and using sabotage.[47] But surely this group would draw the line at sabotaging golden rice, would it not? In fact, Greenpeace has declared, "Golden rice has not been ruled out as a target for direct action in the future."[48] As to the Union of Concerned Scientists, which despite its elevated title comprises mostly lay members, it describes its mission in part as "tenacious citizen advocacy."[49] But many people in the scientific community never bought the line. One close observer described it as a "savvy activist group dedicated to mobilizing public support for political goals that many scientists cannot condone on the basis of sound science."[50]

Golden rice makes agbiotech opponents look like the Luddites they've been accused of being, and it reveals them to be utterly insensitive to the needs of the developing world. But this is not the crop's *raison d'être.* "It is going to be harder for the environmentalists to say they are battling for the poor if they're fighting something that benefits the poor," notes Gary Toenniessen, director of food security at the Rockefeller Foundation, acknowledging the looming ideological rift. "But those groups were naive if they thought Third World countries were on their side to start off with."[51]

The Asian Transgenic Tiger

Apparently there's a strong sentiment among agricultural biotech critics that those outside the West are too ignorant to know when they're being hoodwinked by multinational seed corporations (or, to use their oxymoronic phrase, "American multinationals"). Or the critics believe that developing countries will not be able to compete. They're wrong on both counts.

Recall the New Delhi researchers who are making plants that require much less water. Consider that the Indian Council of Agricultural Research is developing genetically engineered cotton, rice, wheat and pigeonpea that are resistant to pests.[52] Other Indian researchers have created a sweet potato that not only grows faster than the original but has a much higher protein content.[53] India has also created white potatoes that contain far more protein by inserting a gene from grain that grows in the Himalayas.[54] (Westerners tend to eat so much protein it practically oozes out of our pores, so we forget that protein deficiency is a problem for much of the underdeveloped world.) Indian scientists have also taken a gene discovered and studied in Australia and inserted it into India's three highest-yielding wheat varieties to make them resistant to the most widely used herbicide.[55] Because India uses types of wheat that Western nations don't, and has weeds that Western nations don't, it might have been a long time before an American or European company got around to doing this; so the Indians did it themselves.

South Korean scientists have created a strain of bioengineered rice that they claim could boost yields by up to 26 percent. If commercialized, the enhanced productivity will help the nation reduce food imports valued at more than $7.4 billion in 1999.[56] The researchers modified the genetic structure of rice by transplanting a gene called protox from a microbe called *Bacillus subtilis.* The gene involved in photosynthesis (converting sunlight to nourishment) has increased the number of branches and ears of rice plants, boosting crop yields by 20 percent during two-year field tests. "The super rice is a major breakthrough in botanical science as well as a powerful tool to feed the impoverished millions in the Third World," said Professor Choi Yang-do of Seoul National University. He added that the technology could also be applied to other crops like wheat and oats as well as livestock feed. The research team is testing the method with other plants like corn, which has a double-leaf structure, unlike single-leaf rice.[57]

China is a world leader in developing biotech crops. In some senses, it is *the* leader. In 1988, it became the first nation to commercialize a bioengineered crop: tobacco resistant to a plant virus.[58] Since then, Beijing has approved the release of more than a hundred genetically engineered crops, double the number released in the United States.[59] These include over a dozen new strains of rice, potatoes, tomatoes, corn and trees, all of which were developed domestically.[60] Cotton production in China had remained virtually static from 1949 to 1979. Recently it has almost doubled, largely because of *Bt* plants.[61] Indeed, just from 2000 to 2001, the amount of *Bt* cotton planted tripled.[62] It was originally purchased from Monsanto, but China has now developed its own strain.[63] According to a report in *Science* magazine, Chinese farmers of *Bt* cotton have slashed their use of pesticides by about 80 percent and cut overall production costs by over 25 percent.[64]

The Chinese are even experimenting with making cotton infused with rabbit genes, essentially to be able to grow soft fur. This will do more than delight animal rights activists; a test by the China Textile University and the Ministry of Agriculture's Cotton Quality Inspection Center demonstrated that the new fiber is stronger, warmer and 60 percent longer than ordinary cotton.[65]

While some of China's crops are designed to feed a growing population, it's surprising how many are intended to feed a population growing in affluence. "Quality improvement is the first focus, because when you get rich you get picky," said Chen Zhanliang, who was working for Monsanto in St. Louis when his country brought him back to put him in charge of developing transgenic plants other than cotton. He notes that not only are Chinese eating more grain-fed meat but they're also demanding better-tasting grains. "More and more people like to eat rice that's a little sticky, for example, and that's something controlled by a gene."[66] So advanced are parts of their biotech industry that "In less than 10 years, we'll be accessing technology from China," said John Killmer, vice president of Monsanto Far East in Beijing.[67] In 2001, the Chinese government announced its intent to quintuple funding of agbiotech research by 2005, to $500 million annually, and make it the world's greatest investor in the field.[68] The double helix that Watson and Crick discovered a half-century ago has now literally become a Chinese national symbol of progress, displayed in National Day parades and turned into giant sculptures in public parks.[69]

Out of Africa

Between 1990 and 1995 there were 25 field trials of transgenic crops on the African continent, involving a vast array of crops and introduced traits.[70] Of these, 22 were performed in South Africa, two in Egypt, and one in Zimbabwe. African nations have performed relatively few field trials of transgenic crops compared with many other countries, but that's changing. Kenya, Uganda, Namibia and Cameroon have introduced "National Biosafety" laws and regulations, and discussions are being conducted in Mauritius, Zambia, Tanzania, Ethiopia, Nigeria, Ghana and the Ivory Coast.[71] As a result, the number of biotech field trials has increased tremendously; by early 2000 there had been 120 in South Africa alone.[72] South African local potatoes are being engineered with the protein coat (outer jacket) of the potato leaf roll virus, to give the plants immunity.[73] Current research is focusing on transforming potato plant strains important for resource-poor farmers.*

Potatoes South Africa, an industry-funded group in Pretoria, supports the maintenance of the national potato plant collection, which will provide whole plants or genes to future researchers.[74] Corn is also a major source of food in Africa. A collaborative project to engineer fungi-resistant corn began in 1995, using a gene isolated from a bean to fight *Stenocarpella maydis,* a cause of severe yield losses and poor-quality crops in South Africa.[75] The problem of maize (corn) streak virus (MSV) in Africa is also being fought by biotech with funding from

---

*Ironically, the first trransgenic potato was planted in the U.S. and Canada; Monsanto's NewLeaf potato spliced with *Bt* repelled the voracious Colorado potato beetle. But the fast-food giant McDonald's caved in to activist pressure and told its French fry suppliers to stop using the potato. As a result, J. R. Simplot Co., a major maker of French fries, instructed its farmers to stop growing NewLeaf potatoes. (See Carol Ryan Dumas, "Business Interests Force Idaho Potato Growers to Shun Biotech-Aided Spuds," *Twin Falls [Idaho] Times-News,* 9 May 2000.) The world's largest potato processor and Canada's largest wholesaler, McCain Foods, also backed out. "We think genetically modified material is very good science," said chairman Harrison McCain, but "at the moment, very bad public relations." (As quoted in *Canadian Business and Current Affairs* 58:14 (20 December 1999): 1, 4; see also Michael Fumento, " 'Frankenfood' Activists Failed to Scare the Public," *National Post,* 11 June 2002, p. A17.)

the Rockefeller Foundation, Syngenta and numerous other international organizations.[76]

The ARC-Infruitec Division for Plant Biotechnology and Pathology is based in the Western Cape Province of South Africa, near the home of the stone- and pome-fruit growing industry, and in the heart of the wine lands. It works to develop new fruit varieties and research cultivation problems. The Division for Plant Biotechnology and Pathology seeks to genetically improve deciduous fruit and related crop varieties for disease resistance and better shelf life. Among its important recent achievements are:

- Development of sensitive and rapid molecular-based detection methods for bacterial and fungal pathogens of deciduous fruit.
- Development of efficient shoot regeneration from single cells of the leaves of varieties of three types of apples, three types of pears, two types of apricots, and one strawberry type.
- Identification of transgenic strawberry plants expressing genes for herbicide and fungal resistance.
- Development of tolerance against plum pox disease in transgenic apricots.[77]

It's clear that people in Africa and other parts of the developing world, like those in the industrial world, don't want only to subsist; they want to succeed. Biotech alone will not be the answer. First, they need stable governments and market-based economies. That wonderful new Korean rice will do little to prevent North Korea's famines, for instance, because the famine were created by a Stalinist government that steals food from the people to feed its massive army.[78] Bioengineering cannot overcome wicked dictatorships. But it can and will prove an important tool in taking underdeveloped peoples from merely surviving to thriving. "Our position in Kenya is that biotechnology is not a problem. Poverty is," said Shem Adhola, a senior Kenyan official in the agriculture ministry. "The quest to produce nutritious food and agricultural commodities in abundance is our challenge."[79]

Margaret Mellon of the Union of Concerned Scientists declared in 1999, "Genetic engineering will never, ever serve the needs of the small farmer."[80] But there's no substance to the claim that biotech favors large Western farms, especially American and Canadian, over the tiny plots prevalent in underdeveloped nations. While people who use the word "never" regarding technology usually end up chewing and swallowing their words, you'd think they would at least be cautious enough not to make statements that have *already* proven false. For Mellon had it quite backward: large farms lend themselves to efficiency because they can justify huge expenditures for sprays, fertilizers, tractors and other equipment. In fact, virtually every technology discussed in this section of the book has been developed precisely for farmers working small plots. Not surprisingly, more than one-fourth of the global transgenic crop area of the world was grown in the underdeveloped nations—more than 33 million acres.[81]

Happily, at least one fairly recent poll shows that Americans are very upbeat on the idea of using bioengineering to feed people less blessed than they. According to a survey released in October of 2000, hunger and malnutrition are considered an urgent global problem by 75 percent of American adults, compared with the 68 percent who see disease and epidemics as urgent problems, 66 percent who feel that way about pollution and environmental damage, and 43 percent who say global warming. By a margin of more than two to one, Americans support the use of biotechnology in food and agriculture. And nearly three-fourths, 71 percent, believe biotechnology can help resolve problems of world hunger and malnutrition.[82] "The response by Americans in (this) poll is heartening," said Tuskegee University's C. S. Prakash. "It affirms what most agricultural scientists and policy makers have been saying all along. Science and technology can continue to make a positive contribution in alleviating world hunger, and Americans overwhelmingly support initiatives to increase agricultural productivity and the use of biotechnology in addressing concerns of global food and nutritional security."[83]

American and European antibiotech activists may insist on trying to speak for the rest of us, but they do not. Responding to their efforts, Nobel laureate and former president Jimmy Carter declared in a *New York Times* editorial that "the real losers will be the developing nations. Instead of reaping the benefits of decades of discovery and research, people from Africa and Southeast Asia will remain prisoners of outdated technology. Their countries could suffer greatly for years to come. It is crucial that they reject the propaganda of extremist groups before it is too late."[84]

SIX

Frankenfoods?

Back in 1972, Stanford University's Paul Berg became the first scientist to splice together two DNA molecules. He would later win the Nobel Prize for it.[1] Within a year, two other California scientists, Stanley Cohen and Herbert Boyer, pasted frog genes into the plasmids of *E. coli*. It was the first step in learning how to give bacteria the ability to produce animal proteins.[2] Suddenly there was an outcry from environmentalists and some scientists: the idea of combining parts of such disparate organisms seemed both bizarre and somehow dangerous. But safety arguments were carefully considered, a federal oversight committee was established, and research resumed.[3]

Today there are still polls indicating that Americans are wary of this or that type of biotechnology, with the answers generally depending on how the questions are phrased. But in reality, in almost all areas we've become accustomed to the idea of transgenics, to the point that a beverage company satirized the early fears with a series of advertisements about mixed fruit drinks that "Mother Nature never intended."[4] Still, there are those who continue to fight desperately against biotechnology in its various forms in the hope of reversing progress or at the very least placing "permanent moratoria" on certain research. And they will seize on absolutely anything to support their cause.

Monarch Madness

Most Americans still remember something of the furor over a study indicating that monarch caterpillars might be killed by gathering pollen from *Bt* corn. Unto this day, it remains a reminder that with biotech, as with anything, a little bit of knowledge can be a dangerous thing—especially if that knowledge is intentionally distorted. Headlines included: "Monsanto vs. the Monarch" (*St. Louis Post-Dispatch*), "High-Tech Corn Killing Butterflies" (*Chicago Sun-Times*) and "Attack of the Killer Corn" (*U.S. News & World Report*).[5] Even a monarch weighed in on the monarch, with Britain's Prince Charles, long an antibiotech activist, declaring, "If [biotech] plants can do this to butterflies, what damage might they cause to other species?"[6]

Few know that the butterfly brouhaha was based on a mere letter in *Nature*, and that the chief researcher had publicly stated that the experiment quite possibly had no meaning outside of the laboratory.[7] At the time, I interviewed a large number of experts who said the monarchs were at no risk whatsoever.[8] One reason is that whereas the caterpillars in the study were force-fed *Bt* pollen, in real life if they encountered pollen on a leaf they would simply shimmy on over to another one. "They're not interested in eating anything but a nice clean leaf," Warren Stevens, senior curator of the Missouri Botanical Gardens in St. Louis, told me.[9] And while most corn pollen falls onto the field itself, farmers fastidiously keep milkweed, which the caterpillars feed on, out of the rows. So *Bt* pollen could reach the milkweeds only if carried by the wind outside the field. "They can be found along the edge of fields, but are primarily in pastures and old fields," said University of Maryland entomologist Galen Dively, "I don't think much milkweed grows close enough to corn fields where they would receive pollen." Further, *Bt* corn, by reducing insecticide spray use, protects other friendly insects. "Before *Bt* crops, our sweet corn had to be sprayed eight to ten times," said Dively. "Now it's not sprayed at all. That used to kill a lot of lady beetles" as well as butterflies. "That's no longer happening."[10]

A massive amount of research leading to a huge number of studies seems to have lain to rest worries about the threat to beautiful butterflies, especially a set of six studies published together in late 2001 in the prestigious *Proceedings of the National Academy of Sciences.*[11] Those six found that *Bt* corn carried no risk to the insects, except for perhaps one variety of *Bt* that was rarely used and already being phased out.[12] Tragically, just a few months later, in January of 2002, a major storm utterly devastated the monarch population in their Mexican winter homes. While it's entirely possible that *Bt* corn never killed a single monarch, this storm wiped out an estimated 220 to 270 million in just two days.[13] It was a potent reminder that Mother Nature does not honor our notions of beauty and sentimentality.

Up to this point, one gets the idea that opposing agbiotech is the "left-wing position" in the sense that gun control is. In fact, while the greatest opponents of transgenic crops are indeed on the left side of the political spectrum, so are its greatest supporters. These include such heavyweights as Jimmy Carter, former U.N. ambassador Andrew Young, former senator George McGovern, and Rockefeller Foundation head Gordon Conway. McGovern, who perhaps epitomized American liberalism in the 1970s, says, "The most promising weapon in the global war against world hunger is high-yielding, scientific agriculture, including genetically modified crops. Yet, the gene modification controversy and trumped-up fears of 'Frankenfoods' are stepping on the promise of a hunger-free future." These crops, he adds, "must not be stymied by voices raised against the hypothetical, while real disease and starvation threaten millions of people."[14] Incidentally, McGovern isn't just a former senator and the Democratic presidential candidate in 1972. He is the current ambassador to the U.S. Mission of the United Nations Agencies for Food and Agriculture in Rome, and author of a recently published book titled *The Third Freedom: Ending Hunger in Our Time.*[15]

Interestingly, just as there is a left/left split on agbiotech, there is a right/right split on genetic engineering, with conservatives like Francis Fukuyama, Bill Kristol, Dr. Leon Kass and Dinesh D'Souza aligning themselves with radical technophobes like Jeremy Rifkin and *The End of Nature* author Bill McKibben.[16] Libertarian-right "dynamist" Virginia Postrel lucidly explains how these strange bedfellows came to be in her 1998 book *The Future and Its Enemies.*[17]

Old World Attitudes

If you only read North American newspapers and periodicals and hear North American broadcasts, the opposition to agricultural biotechnology on the other side of the Atlantic would stun you. The terms "Frankenfood" and "mutant food" are routinely used not just by European activist groups but even by the mainstream media.[18] This is entirely at odds with European acceptance of other forms of biotechnology. A 2002 public opinion survey of the fifteen European Union nations found that all supported or strongly supported genetic testing and the cloning of human cells. But when it came to growing transgenic crops, only one (Spain) strongly supported it while seven opposed it. As for consuming transgenic crops, residents of only four nations so much as mildly supported it.[19]

"Why is a process that we trust to be injected into our bodies inherently suspect when it's used to make things that go into our stomachs?" Kenyan biochemist James O'Chanda asks rhetorically.[20]

In part, this suspicion reflects a clash between American and European attitudes toward food. "Food is our culture," one Frenchman told me. And this culinary culture is a deeply conservative one. That biotech food is different *in any way* from "natural" food is, for many Europeans, a strike against it, even if the difference has no effect on health or taste. That such food is non-European may be another demerit. One French winemaker graphically expressed such feelings when he declared, "Each bottle of American and Australian wine that lands in Europe is a bomb targeted at the heart of our rich European culture."[21] Christian Verschueren, director general of Brussels-based CropLife International, a pro-biotech global industry umbrella group,[22] confirmed this to me. "The relationship of people to food is different in Europe compared than in other regions of world," he said. "The culture is different and the relationship people have to their food is different. Europeans have a view of culture and tradition concerning food that equates to what they consider 'naturalness.'"[23]

Ignorance is another factor in the European militant opposition to agbiotech. The aforementioned 2002 survey of members of the fifteen EU nations found that 35 percent thought that nonbiotech tomatoes contain no genes, while an incredible 49 percent thought that by eating genetically engineered fruit a person could find their own genes altered. These were the same responses given during a 1996 survey; so in six years absolutely no progress had been made in educating the European public.[24] Meanwhile, a 1999 survey of the United States and Canada found that only 10 percent of Americans and 15 percent of Canadians thought nonbiotech tomatoes contain no genes.[25] A jokester with a point to make set up a "DNA Free Food Society" website, then polled visitors as to how much they would be willing to pay for food without genes. Of the 541 people who responded, about 20 percent said they were willing to pay ten times as much, while 38 percent said they were willing to pay 50 percent more. It's kind of funny—but kind of not.[26]

Overseas opponents of biotech food also shrewdly exploit anti-American sentiment. "Most Europeans think of the U.S. as a colonial power, albeit a very soft and gentle one," said Gian Reto-Plattner, a scientist and a Socialist member of the Swiss Parliament who broke with his party to oppose a nationwide ban on sales of biotech food.[27] The 2002 survey found that a whopping 80 percent of respondents agreed with the sentiment that "multinational companies are too powerful nowadays." And while multinationals can be based anywhere and by definition include more than one country, all too often they're identified with the United States.

Perhaps the strongest strike against agbiotech in Europe is the threat it poses to an ossified status quo. Almost all biotech seed currently sold has one ultimate purpose: allowing farmers to grow more food on less land at lower cost. But the potential for that kind of efficiency and productivity threatens most European farmers, who are much more heavily subsidized than their American counterparts. "Farmers here are embracing organic techniques precisely because they produce far *lower* yields," Roger Bate, executive director of the European Science and Environment Forum in London, told me. "The last thing they want is to be more productive."[28] That would drive prices down. Since even nonorganic European farmers have so much trouble competing, they receive large taxpayer-funded subsidies. Indeed, half of the European Union's entire budget goes toward agricultural subsidies, and the EU accounts for 85 percent of the world's farm export subsidies.[29] These subsidies create a vicious cycle that further discourages efficiency improvements. Now that biotech is making the gap between American and European production even wider, it threatens to bankrupt European subsidy programs. So these governments, with the support of powerful farm lobbies and cynical allies like Greenpeace, use any arguments they can cobble together to slap import restrictions on grain and seed.[30]

This protectionism by no means applies only to biotech. Western European nations have long kept out U.S. beef on the grounds that growth hormones given to American cattle may render the meat unhealthy, despite a ruling of the World Trade Organization that the prohibition was illegal because it was not based on scientific evidence.[31] Most of the soybeans in American feed are now transgenic, and many Europeans have damned them to the point that Greenpeace with wide public support has blocked American freighters trying to dock in European ports. The bitter irony is that while they were fretting about hormones, Europeans were feeding ruminant animal parts, including brains, to their own cattle. Making cows into cannibals was their way of giving them protein. As we now know, the cattle first of Britain and then of several other European countries suffered an epidemic of bovine spongiform encephalopathy (BSE), better known as "mad cow disease."[32] Many of the animals contracted it from eating brains, then a few humans contracted it from eating the beef. Ironically, during one ship-blocking incident, which occurred in December of 2000 after it had become obvious that BSE had infected herds throughout much of Europe, Greenpeace nonetheless declared: "Europeans are being force-fed with genetically engineered animal products by multinational grain traders."[33] If only they *had* been force-fed with animals fed with biotech grain, people and animals would not have come down with BSE, also known as "variant Creutzfeld-Jakob disease," which turns the brain into Swiss cheese and for which there is no treatment or cure.[34]

Clearly one of the most important differences between American and European attitudes is that Americans tend to have much greater faith in our regulatory bodies, even though we have a cottage industry of think tanks, magazines and writers who argue that many of our regulations are too harsh. Europeans, conversely, have been so shaken by such recent developments as the spread of BSE and infection of humans with the disease, and the discovery of widespread dioxin contamination in Belgian livestock and eggs, that many have open contempt for their health regulators, believing them worse than worthless. As a result, environmentalists have achieved "a very high level of credibility," said Walter von Wartburg, the Swiss co-author of *Gene Technology and Social Acceptance.*[35] "We have a regulatory process with lots of directorates coming up with different opinions," he told me. "And if some are for and some are against and Greenpeace says 'This is the ultimate truth' they become a kind of arbiter."[36]

The aforementioned 2002 EU member survey found that highest marks went to "medical doctors keeping an eye on the health implications of biotechnology," which is almost certainly another way of describing medical professionals who have gotten on an antibiotech soapbox. "Consumer organizations," which are overwhelmingly antibiotech, came in second, while groups openly describing themselves as antibiotech came in third. Not surprisingly, industry ranked dead last while government regulators barely fared better.[37]

Still, Christian Verschueren of CropLife International warns against putting too much emphasis on Europeans' safety concerns. "The debate in Europe is not so much on the science," he says. "In the U.S., when you ask consumers whether they believe a product is safe and then inform them it's not only been approved but is already on store shelves, then that increases their acceptance. If you do the same thing in Europe, you alarm people."[38]

Finally, part of the blame lies right in the lap of American companies who mistakenly assumed Europeans were willing to dive into the future of food as readily as Americans. Old World sensibilities are different, after all. When you see how fastidiously Europeans preserve and reconstruct old buildings, even ones repeatedly wrecked during wars, you appreciate their reverence for tradition. When you're trying to sell them food products developed with the latest scientific technology, perhaps you do not.[39]

U.S.-based companies almost certainly tried to go too far, too fast, leaving themselves, as one newspaper headline put it, "Facing the Backlash of Pushing the Envelope."[40] In 2000, Monsanto's CEO, Robert Shapiro, perhaps went even further when he "put on the hair shirt," in the words of *Science* magazine. He publicly stated that his company's attitude had "widely been seen, and understandably so, as condescension or indeed arrogance," and that, "Because we thought it was our job to persuade, too often we forgot to listen."[41]

Japan, which is even more resistant to transgenic crops than Europe, also has a very strong food tradition, powerful agricultural antitrade laws, and an extreme distrust of its regulatory system. In recent years it has been rocked by one food scandal after another, including one that involved 13,000 people falling ill because milk was pumped through dirty pipes.[42] "The government and politicians may have thought they were protecting" the agriculture industry, said Takeshi Domon, a well-known writer on agriculture policies, to the *Washington Post.* "But their [system] has come back to haunt them."[43]

In the United States, regulation works relatively well.[44] Produce a drug and you have to answer to the FDA. Produce a pesticide spray and you'll have the EPA looking over your shoulder. But makers of biotech food products are regulated by the FDA, the EPA *and* the

USDA.[45] For example, to satisfy just the FDA's regulations, Monsanto had to show that its herbicide-resistant transgenic soybeans have no substantial difference from regular varieties. It did 1,800 analyses comparing the two types, looking at fatty acids, proteins and hundreds of other substances.[46] Such extreme proofs are necessary because of American groups opposing agbiotech, which may not be as powerful as their European counterparts but are equally committed and sometimes willing to fabricate evidence.

How, for example, could Margaret Mellon, director of the agriculture and biotechnology program for the Union of Concerned Scientists, believe that "The purpose of biotechnology is to increase the profits of the manufacturers by persuading farmers to use more herbicides," when: A) agricultural biotech is just a portion of all biotechnology; B) farmers are excellent businessmen who aren't persuaded by anything or anybody that doesn't make their job easier or more profitable; and C) 40 percent of American biotech crops are made to resist insects and have nothing to do with herbicides one way or the other?[47]

One type of genetically modified sugar beet under development in Ireland requires only 40 percent of the herbicides needed by traditional varieties. It's sponsored by Monsanto, the very company the Union of Concerned Scientists claims is using biotech to foist more herbicides on farmers.[48] In one raid after another, ecotage opponents of biotech food destroyed the Irish test plots.[49] But they didn't get them all. In 2000, after four years of trials, Monsanto's general manager in the Irish Republic was able to declare, "The results confirm that this new technology will be of significant benefit to growers in Ireland," adding, "those people who had destroyed valid and important research trials in previous years were not allowed to dictate the agenda."[50]

Monsanto researchers also announced they have been able to take a gene from alfalfa and insert it into potatoes to produce plants they believe are as resistant to fungus as those treated with fungicides. "There have been no previous demonstrations of a single transgene [transplanted gene] imparting ... disease resistance ... that is at least equivalent to those achieved through current practices using fumigants," they reported in the December 2000 issue of *Nature Biotechnology*.[51] One of the fungi it protects against is *Phytophthoroa infestans*, the very same that caused *An Ghorta Mhoir*, the Great Hunger that killed a million Irish in the 1840s and forced millions more to emigrate.[52] (Even today, the fungus costs potato growers about $3 billion a year, according to the International Potato Center in Lima, Peru.) The transgenic potatoes were successfully tested in several Midwestern states where blight is a problem.[53]

The Dutch company Proagro has begun testing transgenic mustard plants it believes will revolutionize Indian agriculture because it increases yields by one-fifth over the best Indian variety, yet requires almost *no* chemical fertilizers.[54] Proagro calculates the typical farmer can earn an extra 6,400 rupees ($229) per harvest from the transgenic mustard if he invests 600 rupees ($21.43) in a 6.6-pound sack of new seeds.[55] To an Indian farmer, $229 is a lot of money.

Agbiotech progress is inevitable, but as we've seen, it can and has been slowed down. There continues to be a clamor in Europe for additional moratoria on growing or importing genetically modified products.[56] Test plots all over Europe and more recently in the United States are being destroyed (or, to use the common euphemism, "decontaminated") by the very groups that call for further testing before commercial sales begin.[57] We'll all suffer for this, but those with the most to benefit will suffer the most.

The most serious charge against biotech food is that it is unsafe, might be unsafe, or at some point in the future may be unsafe for human consumption. At a conference on biotech food safety in Edinburgh sponsored by the Organization for Economic Cooperation and Development, I heard a representative of the most powerful antibiotech group in the world, Greenpeace, admit there was no evidence of lack of safety in any currently produced biotech food.[58] It's the *potential* for such evidence, he said, that we must worry about. Greenpeace and its allies might do well to listen to biotech founding father James Watson, who notes that as soon as gene splicing was first implemented in 1973, "There were cries from both scientists and non-scientists that such research might best be ruled by stringent regulations—if not laws. As a result, several years were to pass before the full power of recombinant-DNA technology got into the hands of working scientists, who by then were itching to explore previously unattainable secrets of life." Says Watson, "The moral I draw from this painful episode is this: Never postpone experiments that have clearly defined future benefits for fear of dangers that can't be quantified. Though it may sound at first uncaring, we can react rationally only to real (as opposed to hypothetical) risks."[59]

All the evidence to date is that the agbiotech industry is being very careful in the work it has undertaken. Never mind that researchers themselves and their families will almost inevitably end up eating whatever they've developed. Put it in strictly selfish terms if you want: Biotech companies don't want to spend tens or hundreds of millions of dollars developing a product that may even acquire the taint of unhealthiness, much less face a food recall or ban, much less lawsuits.

Of course, if someone really wanted to, he could make a benign plant poisonous by inserting genes from a poisonous plant. Although it's still not clear, something like this may have been what Scottish researcher Arpad Pusztai did in 1998 when he took a gene from the snowdrop lectin, inserted it into a potato, and fed the potatoes to rats. While lectins in such foods as lentils, beans, peas, tomatoes and potatoes themselves are a regular part of many people's diet, several previous studies had reported that they can bind to human proteins and tissues. It depends on how long they're cooked, how well they're digested, how much you eat and other factors. But the bottom line is that under the right conditions, lectins can be harmful.[60] When Pusztai's rats ate the lectin-added potatoes, they became ill. Or maybe they didn't. It depends on who's interpreting the data. On a popular TV program in August of 1998, Pusztai claimed that "the effect [of feeding the transgenic potatoes to rats] was slight growth retardation and an effect on the immune system." Yet he also told the House of Commons Science and Technology Committee that "no differences between parent and GM potatoes could be found."[61]

Other scientists and scientific bodies such as Britain's Royal Society have said it was the way the experiment was conducted that would have caused harm to the rodents if any was done. For example, rats don't like raw potatoes any more than we do, and when Pustzai realized this, instead of starting the experiment over he simply began boiling the potatoes. The result was that the rats got far less protein than they should have had, which may have contributed to their illness.[62] Another Royal Society concern was that too few animals were used for statistical significance (a mere six). In general, they said, the study was a mess.[63] After this Pusztai was let go by his employer and rebuked by the Royal Society and his fellow scientists.[64] He then took to the lecture circuit on behalf of his newfound allies in the antibiotech movement.[65]

On the whole, though, making new plants through transgenics is safer than by traditional breeding methods. Why? "When you are dealing with biotech crops, you are changing just one or two genes of the plant's structure. With hybridization, there are more genes involved and more uncertainty," said Dr. Bernard Schwetz, then FDA acting deputy commissioner. Thus you have more control, which not only makes the process potentially safer but speeds it up tremendously as well.[66]

While one might think it enough to have three federal agencies monitoring biotech crop development, some people maintained it was unfair that aside from triple oversight, biotech regulations were essentially the same as those applied to other crops. The Union of Concerned Scientists felt this justified its claim that "the Food and Drug Administration has yet to require safety testing on any of the genetically modified foods on the market, relying instead on the industry's conclusions that these foods are safe."[67] The same could be said of any *nonbiotech* food, although some special exceptions made the bar higher for transgenic crops, such as having to prove nonallergenicity when so many foods already in stores are known to have caused allergy attacks, including fatal ones. But prior to the year 2000, biotech crops usually had to meet the same standards. The idea was to regulate the outcome, not the process. Nevertheless, in May of that year, the Clinton administration announced a plan that would increase federal oversight specifically of genetically modified foods. Among the more significant changes proposed, biotech companies would have to notify the FDA four months before marketing a new genetically engineered food, providing the agency and the public with the research results that affirm the new food's safety.[68] Until then, that process had been voluntary. Although this and other proposed changes will be somewhat burdensome to the agricultural biotech industry, the industry seemed to support them.[69] Apparently it saw the added burden as worthwhile if it would help allay public concerns.

The FDA also saw it that way. "FDA's scientific review continues to show that all bioengineered foods sold here in the United States today are as safe as their non-bioengineered counterparts," said one senior Clinton administration official. "We believe our proposed initiatives will provide the public with greater confidence in the safety of these foods."[70] Whereupon the Union of Concerned Scientists slammed the FDA for not going far enough and for creating a smokescreen to reassure a public that has nothing to be reassured about.[71]

But there actually are serious questions concerning agbiotech, both scientific and ethical, which have or at least seem at first glance to have real merit. As we have already seen, these include the possibility of transgenic crops conferring their properties through pollination onto neighboring related weeds, the possibility of transferring genes that will cause allergic reactions, and the problem of insect resistance.

Regarding problems that have yet to be identified—and there *will* be some—recall James Watson's words about postponing work with defined future benefits for fear of dangers yet undocumented. Consider all the problems that cars created, and yet each has been dealt with by improving the technology. Safety problems led to traffic signals, better roads, guard rails, bumpers, turn signals, brake lights, safety belts, airbags, crumple zones and so on. Pollution problems led to unleaded gasoline, catalytic converters and engine improvements that come almost yearly. The result is that automobile emissions have plummeted and continue to decrease.[72]

Some critics, for example, find it worrisome that "markers" comprising antibiotic-resistant genes usually are inserted into a plant along with the gene or genes that will eventually be used to modify the plant, such as *Bt.*[73] The young sprouts are then exposed to antibiotics that kill plants without the markers, hence revealing plants to which the genes weren't properly transferred. The fear is that this practice will promote antibiotic resistance in humans who eat the crops. Scientists, including those of the EPA and the FDA, downplay this, noting in part that this type of antibiotic is completely different from any that are used on humans and animals.[74] "We have carefully considered whether the use of antibiotic resistance marker genes in crops could pose a public health concern and have found no evidence that it does," said President Clinton's FDA commissioner, Jane Henney.[75]

Whether or not such worries are groundless, researchers are nevertheless developing several new technologies to excise marker genes from transgenic plants.[76] For example, instead of inserting an antibacterial marker, researchers at Rockefeller University in New York and the University of Singapore have devised a method in which a hormone is produced that will allow plant shoots to develop only if they have successfully received the gene or genes inserted into them. If the DNA doesn't "take," no plant grows. The hormone is emitted only long enough for shoot formation, and not long enough to cause abnormalities in the mature plant.[77] While some of these earlier methods of avoiding marker genes weren't very reliable or couldn't be used in crops grown with seeds, researchers have now surmounted these obstacles.[78]

"Super Bugs"

There has also been great fuss over the possibility of insect-resistant crops causing insects to develop immunity to the pest-resistant gene in the plant. "A growing body of scientific evidence points to the likelihood of creating 'super bugs' resistant to the new pesticide-producing genetic crops," claims Jeremy Rifkin, widely considered the nation's premier technophobe.[79] But that's been a problem with insecticides for decades. Why must biotech be superior in this regard in order to be accepted?

Actually, though, it *does* appear to be superior. A popular insecticide among organic farmers is made from *Bt.* These farmers consider it "natural" and hence organic because before it was engineered into a spray, its source was a common bacterium. Insects were showing resistance to *Bt* as a spray long before the first biotech *Bt* crops were commercially grown, but the information on various types of *Bt* plants is quite encouraging. Resistance did build in insects initially, but a 2000 study in the *Proceedings of the National Academy of Sciences USA* showed it had leveled out at a low rate.[80] "*Bt* cotton has helped to reduce insecticide use in Arizona cotton to the lowest levels in the past 20 years," said Larry Antilla, one of the study's authors. "This benefits the public, farm workers, and the environment."[81]

Nor is this simply an academic argument. Insect resistance to pesticides has seemingly become a major cause of suicide in India. *St. Louis Post-Dispatch* reporter Bill Lambrecht, writing from India in 1998, told of a farmer named Damera Shanker who "walked the red dirt road from his village to check his two-acre plot. He returned at 1:30, mixed a cocktail of insecticides and drank. By 5:30, he was dead. For an hour, Shanker had watched a voracious green caterpillar called the American bollworm devour his cotton and vegetable crops. At age 23, he owed

$17,000—a fortune in India—for pesticides that had stopped working." Lambrecht went on to report that "more than 700 farmers have committed suicide in four Indian states since the beginning of last year. Many drank poison from a bottle and then lay down in their doomed fields to die; some hanged themselves or jumped down wells."[82]

Yet although Lambrecht up to this time had been his own cottage industry of negative reporting on the agbiotech programs of his hometown company, Monsanto, he wasn't negative this time. "The bollworm that is creating despair on India's farms is the target in 40 farm-field tests that Monsanto is conducting across India," he wrote. Lambrecht quoted Monsanto scientist T. M. Manjunath as saying, "It is this pest that is driving farmers to commit suicide. If using transgenic seeds would mean spraying two or three times rather than 12 or 15, it would be a great contribution."[83]

Despite the potential superiority of biotech crops in defeating insect resistance, special precautions are still being taken to prevent such resistance from developing. One EPA-mandated measure is that a specific portion of the crop must be set aside as a "refuge" in which no *Bt* crops are planted.[84] This helps prevent resistance because resistance is a recessive genetic trait that can be passed on to offspring only if both parents have the gene. The refuge reduces the chance of two such bugs meeting and mating.[85] Another way of defeating resistance is stacking two different insecticide genes into one plant, as does a new type of cotton named Bollgard II from Monsanto.[86] That means an insect predator would have to develop mutations conferring resistance to both insecticides, then mate with another insect that likewise did. It's not a difficult matter to insert yet a third and fourth insecticide gene into cotton or other crops, each time decreasing the possibility of resistance developing and being passed on.[87]

"Genetic Pollution"

Assuming that Jeremy Rifkin simply didn't choose the best example he could have, just how real is the danger of "genetic pollution"? In the broadest sense, it's quite real because it's completely natural. Nature creates new plants from old, while old ones become extinct.

When Europeans established new trade routes to the East in the Middle Ages, they brought back all sorts of wonderful spices and textiles.

They also brought back a tiny package of genes called *Yersinia pestis.*[88] *Y. pestis* traveled inside another package of genes belonging to an organism of the order *Siphonaptera,* better known as the flea. This in turn made its way to Italy and then the rest of Europe on the back of a package of genes from species of the order *Rattus rattus.* This genetic pollution led to the wave of bubonic plague called the Black Death, which took one-third of Europe's population in four years.[89] It was genetic pollution, and it occurred a rather long time before James Watson and Francis Crick happened on the scene.

Were you to ask the experts to name one notorious plant, animal and insect pest, then chances are that near the top of each list would be kudzu, rats and Africanized or "killer" bees. None are native to North America.

Kudzu was brought to the United States in 1876 as a gift of the Japanese government. During the Great Depression the Soil Conservation Service promoted the plant for erosion control. Farmers were *paid* to plant fields of the vines. Now it covers seven million acres of the South and wipes out any plant that dares grow in its path.[90]

The rodent that plagues inner cities and farms, the "Norway rat," probably originated in Central Asia. It's widely thought the rats immigrated to the United States aboard the ships of Hessian mercenaries. Consider them revenge for Valley Forge.[91]

The tale of the Africanized bee is the most appropriate for our purposes because it shows the dangers of old-fashioned crossbreeding. Brazilian honey farmers sought to mate the hardiness of the African bee to the productivity of the European honey bee. Instead, they got superaggressive bees that consume honey as rapidly as they produce it, leaving nothing for beekeepers to collect.[92] The insects then began a trek north that has resulted in dozens of dead people each year and jeopardizes the North American honey industry because it drives out the gentler honey bees.[93]

A federal study that looked at a selection of just seventy-nine harmful species and went back only a century found that these unwelcome migrants have cost the nation a cumulative total of $96 billion.[94] A 1998 article in *Bioscience* put the cost of bio-invader damage plus control at $136 billion a year. Yet every penny of this was spent on so-called "natural" species, not on biotech ones.[95]

Genetically engineered organisms have an inherent advantage in that biotechnology allows the selective transfer of a single gene or set of genes for a specific trait. If the sought-after trait is expressed, eureka! If not, the experiment ends.

Genetic Erosion

What about the potential of biotech to destroy biodiversity? Loss of biodiversity, also known as "genetic erosion," is well documented. But it started long before biotech. In Mexico, four out of every five varieties of corn that existed in the 1930s are already gone. So if it's true that "Loss of genetic diversity is leading us to a rendezvous with extinction, to the doorstep of hunger on a scale we refuse to imagine," as Cary Fowler and Pat Mooney assert in their book, *Shattering: Food, Politics and the Loss of Genetic Diversity*, don't blame biotechnology.[96] (*They* didn't; their book came out in 1990, five years before the first U.S. biotech row crops were planted.) In any case, no such worst-case scenario will happen. One thing preventing it is that both the USDA and the U.N. have libraries of genetic material, known as germplasm, taken from plants around the globe. The USDA's is named the National Plant Germplasm System and provides the material free to researchers around the world, while the U.N.'s Food and Agriculture Organization (FAO) and the Consultative Group on International Agricultural Research have collected donations to maintain "in perpetuity" sixteen international crop-research institute seed banks plus many independent ones. These will help to ensure that plant breeders have unrestricted access to seeds anywhere in the world.[97] In November 2002, the United States also signed an international treaty to preserve plant genetic resources.[98]

Far from being an enemy of biodiversity, biotechnology will allow it to increase. One reason, noted Green Revolution "father" Norman Borlaug, is that "if we grow our food and fiber on the land best suited to farming with the technology that we have and what's coming, including proper use of genetic engineering and biotechnology, we will leave untouched vast tracts of land, with all of their plant and animal diversity." He pointed out that "It is because we use farmland so effectively now that President Clinton was recently able to set aside another 50 or 60 million acres of land as wilderness areas."[99]

With biotech we'll even have more diverse crops, as current ones are modified to grow better in different climates and soils and with different forms of pest resistance. The trend toward one-size-fits-all crops is already being reversed, not only because we've learned so much about how differences in rainfall, aluminum content in soils and temperature can affect yields, but because we're finally able to do something about it. "Four or five years ago, you could count the number of canola hybrids [made with both traditional crossbreeding and transgenics] on one hand," explained Barry Coleman, executive director of the Northern Canola Association, based in Bismarck, North Dakota. "Today, there are about 150 to choose from."[100] Argentina alone is already using seven different strains of transgenic canola.[101]

This hardly portends a *Last Harvest,* the title of science writer Paul Raeburn's book warning of the perils of losing crop diversity. The subject of this book is not biotech but rather the older crossbreeding technology that produced crops with much higher yields but gave them certain disadvantages when it comes to survival. Still, Raeburn appears disposed to the belief that transgenic crops will help reverse the trend toward genetic erosion.[102] He even quotes *Population Bomb* author Paul Ehrlich, perhaps as pessimistic a man who has ever walked the earth, as saying that genetic engineering "could also play an important role in maintaining the genetic diversity of crops, since it permits the simultaneous introduction of a given useful trait into all varieties," and that "locally adapted varieties could be genetically enhanced while remaining in production."[103]

Biotechnology isn't making us more vulnerable to crop disease, but less so. Instead of combing through a bunch of potato or wheat strains and hoping to find one that by itself has resistance, or crossbreeding a couple of plants to make a new disease-resistant one, we can now quickly screen genes that appear to have resistance potential and insert them into the threatened plant. In little more than the time it takes to grow sprouts, we can determine which of any number of the new strains can fight off the predator, be it a fungus, a virus, a bacterium or an insect. The gene for the new plant may be from a close relative of that plant or it may be taken from something that could never be crossbred with it, maybe putting a gene from that wheat plant into the potato plant or vice versa. Or the gene could come from the plant itself. Researchers can mutate a potato, shuffle the genes, and within a day have thousands of variations of the potato of which dozens might be resistant to *Phytophthora infestans.* The only time-consuming aspect is growing the seedlings long enough to expose them to the fungus and ensuring that what worked in a test tube works in real life. If that sounds in the least bit far-fetched, you should know that already genes from an ancient sunflower variety cultivated by the Havasupai Indians of the southwestern United States have been used to fight a form of disease-causing fungus called a rust that threatened crops grown in the U.S., Australia, China, South Africa and India.[104] This was done using traditional crossbreeding. But biotech methods would have speeded up the process. And remember the transgenic papaya that saved the farmers in Hawaii?

Then there are the complaints against biotech crops that essentially come down to: They work *too well.* One newspaper charged that if you kill weeds too effectively, you'll kill birds. How? "Although Roundup [Ready] does not directly kill the insects, they die anyway, because they no longer have the seeds from the weeds to eat," declared the *London Evening Standard.* "Song thrushes, for example, depend on sawfly larvae from weeds for their existence, and they are already under threat from conventional spraying methods."[105] Thus, the reporter made a case for weedy fields, even as he admitted that the "problem" of not having weeds was there long before biotech. In any case, as a writer for *Nature Biotechnology* has pointed out, herbicide-tolerant crops "allow farmers to maintain weeds longer (because they can be treated after crop emergence, not before) and the possibility that greater efficiency of [biotech] crops may allow more land to be 'set aside' for wildlife."[106] Regarding insect-resistant biotech crops, since they kill only insects that feed on the crop rather than any that happen to be caught in an insecticide spray, they also leave more bugs for birds to feast upon.

In yet another of the myriad double standards that activists employ, we are often told that our regulators allow the companies to do their own testing. But that's virtually always how the EPA and the FDA operate. Chemical manufacturers test their own new products and submit the results to the EPA; pharmaceutical companies test their own new drugs and submit the results to the FDA. That the system seems to work fairly well is indicated by how often the FDA rejects new drug applications, or sometimes stretches out the approval process for a decade or more.[107]

Even when the FDA was using the same overall standards for biotech and nonbiotech foods, the agency had some exceptionally tough rules for transgenic crops. For example, transgenic potatoes must be tested for a natural toxin called glycoalkaloid solanine because it has been detected at harmful levels in some new potato varieties that were developed with *conventional* genetic techniques.[108] Likewise, both the EPA and the FDA demand special documentation for potential allergenicity caused by biotech crops.

Accidentally inserting allergens into a transgenic plant is indeed a potential problem, as shown by an experiment that successfully put an allergy-causing Brazil nut gene into soybeans. But that was the *purpose* of the research, published back in 1996.[109] The plant was not caught by industry or regulators at the last minute, as some activists would have us think.[110] The activists have converted a good thing—researchers being on the lookout for health problems—into a scary thing in the public's mind. Agbiotech companies are very concerned about the potential for transferring proteins into new crops that may cause allergies, and have developed a multitiered system to prevent it from happening.[111]

"It's a fairly lengthy process we go through," explained Roy Fuchs, a regulatory science specialist at Monsanto, "and a several-pronged approach. First, we typically select genes from sources that have a history of safe use. Over 90 percent of human food allergies are in just five food groups as sources of genes, so by avoiding those groups you can significantly reduce the likelihood that your protein is an allergen." For example, "With our [herbicide-tolerant crops] we use a class of genes that's found in all plants. Then we look at the characteristics of the protein we're going to be transferring to see if it will be rapidly digested. That means it's more likely to be broken down in the gut and not make it into the bloodstream." An allergen can't cause a reaction if it's destroyed during digestion. After this, "We search to see if the amino acid sequence in the protein has any relationship to known allergens, either in food or in pollen." If they were to consider transferring genes from a food known to cause allergies, they "could obtain blood serum from people known to react to that plant and test the protein to see if it causes a reaction." In any event, said Fuchs, "Any gene from an allergenic source is considered to produce an allergen until we establish otherwise. That said, only five to seven genes of the forty to sixty thousand genes in any allergenic food crop would actually encode an allergen."[112]

It was the first part of this analysis that was concerned in the huge Taco Bell flap in late 2000, which led to such headlines as: "Allergy-Causing Taco Bell Taco Shells Found in Groceries" (UPI), "Banned GM Corn Reported in Taco Snack" (London's *The Guardian*) and "Alarm at 'Harmful' GM Corn in Snacks" (London's *Daily Mail*).[113] But it was all a tempest in a taco shell.

The corn was produced by Aventis CropScience, which later merged into Bayer CropScience. Its "StarLink" corn contained a *Bt* protein that killed corn borers. But, unlike all the other *Bt* variants in our biotech crops, it was not a rapidly digested protein such as Fuchs described. Thus Aventis did some additional allergenicity testing, but the results were not conclusive according to the EPA and to a scientific expert advisory panel that the EPA called together to provide advice. Since there was no clear answer yet to the allergy question, EPA approved the product for use in animal feed but not yet for human food. But apparently some farmers didn't get the word that the crop was supposed to be kept separate from other corn. It ended up in Taco Bell shells sold both in restaurants and in stores, as well as in a few other food items, although always in amounts too small to provoke an allergic reaction even if it *were* an allergen. Only 300,000 acres of StarLink corn were grown in the United States, about one-half of 1 percent of the corn planted each year. Further, only 1 percent of the Taco Bell shell sample tested contained the meal from StarLink corn.[114] When I checked the FDA voluntary recall page that listed Kraft Foods as having recalled the Taco Bell shells, I found no fewer than fifteen other such recalls in the previous sixty days (none biotech), including four that were expressly because of possible *real* allergen contamination.[115] I had heard about none of these in the news, and I suspect virtually no one other than store and restaurant managers had either. Ben & Jerry's, maker of tasty, politically correct ice cream and supporter of antibiotech groups, had to recall its Peanut Butter Cup pints because they contained tree nuts that may provoke allergic reactions.[116]

Aventis announced it was stopping the sale of StarLink seeds, and any foods that might contain even the tiniest amount of the questionable grain were recalled.[117] But naturally, this became the rallying point for another attack against biotech food as a whole. Nor was it a coincidence. The group that had done the testing of twenty-three different products before finally finding what it was looking for in the taco shells was Friends of the Earth, a member of a coalition named Genetically Engineered Food Alert, which opposes all biotech food.[118] It seized upon this to push its agenda.[119] Individuals claimed that exposure to products they believed contained StarLink had made them ill, and this drew further attention. While a CDC report in June 2001 dismissed this possibility and got wide media attention, the StarLink flap remains a good "war story" for biotech bashers.[120]

Further, there's a powerful flip side to the biotech allergenicity issue. Each year, food allergies cause 2,500 emergency room visits and 135 deaths.[121] It's estimated that over one hundred different foods or food components have produced documented allergic reactions, and that up to 5 percent of children and 2 percent of adults suffer from food allergies.[122] Some children are so incredibly sensitive to peanuts that the mere whiff of a peanut butter and jelly sandwich can send them to the emergency room.[123]

Biotechnology can actually be used to de-allergenize these foods. I asked Roy Fuchs about work in this area. "There are several approaches," he told me. "One is antisense technology, in which the production of the offending protein produced from a gene is either greatly reduced or shut off. Antisense technology can be applied to rice, peanuts, soybeans, other foods, and even pollens."[124]

A team of scientists from the Agricultural Research Service, Pioneer Hi-Bred and elsewhere have silenced a gene in soybeans that's thought to cause about two-thirds of all allergenic reactions associated with the bean.[125] That's truly important because so many people are allergic to soy and because soy is such a common ingredient in our food. Some products that could potentially benefit from the improved soybeans include baby formulas, soy milk, cereals, flour and pet food.[126] "You're never going to make a completely allergy-free soybean plant because you're not going to be able to eliminate all the proteins in it," said one of the investigators.[127] But by turning off that one gene, they could reduce the allergenicity to a level that's not even noticeable for most people who now must completely avoid soybeans.

"A second process is to modify the amino acid sequences in allergenic proteins," said Fuchs. "Changing these so that they can no longer bind to the antibodies that cause allergy could greatly or completely de-allergenize a food."[128] Finally, "you can engineer into plants genes such as thioredoxin, which increase the digestibility of food allergens." This approach is being tested at three University of California campuses with new wheat and barley varieties as well as milk.[129] The University of California researchers, working together, have reduced an allergy-provoking protein in milk by 99.7 percent by changing the structure of the protein. This also made the milk more digestible.[130]

Unfortunately, one subtle stab at biotech comes from product distributors, stores and restaurants that see it as a way of gaining market share. They know that a consumer with the least doubt about biotech food is likely to be drawn to—and quite possibly pay extra for—something labeled "biotech-free." So they label the product, and the label itself becomes a tiny reinforcement that there is something wrong with biotech. After all, why else the special label? Sadly, if I ran a grocery store and put up signs that said all my products were "DHL-free," it would probably improve my business. No matter that DHL is an overnight delivery system.

In April 2001, the *Wall Street Journal* reported that it had tested twenty different U.S. food products labeled "GMO-free" ("GMO" meaning "genetically modified organism"). Of these, "11 contained evidence of genetic material used to modify plants and another five contained more substantial amounts"; thus only four were actually "GMO-free." In one case, for example, Yves Veggie Cuisine placed "non-GMO" on its entire line of vegetarian products; but when the *Journal* sent a package of its Canadian Veggie Bacon to a lab, 40 percent of it had DNA from biotech crops. Everybody told the *Journal* that the presence of biotech ingredients was the result of some sort of individual mix-up, rather than a systematic violation of their own claim. Assuredly some really believed their products to be biotech-free, but even these didn't conduct the lab tests that would substantiate this. *All* were in violation of FDA law forbidding false claims on food packages, and *all* took advantage of consumer fears and helped spread those fears even as they also got the benefit of using biotech food products.[131]

Okay, so fear sells. But really, can you blame these stores and chains for trying to make an extra buck by playing on people's doubts and worries and muddying the name of a safe technology? Absolutely.

We need gadflies in all areas of policy, from national defense to energy policy to education. We need biotech gadflies, too. My dictionary defines "gadfly" as a critic who *either* stimulates or annoys. To fulfill their role, gadflies need to be stimulating, intelligent and informed. They need to learn how to say something more than "no."

Yet many of the arguments against agricultural and some other forms of biotech really aren't arguments at all. Britain's Prince Charles, who fancies himself an organic farmer, has repeatedly stated his claim that biotech "takes mankind into realms that belong to God and to God alone."[132] Speaking for God is always an iffy proposition, in part because there's always somebody else out there to claim his connection is clearer. Indeed, the Vatican has said, "We cannot agree with the position of some groups that say it is against the will of God to meddle with the genetic make-up of plants and animals."[133] The Church of England, of which Charles will be titular head if he becomes king, has concluded: "Human discovery and invention can be thought of as resulting from the exercise of God-given powers of mind and reason" and "in this respect, genetic engineering does not seem very different from other forms of scientific advance." A March 2000 report commissioned by the Anglican Church specifically considered "the view that such manipulation represents a usurpation of divine privilege and a violation of the natural order of God's creation," but responded that "human intervention has been pivotal in pursuing scientific and medical revelation over time; discovery and invention are the result of exercising God's gifts of mind and reason."[134]

The "It's not God's will" argument is closely aligned with the "It's not natural" view, the advantage of the latter position being that it can be used by atheists.

While biotech food is not completely natural, neither is virtually anything else we consume. A New Zealand restaurant owner who declared to a reporter that "Nature has provided an exceptional range of natural products that do not need human tampering" was flatly wrong.[135] Five thousand years ago in Peru, potatoes were grown selectively.[136] Plants that produced potatoes with desirable characteristics, such as higher yields, were used to propagate future plants. There's evidence that what we call corn was developed in what is now Mexico by farmers—the first biotechnicians—who interbred two different wild grasses. Again, that was about five thousand years ago.[137]

In modern times, more than two thousand types of plants—many of which adorn our dinner tables—have been genetically modified by irradiation. While biotech products are sometimes derisively called "mutant foods," the whole purpose of this particular form of irradiation is to cause genetic mutations. The breeders then pick out the plants with mutated traits they like and use that seed in future plant generations. Yet almost nobody in either the United States or Europe seems even to know about this.[138]

Did nature provide that New Zealand restaurateur's beer and wine? Did nature provide the type of beef or lamb he sells? Did nature provide the bread, vegetables and fruits he sells? Quite possibly every item on his menu came from intentional genetic manipulation by the crossing of whole plants or animals, or the making of yeast cultures. Virtually everything we eat that doesn't have the word "wild" in it, such as "wild game," has been modified by man.

Jared Diamond, in his book on social development, *Guns, Germs, and Steel,* chronicles the development of wild crops into those similar to what we eat today. "Supermarket apples are typically around three inches in diameter, wild apples only one inch," he writes. "The oldest corn cobs are barely more than half an inch long, but Mexican Indian farmers of A.D. 1500 already had developed six-inch cobs, and some modern cobs are one and a half feet long."[139] He also notes that "wild squashes and pumpkins have little or no fruit around their seeds."[140]

Numerous useful mutations involved the reproductive system. "Some mutant [plants] developed fruit without even having to be pollinated, resulting in our seedless bananas, grapes, oranges, and pineapples," writes Diamond. "Some mutant hermaphrodites lost their self-incompatibility and became able to fertilize themselves—a process exemplified by many fruit trees such as plum, peaches, apples, apricots, and cherries."[141] Wild wheat, barley and peas grow in pods that normally shatter to spread the seed to the winds. That's hardly beneficial to us. The ones that could be used were mutant plants whose pods stayed closed until the food inside was edible. These were gathered, with some eaten and some sown. "Thus, human farmers reversed the direction of natural selection by 180 degrees," Diamond says. "Over 10,000 years ago, that unconscious selection for non-shattering wheat and barley stalks was apparently the first major human 'improvement' in any plant."[142] "Mutant hermaphrodite" foods are commonplace in New Zealand restaurants—and everywhere else.

Norman Borlaug also notes that "Mother Nature has crossed species barriers, and sometimes nature crosses barriers between genera—that is, between unrelated groups of species." Says Borlaug:

> Today's modern red wheat variety is made up of three groups of seven chromosomes, and each of those three groups of seven chromosomes came from a different wild grass. First, Mother Nature crossed two of the grasses, and this cross became the durum wheats, which were the commercial grains of the first civilizations spanning from Sumeria until well into the Roman period. Then Mother Nature crossed that 14-chromosome durum wheat with another wild wheat grass to create what was essentially modern wheat at the time of the Roman Empire.[143]

When scientists finished mapping out deadly *E. coli* O157:H7, they made a startling discovery. Part of the reason this bacterium is such a bad actor and so hard to treat is that parts of it have come from different viruses.[144] Meanwhile, it appears that thousands of viruses have incorporated themselves into our own genome.[145] Ma Nature was stirring hard in her mixing bowl long before scientists came along. Gene splicing and recombining allows *us* to do things *we* couldn't do before. As Steven Pueppke of the University of Illinois writes, the chief difference between the classical tools and biotechnology is not what most people think of, which is the crossing of species boundaries, but rather *precision.*[146]

In one real sense, biotech opponents are completely right in their critique of new innovations as "unnatural." Throughout most of history, surviving to what we call old age has been unnatural, whereas infant and child deaths have been quite natural. Multiple sclerosis is rare, but nonetheless quite natural. *Treating it* is unnatural. That said, it's a good bet that persons with the disease aren't too keen on Mother Nature's handiwork in this particular case. Yet biotech opponents in New Zealand have succeeded for years in preventing cows from being given genes that could allow them to produce milk that would treat MS.[147] Since civilization began it has been natural for humanity to try to improve its lot. It has been natural to attempt to end starvation, eliminate disease, increase the length and quality of life, and live in more pleasant surroundings.

It can be said there's no "need" for biotech, in the sense that the fate of humanity doesn't hang on it. As a species we have survived and flourished without running water, canning or microwave ovens. We made it without all the wonderful food varieties that have been created in the last century through crossbreeding. We even survived without pizza. In a real sense, we don't "need" any of these things that keep nature from taking its course. But all of them improve the quality and length of life. That's what biotech is beginning to do, with rapidly accelerating progress.

PART FOUR

Biotech Brooms—Letting Nature Clean up Man's Messes

ONE

Tackling Toxic Waste

While the number of toxic waste sites in the United States and around the world seems enormous, it's actually a sign of progress that we're even using such terminology. "Pollution," believe it or not, first came into English as a term for masturbation.[1] It's not that people in the past didn't pollute in the modern sense; they just didn't think about it much. Likewise, while dumping junk into the air and open water *were* long ago considered pollution, dumping it into the ground where it could eventually make its way into the water table was not. In the U.S., federal laws mandating systematic monitoring, prevention and cleanup of toxic waste sites weren't enacted until 1980 and weren't really given teeth until six years later.[2]

Cleaning up these sites will cost a fortune. Air pollution disappears almost as soon as you cut off the source. Water pollution also can quickly vanish if effluent discharges stop, depending on whether the water is stagnant or moving. But this isn't true of solid waste, be it toxic, hazardous or radioactive.*

And there are lots of toxic waste sites. Over 1,200 had been placed on the EPA's "National Priority List" as of July 2003, with new ones being added regularly.[3] Then there are brownfields, meaning property that can be left alone for the time being but the "expansion, redevelopment, or reuse of which may be complicated by the presence or

---

*Toxic and hazardous are usually used interchangeably, meaning solid waste that poses a risk to the safety or health of people or the environment. "Toxic waste" will apply to both in this book. Radioactive waste is just that. A mixture of toxic and hazardous waste and radioactive waste is called "mixed waste."

potential presence of a hazardous substance, pollutant, or contaminant."[4] There are about 450,000 brownfields in the United States with 100,000 to 200,000 being underground storage.[5] The cost of cleaning up any type of toxic waste site varies tremendously, depending on more factors than you could possibly care about. But a total of something in the hundreds of billions is clearly indicated.[6] There are an estimated additional 400,000 contaminated sites in Western Europe that could cost as much as $400 billion to clean up over the next couple of decades. The total world market for remediation in 1998 was $15–$18 billion.[7]

How clean a site should be, how quickly it should become clean, and whose responsibility it is to clean it up are policy decisions. Policy decisions are part science, part economics and part politics; as such, they cause arguments. But it seems that we should be able to agree that to the extent somebody does clean them, it should be done in the most efficient manner. And efficient doesn't just mean "cheapest." The traditional cleanup process of "dig and cart away" or "pump and treat" can itself be a messy affair. A botched cleanup job can be worse than none at all.[8] Bioremediation could change all that. At this point in the development of the technology, bioremediation is rarely advanced enough to clean a toxic waste site completely without some use of the old-fashioned, expensive techniques. But as with all things biotech, it advances by leaps and bounds. And for all the talk about "natural" and "unnatural" regarding biotechnology, in this case the biotech solution is clearly a lot more natural than state-of-the-art remediation.

"Bioremediation" means using living organisms including bacteria, plants or fungi to do a cleanup job.[9] It can often eliminate the need for messy, expensive chemical treatments, digging, incineration and landfilling. Further, rather than simply move contaminated material from one place to another, bioremediation can actually eliminate it.

Bioremediation first claimed the spotlight over a decade ago, when the supertanker *Exxon Valdez* ran aground in Alaska's Prince William Sound. A huge gash in the ship's side spilled more than 11 million gallons of crude oil into the water, the worst tanker accident spill in U.S. history. Biotechnology in the form of bacteria engineered to "eat" oil took the edge off it. Although older methods were used in conjunction with the bacteria, such as adding a form of soap to the spill, more than seventy-four miles of shoreline were cleaned far faster than through conventional reclamation methods, such as incineration, landfilling and chemical treatment.[10]

It's important to understand that while both plants and microbes can break down chemicals and petroleum into harmless products, breaking down metals is far more difficult and thus more rarely done. This includes radioactive material. Rather, with metals the effort is usually to "bind" or "immobilize" or "sequester" these so that if they are in water they stay put and don't leach throughout the ecosystem. If the immobilizing plants or micro-organisms are on land, they collect the metal so that it can be carted off or sometimes released through the leaves into the atmosphere in harmless amounts.

"Current engineering-based technologies used to clean up soils—like the removal of contaminated topsoil for storage in landfills—are very costly," says Agricultural Research Service (ARS) plant physiologist Leon Kochian. The EPA estimates that phytoremediation, using plants to break down or isolate undesirable material, could save 50 to 80 percent of conventional treatment costs.[11] During his thirteen years at the ARS lab in Ithaca, New York, Kochian has become an international expert on making plants do our dirty work. Just because the United States is the world's largest user of resources and hence in many ways its largest polluter, Americans shouldn't forget that "contaminated soils and waters pose major environmental, agricultural, and human health problems worldwide," he says.[12] The difference is that in the U.S. these rarely kill anyone, whereas they sicken and kill countless persons in less fortunate countries.[13] According to the CIA, Nigeria's Niger River Delta alone, a wetlands area of 42,000 square miles with 7 million inhabitants, has over 2.5 million barrels of oil pooled in spill sites. That's equal to more than ten *Exxon Valdez* disasters.[14] And it's getting worse.

Bioremediation can be used at the source of the pollution. Offshore, bilge water from cargo, cruise and military ships frequently contaminates delicate ecosystems. By law, it must be treated before being discharged. But current treatment systems are poor and often large amounts of oil effluents go straight into the water. That may become a thing of the past as a result of the recently approved PetroLiminator, a cleaner that works automatically and around-the-clock. The maker, EnSolve Biosystems of Raleigh, North Carolina, uses naturally occurring bacteria from oily, brackish water environments to digest and break down the bilge water into less than one part per million, far below federal standards.[15]

Another form of "bilge," if you will, is human. The use of microbes in septic tanks and municipal sanitation goes back decades. What's new is using them in confined areas, such as train toilets. When I was a kid, I thought it great good fun to use these toilets because you could see the tracks flying by as you flushed. But this caused real problems when riders ignored the signs about not using the toilet in the station. This system was replaced with an enclosed one, but these are prone to blockage and require frequent, expensive emptying. So engineers from QinetiQ, the commercial arm of the U.K.'s defense research agency, used technology developed for warships to design a cheaper, cleaner, more environmentally friendly train toilet.[16]

Something called a macerator chews up waste and feeds it into an aerated tank containing membranes coated with muck-munching bacteria. The solids are broken down primarily into carbon dioxide and water, while the gas is pumped off and bacteria-free water passes across the membrane. Some of the water is sterilized with ultraviolet light and returned to the flushing tank. The rest goes through a reverse-osmosis device that filters out the remaining chemicals, such as proteins and urea, so that the water is entirely microbe-free and can be used for hand washing. This way the system needs servicing only once a month to remove built-up sludge. Further, it provides a full six-liter flush as opposed to the small squirts in current systems. The technology could also decrease waste processing, reduce the need for clean flushing water in homes, and cut the huge cost—both financial and environmental—of wastewater treatment, especially in densely populated cities.[17]

A growing number of businesses and industries are looking to microbes and fungi to remove pollutants from their air emissions. The technology is already well established in Germany and the Netherlands. The basic approach involves establishing a colony of microbes and fungi that can biologically absorb and digest airborne pollutants. These have advantages over the oxidation and carbon filtration technologies, including being cheaper and using less energy than the high-temperature oxidation methods.[18]

For example, at the U.S. Department of Energy's Brookhaven National Laboratory in Upton, New York, chemists Mow Lin and Eugene Premuzic are experimenting with using bacteria to strip coal of sulfur, heavy metals and other toxic impurities. They've identified strains that could survive at high pressure, acidity and concentrations of toxic metals and at temperatures up to 185 degrees Fahrenheit, while digesting coal slurry. The microbes break the complex organic molecules in coal into simpler ones that burn more efficiently than normal coal or that can more readily be converted into coal gas or liquid fuel. At the same time, the bacteria remove sulfur and other contaminants, making the resulting fuel cleaner.[19]

TWO

Phytoremediation and Mighty Microbes

At this early point in the career of bioremediation it's much easier to use organisms that have already been discharged—itself no mean feat. One Department of Agriculture agronomist estimates that the cost of using plants to clean polluted soils could be "less than one-tenth the price tag for either ragging up [scraping or digging] and trucking the soil to a hazardous waste landfill or making it into concrete."[1] Indeed, the USDA has been at the forefront of using nature to clean up man's messes, with many of its best ideas coming from plant physiologist Leon Kochian.

"Phytoremediation" takes advantage of the need of certain plants—including trees, grasses and aquatic plants—to remove, destroy or sequester hazardous substances from the environment.[2] The most common types of phytoremediation are:

- Rhizofiltration: A method of cleaning water through the uptake of contaminants in roots.
- Phytoextraction: Taking contaminants up from the soil and storing them in either the roots or the shoots.
- Phytotransformation: Involves degradation of contaminants through the plant's metabolism, and can take pollutants from either soil or water.
- Phytostimulation: Using plants to stimulate bacteria and fungi to degrade pollutants.
- Phytostablization: Using plants to reduce the mobility of contaminants in the soil, locking them in place.[3]
- Phytovolatilization: Refers to the uptake and transpiration of contaminants, primarily organic compounds. The contaminant, present

in the water taken up by the plant, passes through the plant or is modified by the plant, and is released into the atmosphere where it evaporates or vaporizes.[4]

Obviously, these different methods are not exclusive of one another. Further, they can be combined with other bioremediation methods such as the use of microorganisms. Indeed, except in very shallow bodies of water or waste sites that don't reach too deeply into the ground, the major drawback of phytoremediation is that it can remediate only as far down as the roots go. Thus a plant-microbe combination can be an effective method of cleanup, especially since plant roots often add nutrients into the ground or water that help the microbes thrive.

Currently there is little bioremediation making use of transgenic plants, in part because it's a relatively new area and in part, unfortunately, because of antibiotech activist efforts.[5] But phytoremediation's future lies with the transgenics. In their normal state, plants may process pollutants too slowly for our needs. But in what is one of many cases of bioengineering piggybacking on other bioengineering, these plants are being transgenically induced not only to grow faster than normal but to consume waste more quickly as well.[6] David Glass, an independent analyst based in Needham, Massachusetts, predicts phytoremediation in the U.S. will grow to a $235 million to $400 million industry by 2005.[7]

The Salad That Saved Leadville

Nobody has asked the Produce Marketing Association how they feel about it, but a pilot study using a whole salad's worth of plants, including mustard, spinach, broccoli, radishes and carrots, is being carried out in a former mine drainage tunnel in Leadville, Colorado.[8] Leadville got its name from the lead smelting operations that made Andrew Carnegie rich, but the Rocky Mountain town has played host to individuals and companies extracting silver, molybdenum and numerous other metals since the Spanish arrived from Mexico in the sixteenth century.[9] Most of the companies have long since disbanded, leaving tailing piles and contaminated water and soil. The EPA declared twenty-two acres of the town a Superfund site without discussing matters with the local residents. This rather ticked them off, because they had already been living with the pollution all their lives and didn't care much for the idea of their town being ripped up like a sod field.[10] So Frank Burcik, a former mining engineer and president of Water Treatment and Decontamination International in Lakewood, Colorado, went to the government and said he had a better way, namely growing specific plants over the contaminated area that would suck up the metals through the roots.[11] "Our investigative work has already proved the ability of (our) proprietary phytoremediation system to remove 80 percent to 95 percent of toxic trace elements and other contaminants from mine water in a range of conditions," Burcik said.[12] "I'm so excited because this works. This is a magic thing."[13]

Burcik believes he can profitably recycle metal from the phytoremediation plants after they're dead, while the remaining plant material can safely be used to make such products as particleboard. "While plants found in wetlands have been used with only modest success in other clean-up efforts, my secret has been to harvest the saturated plants and reseed new ones," Burcik said.[14] According to Bill Holdsworth, an expert on bioremediation techniques, Burcik's deep-mine "greenhouse" may be a catalyst for reviving the fortunes of an ailing town, creating new jobs and cleaning up its reputation.[15] The EPA and the U.S. Bureau of Reclamation are interested in transplanting Burcik's process to help clean the California Gulch Superfund site, while World Bank officials say they find intriguing the possible applications in underdeveloped countries.[16]

Scientists from Colorado State University have also become involved in the project, using different plants. They are conducting tests to extract toxins from runoff water via plant roots using quinoa (a Latin American grain), Indian mustard and yarrow, a flowering herb that American Indians and settlers used to help staunch wounds.[17] Flow rates are varied through the channels, while all have constant levels of fertilizer and light.[18] Perhaps soon Leadville will have to change its name to just plain "Ville."

Tobacco against TCE

In mid-June of 2000, researchers at the University of Washington announced their successful insertion and implementation of a gene from a mammalian liver enzyme into the tobacco plant. The enzyme breaks down a variety of toxic chemicals, including the most widespread groundwater contaminants in the U.S., known as chlorinated solvents. These are various types of extremely useful agents that can break down substances in order to produce various materials or clean products and machinery. The most common is trichloroethylene (TCE), a solvent used to clean greasy machinery. Others include dibromide, carbon tetrachloride, vinyl chloride and chloroform. The transgenic plants were extremely fast at absorbing TCE through the roots—as much as 640 times faster than normal tobacco plants.[19] "The plants take [the pollutants] up and degrade them completely," said Lee Newman, a professor in the School of Forest Resources at the University of Washington and a co-author of the study.[20]

Because chlorinated solvents are such common pollutants, it's hard to exaggerate the usefulness of this gene. Approximately 400,000 soil and groundwater sites are considered contaminated with the chemicals.[21]

In the world of phytoremediation, certain plant species are known as hyperaccumulators. They "have the ability to extract elements from the soil and concentrate them in the easily harvested plant stems, shoots, and leaves," explained Leon Kochian. "These plant tissues can be collected, reduced in volume, and stored for later use." And even as they "vacuum, they must be able to tolerate metals that would kill other plants."[22] Amazingly, over four hundred different metal hyperaccumulating plants have now been discovered.[23] One metal hyperaccumulator that Kochian and his colleagues have been working with is *Thlaspi caerulescens,* commonly known as alpine pennycress. "*Thlaspi* is a small, weedy member of the broccoli and cabbage family," Kochian said. "It thrives on soils having high levels of zinc and cadmium."[24] His lab has been trying to discover the underlying mechanism that enables the plant to accumulate high amounts of these metals.[25]

"Hyperaccumulators like *Thlaspi* are a marvelous model system for elucidating the fundamental mechanisms of—and ultimately the genes that control—metal hyperaccumulation," explained Kochian. "These plants possess genes that regulate the amount of metals taken up from the soil by roots and deposited at other locations within the plant," he said, adding that "There are a number of sites in the plant that could be controlled by different genes contributing to the hyperaccumulation trait."[26] This is the rhizofiltration process described earlier.[27] Kochian and his colleagues have learned how *Thlaspi* accumulates these metals at terrific levels. "A typical plant may accumulate about

100 parts per million (ppm) zinc and 1 ppm cadmium," he said. "*Thlaspi* can accumulate up to 30,000 ppm zinc and 1,500 ppm cadmium in its shoots, while exhibiting few or no toxicity symptoms. A normal plant can be poisoned with as little as 1,000 ppm of zinc or 20 to 50 ppm of cadmium in its shoots." To find out what makes this super-plant tick, he and his team cloned a zinc transport gene. The breakthrough allowed them to discover that zinc transport is regulated differently in normal and hyperaccumulator plants. In essence, *Thlaspi* is always working at full horsepower, while other plants try to ramp up their power to meet the oncoming rush of zinc. For *Thlaspi,* "This results in very high rates of zinc transport from the soil and movement of this metal to the leaves."[28]

Down south, Richard Meagher of the University of Georgia, Athens, is coaxing *Arabidopsis* plants into cleaning up dangerous mercury from hazardous waste sites. When mercury is dumped, microbes in the soil and water turn it into methyl mercury, which then accumulates in the food chain to levels that could cause human neurological problems.[29] But Meagher isolated a bacterial enzyme, called mercuric ion reductase, that converts the metal into its least toxic form. When he placed this gene into *Arabidopsis,* it allowed the plants to grow in soil heavily contaminated with mercury and release it into the air, using the aforementioned process of phytovolatilization. Methyl mercury never gets a chance to form. To increase the cleanup efficiency, Meagher gave the plant a shorter lifespan so it can be grown and harvested six or seven times a year, and he gave it more height so it can absorb considerably more mercury during each of those lifespans.[30]

"The results were astounding—far better than we expected," Meagher said. "The amount of mercury that my plants converted in the laboratory was 10 nanograms per milligram of plant tissue per minute by weight of the plant tissue." (A nanogram is a billionth of a gram.) "That may not sound like much, but remember that's per plant per minute. It may be less efficient in the field, but the power is there to do something remarkable." The ultimate aim of the Georgia team's work will be getting those same transgenic properties working in trees and in other small plants, including grasses. "Meeting that goal may take two tries," said Meagher, "but I don't think it will take 10."[31]

While having something poisonous removed from the soil and released into the air might not sound particularly reassuring, Meagher says that compared with the background mercury levels already in the air, the amount released from contaminated sites would be minimal. "There would be no measurable increase in the atmospheric levels if plants were used to clean every site in the U.S. over the next 20 years," he said.[32] In any case, in his next stage of research, he and his colleagues employed a new strategy in which they inserted a second modified bacterial gene that enabled the plants to break down methyl mercury directly. This means it would release an even more benign form of mercury. They hope the technique will be tested in trees, shrubs and aquatic plants.[33] "Our working hypothesis," Meagher explained, "is that the appropriate transgenic plant, expressing these genes, will remove mercury from sites polluted by mining, agriculture, and bleaching, for example, and prevent methyl mercury from entering the food chain."[34]

Poisoning by the Millions

In yet another blow to those who believe in a benign Mother Nature, it happens that one of the world's worst poisons is the 100 percent natural element called arsenic. It's not that it's so lethal in tiny concentrations, but rather that it's so commonly found at extremely high concentrations. Usually it comes from the dissolution of minerals and ores. "The contamination of groundwater by arsenic in Bangladesh is the largest poisoning of a population in history, with millions of people exposed," according to the *Bulletin of the World Health Organization.*[35] It is estimated that of the 125 million inhabitants of that country, between 35 million and 77 million are at risk of drinking water naturally contaminated with arsenic.[36] But believe it or not, it gets worse. Recent studies startled researchers by showing that arsenic in irrigation water pumped from contaminated wells works its way not only into the roots and stems of the rice plants but into the very grain itself—and three-fourths of the calories in the Bangladeshi diet come from rice.[37] Considering that this poor nation is routinely swept by typhoons and floods and that most of its citizenry aspire to achieving *mere* poverty as opposed to *grinding* poverty, the arsenic is the icing on the cake from hell.

Arsenic can cause direct poisoning, like you see in the movies, but it has also been linked to a number of different cancers, including but not limited to lung, skin, liver, kidney and bladder.[38] Further, while natural concentrations are often enough to kill people, throughout the world man has made the situation worse. That's because arsenic is widely used in wood preservatives, along with paints, dyes, metals, drugs and soaps. It can also be released into the atmosphere from burning fossil fuels, producing paper and making soap.[39] Extracting metal from ore also releases arsenic, harming miners and perhaps others.[40] In a move that was controversial because many perceived it to have an extraordinary cost while conferring little benefit, the EPA in 2001 cut by 80 percent the amount of arsenic allowed in U.S. drinking water.[41] Water supply agencies had complained that initial compliance costs alone would be about $5 billion.[42]

But Richard Meagher has a trick for that, too. He and his University of Georgia colleagues inserted two unrelated *E. coli* genes into *Arabidopsis,* allowing it to remove arsenic from the soil and transport it to the plant's leaves. The transgenic plants accumulated seventeen times as much fresh shoot weight and two to three times as much arsenic per gram of tissue as nontransgenic plants.[43] "One of the most important aspects of the research is that this new system should be applicable to a wide variety of plant species," said Meagher. He added that a colleague of his is already "working on putting the genes into cottonwood trees, which have a large root system and could be useful in the phytoremediation of arsenic."[44]

The brake fern may also come to the rescue. This plant grows naturally in the southeastern United States and in California, as well as in other countries.[45] Researchers at the University of Florida at Gainesville systematically screened fourteen species from an abandoned lumber yard contaminated by arsenic.[46] Their tests showed that the brake ferns growing there were literally suckers for arsenic. They concentrated in their cells up to 200 times the arsenic level found in the soil. It also came as a surprise to Florida researchers that their ferns grown in pots containing arsenic at 50 or 100 parts per million soil particles thrived even more than in clean soil. At these higher exposures, the brake fern took up so much of the metal that it was 2.3 percent arsenic, so by EPA standards each of these plants was its own little toxic waste site.[47]

Unlike many ferns, this one enjoys basking in the sun. It could potentially be cultivated in water and thereby act as a natural arsenic filter. And while the fern grows all over the planet, for places where other plants may grow better, the appropriate brake fern genes could be spliced into them. Obstacles such as mass-producing the ferns in different climates and developing the technology to achieve results in real-world settings still pose challenges. But the method could prove useful in sites contaminated with arsenic from herbicides, pesticides and smelting wastes. That includes the potential to help clean up arsenic contamination in shallow bodies of water in countries like India and Bangladesh.[48]

Nuclear Waste

Near the Chernobyl nuclear plant in the Ukraine, sunflowers have been used to absorb deadly radioactive elements in groundwater, cleaning up 95 percent of a pond in a mere ten days.[49] Researchers from what is now EdenSpace Systems Corporation of Reston, Virginia, and elsewhere knitted together rafts, each of which held twenty-four sunflowers, and dotted a small pond about half a mile from the reactor.[50] The flowers sucked up both the radionuclides cesium 137 and strontium 90.[51] EdenSpace and Department of Energy (DOE) scientists have also been using sunflowers to remove uranium from contaminated springs at the Oak Ridge National Laboratory in Tennessee.[52]

Cleaning up soil contaminated with radioactive cesium is also one of Leon Kochian's projects. "Although the [main] cause of cesium-137 contamination—above ground nuclear testing—has been reduced, large land areas are still polluted with radiocesium," Kochian said. "Cesium is a long-lived radioisotope with a half-life of 32.2 years. It contaminates several U.S. Department of Energy sites in the United States. The projected costs of cleaning up these soils are very high—over $300 billion."[53] In their initial studies, Kochian's people found that the radioactive element was present in a form that plants couldn't absorb. So systematically they began adding things to soil in test patches to see if they could give the plants a boost. They hit the right formulation with ammonium ions. Then Kochian and a DOE scientist at Brookhaven National Laboratory in New York tried different plant varieties to make even better use of the formula. "One species, a pigweed named *Amaranthus retroflexus,* was up to 40 times more effective than others tested in removing radiocesium from soil," he said. "We were able to remove 3 percent of the total amount in just one three-month growing season. With two or three yearly crops, the plant could clean up the contaminated site in less than 15 years."[54] DOE is currently performing pilot studies at Brookhaven using this technology.[55]

Peter Goldsbrough, a genetics professor at Purdue University in Indiana, is working on plants to increase their rate of phytoremediation of metals, and to *decrease* it, and to get the plants to store the toxins in the right place. "Obviously if you're going to use plants to clean up toxic sites you'll want to increase the absorption rate," he explained. "But in many cases you want to decrease it. For example, if you are growing a vegetable crop in Poland or elsewhere in Eastern Europe which has suffered from pollution, you will want to decrease the uptake of metals." As to where the material is stored, if the plants are to be used for food, you'd want it somewhere that doesn't get eaten. With grain, you'd want it in the roots; with tubers, in the leaves. But "in the case of phytoremediation plants, you would want the metal stored in the shoots, where it can be easily harvested." To achieve the ultimate phytoremediator, "You would need to engineer a plant genetically that has a high biomass, i.e., it would have to be very large." Further, said Goldsbrough, "it would also need to be sturdy and capable of growing in contaminated sites. And it would have to be able to absorb lots of metals in its shoots, above ground, so that it could be harvested and, ideally, the metals extracted and recycled."[56]

Yes, recycled. The same aspect of heavy metals that makes them so pesky—they're hard to get rid of—makes them ideal for recycling. That's why DuPont and other companies are looking into phytoremediation not just to clean up metal but to reuse it.[57] The material from a battery you threw away three years ago may show up in a battery you use three years from now. "At the moment, it still is not a widely accepted technology, but that's likely to change because the cost factor is so favorable," said Goldsbrough. "And when you take into account the recycling potential, the possibilities are staggering."[58]

The range of plants that can be used for phytoremediation is vast, and many are quite common. A Canadian researcher found that by adding a common bacterium to prairie grass seeds he could boost their ability to clean up toxins in soil from pesticides and polychlorinated biphenyls (PCBs) from 30 percent to 50 percent. Steve Siciliano at the University of Saskatchewan and his thesis supervisor, Jim Germida, showed that the microbes not only improve the existing cleansing power of some plants but even stimulate some grasses to produce a previously unknown soil-scrubbing enzyme.[59] "Typically, we screw up the environment and then screw it up some more trying to fix it," said Siciliano. "Phytoremediation is a natural solution, using living resources to restore environments."[60]

Green Recycling

Poplars and other trees have an advantage over most vegetation as phytoremediators for several reasons. One is that they are large, with extensive root systems through which to process water. Another is that the roots can go quite deep. Still another is that the roots secrete carbon-rich exudates, which can act as food for soil microorganisms and increase pollutant absorption. Poplar trees have already been used to remove chlorinated solvents and nitrates from groundwater and heavy metals from soil.[61] Not content to call them simply poplars, some in the phytoremediation business refer to a stand of the trees as a "self-assembling solar-powered pump-and-treat system."[62]

Phytokinetics of North Logan, Utah, is one of a growing number of companies using trees to clean up waste sites.[63] One of the company's current projects is cleaning up the Bofors-Nobel Superfund Site in Muskegon, Michigan. Sludge lagoons and groundwater at the site are highly contaminated with a number of chemicals including dichlorobenzidine, which the EPA classifies as a probable human carcinogen.[64] Five native species of trees along with willows and poplars are being used. Ultimately, the barren twenty-acre site will be forested with 13,000 trees, all sucking away at the soil. Another one of Phytokinetics' projects is the New England Superfund Site in Southington, Connecticut. Beginning in 1998, the company planted 1,000 poplar and willow trees at the site to replace a mechanical pump that had previously been installed, greatly reducing maintenance costs. At the peak of the growing season, according to Ari Ferro, a biochemist and president of Phytokinetics, each tree can draw up more than fifteen gallons of water each day. "The trees are pumping like crazy," said Ferro. "The contaminants get sucked up into the root zone and biodegraded."[65]

But genetically engineered poplars in experimental plots will do even better. University of Georgia researchers took a gene from a strain of bacteria that enables it to tolerate high levels of toxic mercury, modified it and spliced it into the yellow poplar genome. Although too much mercury can kill a poplar just as it can kill an animal, tests showed that the transgenic trees had a tenfold increase in their ability to withstand it.[66]

So far, mustard has proven to be one of the most effective phytoremediators. Leon Kochian's ARS colleague David Ow in Albany, California, is searching for genes that increase mustard's tolerance to toxic chemicals. Ow and his fellow researchers first moved genes from the Indian brown mustard plant into yeast so they could be readily copied and studied.[67] Currently they are in the process of moving about fifty genes into another plant, *Arabidopsis,* in hopes of increasing the plant's ability to survive taking in large amounts of metals.[68]

Even nontransgenic mustard can be useful. "Researchers have found the Indian mustard plant to be particularly suitable for phytoremediation for several reasons," said Elizabeth Pilon-Smits, a Netherlands native and now a biologist at Colorado State University.[69] "It has a large biomass; it is relatively insensitive to pollutants; it accumulates trace elements, such as selenium; it is a tolerant plant, capable of thriving from freezing point to tropical temperatures; and it lends itself to transgenic alteration."[70] But while wild types of mustard can reduce selenium levels in soil by around 50 percent, her transgenic mustard plants are far more effective yet.

Selenium is another metal that can pose real problems. "In very small doses, selenium is a trace nutrient, vital for good health," said Pilon-Smits. "In larger doses, selenium exposure can result in toxicity, profound physical deformities, and death."[71] Selenium contamination can be purely natural. Soils that were developed from ancient marine sediments, particularly those in the western United States, are laden with the mineral. When grazing plants take it up from the soil, it can harm livestock. Irrigation can also cause selenium to concentrate in dangerous amounts. As the water goes through fields, it pulls the mineral out of the soil and sends it into streams and lakes where it accumulates in fish and other wildlife.[72] Central California's Kesterson National Wildlife Refuge made headlines in the 1980s when migrating waterfowl suffered bizarre deformities, such as twisted beaks and bulging brains, after nesting at the reserve's ponds, which were filled with selenium runoff from nearby farms.[73] In China, natural concentrations are so high they have caused serious health effects in humans, including heart damage and disfigurement of teeth and fingernails.[74]

Selenium is also found in coal and oil, so wastewater from power plants that burn selenium-rich fuel is often contaminated.[75]

But some vegetation is "very effective at removing selenium from contaminated soils," said Norman Terry of the University of California at Berkeley.[76] They not only absorb it; they also convert some of it into dimethyl selenide gas. That's good, because this gas has only about 1/700th the toxicity of the form selenium normally takes in the soil.[77] "I really think it has a great chance of succeeding as a way of cleaning up [agricultural] drainage water," said Terry, adding, "If it works, the whole western United States would be able to benefit from the technology."[78]

During research at a Chevron oil refinery near San Francisco, Terry and other researchers used cattails and another wetlands plant, the bulrush, to remove about nine-tenths of the selenium from 10 million liters of wastewater pumped out daily. In another demonstration in Corcoran, California, they reduced selenium levels from agricultural drainage water by 80 percent.[79] A problem they ran into was that some migratory birds were becoming poisoned from eating the very plants that were removing the selenium from the water and then storing it. But Terry said as much as 10 to 30 percent of the mineral is released harmlessly into the air and that he's continuing to investigate which plants are best at releasing selenium into the air or, alternatively, genetically engineering plants to release more. "We need to test it," he said. "If at the end of the day we're killing birds, then we've lost. But if it works, it's a win for everyone."[80]

Bioremediation has already proven a great success at reducing lead contamination, too. EdenSpace Systems bulldozed the top four feet of soil on a parcel of land belonging to DaimlerChrysler in Detroit, then carted it to a nearby area and planted it with sunflowers and Indian mustard. Both are powerful lead accumulators. The lead concentration in the soil dropped by over 40 percent, bringing it below federal and state limits. The project cost $900,000, but that was at least $1 million less than it would have cost to cart the 5,750 cubic yards of soil to a hazardous waste landfill. Instead, only a few cubic yards of tainted plant material had to be disposed of.[81] "We avoided the problems that go with hauling the waste to a landfill," said DaimlerChrysler senior manager Greg Rose. "We re-used the soil on the site—now safe and usable for future projects—and we saved money."[82]

In India, scientists are experimenting with several aquatic plants to remove chromium pollution produced by tanneries.[83] Ohio State University researchers have engineered a form of algae that exists throughout the world to make it extract copper, zinc, lead, nickel, highly toxic cadmium and mercury, and other metals from contaminated water.[84] It's named *Chlamydomonas reinhardtii.* The algae are cheap and common, according to Richard Sayre, the lead researcher. "This organism occurs everywhere on the globe . . . and it's innocuous," he said.[85] "It grows in Antarctica, it grows in the ocean. It naturally grows in soils in the back yard. It's called the cockroach of the algae world."[86] Sayre's team discovered that the genetically altered algae could be turned into heavy-metal sponges, either living or freeze-dried. Some metals bonded to the altered algae, while other nontoxic metals remained in the water.[87] "The material will sequester 20 percent of its weight as metal, and there's very little interference by nontoxic metals," Sayre said. Specifically, he saw that calcium and potassium, which we want to remain in any cleaned-up waters, did not interfere with attempts to get cadmium to bind to the algae.[88]

Recently, scientists in Spain got the idea of taking genetic material from mice and attaching it to a type of bacteria that was already resistant to heavy metals. Even by biotech standards, this mouse-and-microbe combination was a strange one, but it worked.[89]

Bugs with Bite

Microbiologists estimate that our planet has five million trillion trillion bacteria (a 5 followed by 30 zeroes).[90] Bacteria are the most common life form on earth. And now scientists are realizing that there is a bacterium that will eat nearly every kind of toxic waste. "There are some manmade compounds—like Styrofoam cups—that bacteria haven't figured out how to degrade yet," says Mike Nelson, a professor of mining engineering at the University of Utah. "But, give them time," he jokes, "and I'm sure they'll figure it out."[91]

There are essentially three ways for microbes to clean up toxic waste sites. The cheapest is to let them go about it themselves. As Nelson indicated, for almost every pollutant there is a bug that will break it down. So long as no new contamination is added, over time a site will become naturally cleaner until eventually it no longer qualifies as toxic. But that can take decades. Sometimes the land must be cleaned much more quickly by the bugs, and there are two ways of accomplishing that: biostimulation and bioaugmentation.

Biostimulation means adding specific nutrients so that naturally occurring microbes will thrive. It relies on Nelson's assertion that there probably already are naturally occurring bugs in the waste, but they need a boost, such as being fed nitrogen and oxygen or hydrogen. Alternatively, one can use bioaugmentation, which involves adding specific microorganisms. Different sites require different methods, but often either or both can be used.[92] And as one can imagine, different companies favor the methods they're best at.

Regenesis Bioremediation Products in San Clemente, California, is the world's largest bioremediation company, specializing in getting either oxygen or hydrogen down to the microbes at just the right levels for them to thrive.[93] One of their products, Hydrogen Release Compound (HRC), stimulates bacteria that break down degreasers, pesticides and other widely used industrial and agricultural chemicals that are notoriously difficult and costly to treat. According to Stephen Koenigsberg, vice president for R&D at Regenesis, these chemicals can penetrate deep into soils and bedrock, making them difficult to remove through conventional means. HRC is a honeylike formulation applied to soil and groundwater through a common (and cheap) process known as direct-push injection. This is a truck-mounted hydraulic device that drives a pointed rod into the ground, like a hypodermic syringe. Typically it can inject substances 50 feet into the ground, though in loose soils it can push down over 100 feet.[94] Once it's in place, it interacts with anaerobic microbes so there's a slow, steady release of hydrogen. These indigenous microbes flourish in hydrogen, allowing them to convert contaminants into harmless end-products. An application of HRC typically induces reducing conditions in the aquifer for an average period of 18 months, while the extended-release product HRC-X can keep the process going for 36 months or more. Studies have shown that the product can reduce concentrations of the acutely toxic PCE (perchloroethylene) by about 90 percent in merely eight months.[95]

Part of the bioremediation process can involve monitoring microbes. First, you want to see if the right ones are there, and if not, apply them in a process called bioaugmentation. If they *are* there, then "we ask the bacteria what they're doing and if they're happy or sad," said David White, director of the University of Tennessee/Oak Ridge Laboratory's Center for Biomarker Analysis (CBA).[96] "So we make a diagnosis, write a prescription, and check to see if it works," he explained. "Primarily we monitor wells during the bioremediation process, including sediments and groundwater." One of the biosensors they use employs beads about three millimeters in diameter with powdered, activated carbon inside. "We purge them of bacteria and put them in solid-state devices and lower them into wells where the bacteria rapidly colonize them. Within a month you can pull them up to see if the bioremediation is working," he said. "For example, if there's not enough nitrate or phosphate down there, it will tell you that." CBA's usual jobs include testing for the potential of compounds such as HRC to immobilize uranium (in conjunction with the right microbes) for the Department of Energy at its various former nuclear weapons facilities. One bacterium they commonly use is *Geobacter.*[97] "It stops uranium in its tracks so it doesn't make it into rivers or other bodies of water. Geobacter can keep uranium in place for over ten thousand years."[98]

Biotech bugs may also help solve the problem of mixed waste from nuclear weapons facilities. Unfortunately waste doesn't sort itself out the way we separate cans from newspapers when we recycle, so mixed waste is common. Indeed, in the United States about one-third of the reported three thousand waste sites created by nuclear weapons production between 1945 and 1986 are radioactive, mixing material such as uranium 235 with heavy metals such as mercury and solvents such as toluene.[99] The cleanup cost of these waste sites by traditional methods has been estimated to be about $265 billion.[100]

One problem with trying to use bioremediation against the metals or the solvents in mixed waste is that bacteria that might be able to break down or sequester this part of the waste will be zapped by the radioactive part of the mix. Enter recombinant *Deinococcus radiodurans,* discovered at the University of Oregon in Corvallis.[101] *Deinococcus radiodurans* was originally found living in a can of irradiated meat in the 1950s. University of Michigan scientists spliced it with genes from a strain of *E. coli* bacteria that can break down mercury, toluene and chlorobenzene, which are commonly used as carrier fluids for radioactive materials. Indeed, it appears that *Deinoccocus* can play host to genes from any bacterium that breaks down waste products.[102] It does *not,* unfortunately, neutralize radioactive material. But by helping to convert mixed waste to only radioactive waste, it can vastly reduce the overall mass that has to be sequestered for many thousands of years and save a huge amount of money. *Deinococcus* does not have a super ability to fend off radiation, but the bacteria "thrive because they have tremendous repair mechanisms," said Larry Wackett, the Michigan biochemist who led the study.[103] Because of this, they can survive 1.5 million rads of gamma radiation, or about 3,000 times the lethal dose for humans.[104]

We Don't Need No Steenking Sewers!

Sewage systems have to be one of the most underrated marvels of the modern age. There are few things fouler and more capable of transmitting horrible diseases than what goes down the toilet. If you don't believe that, there are still lots of countries you can go to where people draw their water downriver from where other people "do their business" and pick up any number of dreadful bacteria, viruses, amoeba and parasites. According to the World Health Organization, two-fifths of the world's water sewage is untreated, resulting in about 2.2 million diarrhea deaths a year, 6 million cases of blindness from trachoma (caused by bacteria), and 10 percent of the underdeveloped world's population suffering from intestinal parasites.[105] Alternatively, you can just read about how sewers were in the industrialized nations until quite recently, in Jonathan Swift's "A Description of a City Shower," about the open sewers of London after a rain:

> Sweepings from butchers' stalls, dung, guts, and blood,
> Drowned puppies, stinking sprats [small herrings], all drenched in mud,
> Dead cats and turnip-tops come tumbling down the flood.[106]

It was only a few decades ago that the river Thames lost its reputation as a sewer spanned by historical bridges. Yet another British river was described in Victorian times as being "full of refuse from water closets, cesspools, privies, common drains, dung-hill drainings, infirmary refuse, wastes from slaughter houses, chemical soap, gas, dye-houses ... pig manure, old urine wash" and dead animals, including "occasionally a decomposed human body."[107]

To keep our water clean at affordable prices even as urban areas become ever more congested, technology must improve. The cost of upgrading and replacing wastewater systems to meet the mandates of the federal Clean Water Act will reach nearly $1 trillion over the next twenty years, according to a report by the Water Infrastructure Network, a coalition of municipal agencies.[108] That's probably overstating the price, a typical tactic of advocacy groups. But the cost will also be less for another reason, as well: bioremediation. In some places, it's already saving money.

Until recently, Tampa, Florida, had a real odor problem. Being subtropical, Florida plays host to all sorts of creatures large and small that other states just don't have. Or other states have them in lesser numbers because the winter cold kills many of them off. Among the especially abundant Florida life forms are bacteria that were stinking up Tampa's sewage system with the rotten-egg smell of hydrogen sulfide. So the authorities decided to fight one bug with another. They called upon In-Pipe Technology, a company based in Wheaton, Illinois, which placed small units in strategic sites in the sewage collection system.[109] These boxes squirt doses of 10 to 14 different bacteria into the pipes. They eat up the "food supply," as it were, before their odor-emitting brethren can. Soon the foul-smelling bacteria give up. "All our bugs are native to sewage, but only in low numbers," said Dan Williamson, In-Pipe's founder. "We're reinforcing what's already there so that they dominate the bad boys."[110] The only byproducts of In-Pipe's secret recipe are water, nitrogen and carbon dioxide.

But there's a lot more than olfactory aesthetics at stake. Hydrogen sulfide doesn't just stink; it corrodes pipes. "The sulphide dissolves to form sulphuric acid, and so you get corrosion as well," explained Williamson.[111] Because they stop hydrogen sulfide from building up, the added bacteria also halt corrosion. And there's yet another advantage: by the time wastewater reaches a treatment plant, fecal matter and other suspended solids have been reduced by as much as 60 percent.[112] But what's the bottom line for enlisting these magnificent microbes? Compared with the previous system, which only masked rather than prevented the problem, it's saving Tampa about $175,000 a year.[113]

Bacteria have been altered to destroy natural chemicals such as toluene, a common gasoline additive, and manmade ones such as TCE, which often pollutes the ground and waterways.[114]

Earlier we saw how people search places with wretched conditions, like hot springs or volcanoes, for superhardy bacteria known as "extremophiles," which can then be spliced into plants to transfer that toughness. But hunting for extremophiles is also important to the bioremediation industry. Oregon State University veterinary professor Morrie Craig, for instance, went whale hunting, figuring that since whales take in so much pollution with their food, looking inside the intestines of one might reveal all sorts of extremophiles.[115] (One wag referred to this as Craig's "gut instincts.")[116] Craig began investigating the thousand or so species of bacteria that live inside the leviathan's digestive system and found a smorgasbord of strange creatures. Some bacteria break down carcinogens like naphthalene and anthracene; others chew happily on PCBs. Moreover, the whale bacteria are anaerobic, capable of eating without oxygen. This would make them particularly useful in going after underground pollutants.[117]

Another hazardous waste with an extremely long name is tetrachloroethylene, better known as perchloroethylene, which is thankfully shortened to PCE. It's commonly used by dry cleaners. Also, "it's widely used as a degreaser," according to chemical engineer Thomas Wood of the University of Connecticut main campus at Storrs. "PCE is used a lot in military bases to clean aircraft engines and industrially to clean metal," he said. "It's been around since World War II."[118] Considered one of the five most common groundwater pollutants, it is classified as a possible carcinogen.[119]

Trying to break down chemicals like PCE with bacteria or plants is no easy task because there's nothing like it in nature. Chemicals with a major natural component, like gasoline from petroleum, are much easier to deal with because petroleum has been around practically forever and there's been plenty of time for bacteria to evolve that feed on it. Scientists can alter these to make them much more efficient at the job, but something so removed from nature as PCE is a tough artificial nut to crack. Further, you need something that will break it down in the presence of oxygen. If it breaks down without oxygen, PCE actually becomes *more* toxic as it degrades until it finally loses enough molecules.[120] This is true for some other chemicals as well.[121]

So instead of looking in a natural source, Wood and his fellow researchers looked in an *unnatural* one, a municipal wastewater treatment plant in Italy. There they found a bacterium with an enzyme named ToMO.[122] "We were able to show the ToMO enzyme is capable of degrading PCE by cloning the genes into *E. coli*," Wood explained, whereupon "the *E. coli* then had the ability to degrade the PCE."[123] To get the bacteria down to the water table to do battle with the PCE, Wood has created "a system where we take the bacteria and marry them to the roots of poplar trees—which grow about 10 feet a year, and put down a very extensive root structure. Wherever the root goes, the bacterium goes. So it cleans the soil and can go all the way down, cleaning the underground aquifer."[124]

Remember that according to the EPA, there are some 100,000 to 200,000 leaking underground storage tank sites. Many contain petroleum hydrocarbons. A 1998 experiment was remarkably successful in cleaning up the ground at a New Jersey gas station, where the installation of new underground storage tanks prevented the use of the traditional method of excavating the soil and hauling it away. Mighty microbes were injected into bore holes. Samples taken sixty days later showed contamination decreased by more than 99 percent, with improvement in nearby groundwater as well. The entire remediation took less than three months. A further advantage of using bioremediation in this case is that the bacteria will destroy not just petroleum but a common byproduct of making petroleum into fuel, the carcinogen benzene.[125]

Making Light of Pollution

Biotechnology techniques are used to degrade waste from breweries, paper manufacturing plants and the production facilities of many synthetic organic chemical materials. British researchers are creating tobacco plants that can defuse literally explosive waste sites by degrading TNT (2,4,6 trinitrotoluene) into harmless compounds, which is important because TNT isn't just explosive but also toxic to fish and algae and is considered a possible human carcinogen.[126] The mushroom *P. chrysosporium* has also shown the ability to break down TNT. "The enzyme can break down organic toxins, leaving other harmless compounds that other microorganisms use as food," says Tulane molecular mycologist Joan Bennett.[127] Tobacco spliced with the bacterial nitroreductase gene also appears promising. The half-life of TNT in water surrounded by the transgenic plants was a mere few hours, while normal tobacco took weeks to reduce the TNT level by just 20 percent.[128]

Another way biotech can contribute to cleaning up waste is not just illuminating the problem in a metaphorical sense but also in a literal one. If it's true that you can't fight what you can't see, how can we combat waste products that produce harmful agents so small as to be essentially invisible? Remidios of Aberdeen, Scotland, has modified bacteria so they light up to different intensities based upon factors such as how well they're respiring.[129] When mixed with dissolved soil samples, the bacteria respond to toxicity. Their luminescence shows whether volatile organic, nonvolatile organic or inorganic contaminants are causing the bacteria to sicken. Workers can then determine the site's suitability for bioremediation in the first place, or monitor it after it's begun.[130] Remedios used these glowing "biosensors" in a three-year test of a thirty-acre site of a century-old paint-making plant in Dusseldorf, Germany. Plant officials said the probe found contamination by toluene, xylene and trichloroethene dating back six decades. The company is now using bioremediation and other strategies to clean up the chemicals, at a cost of $20 per ton of soil versus their original estimate of five times that for chemical treatment. Conventional monitoring and use of biosensors cost about the same initially, said Remedios technical director Anne Glover. "The difference is that we provide an immediate assessment of the site without having to revisit" it for further sampling and testing.[131]

As rapidly as the field of bioremediation has been moving, it may soon go into warp drive. One reason is that Russell Hill, an associate professor at the University of Maryland Center of Marine Biotechnology in Baltimore, found an unusual bacterium in a heavily polluted part of the Chesapeake Bay.[132] It proved capable of withstanding incredibly high levels of mercury, which it eats like popcorn. Hill was delighted to find that this penchant for mercury-chomping could be readily transferred from one strain of bacteria to another. "That means it will be easy to transfer these genes to another strain of bacteria that can, say, degrade hydrocarbons," he postulated.[133]

Scientists at the Department of Energy's Joint Genome Institute have released a draft gene of a toxic-tolerant bug called *Ralstonia metallidurans.* It was discovered in a settling tank in Belgium that was filled with pollution from heavy metals. Examination of *Ralstonia* showed that it has two large plasmids that house genes making the bug resistant to the harmful effects of a wide array of heavy metals including zinc, cobalt, cadmium, mercury, lead, chromium and nickel.[134] "Having a draft sequence of the *Ralstonia* genome, which contains some 3,000 genes, will make manipulation of these naturally existing resistance factors much more practicable," said John Dunn, a scientist at DOE's Brookhaven National Laboratory, who helped decipher it.[135] "Eventually, we'll want to understand how these genes are regulated under a variety of growth conditions and in different environments to see how they might be applied in bioremediation."

Scientists could either transfer the heavy-metal-resistant genes from *Ralstonia* into other microbes that break down organic pollutants, or genes from other microbes could be spliced into *Ralstonia.* Either way you end up with a microbe capable of any number of antipollution activities. *Ralstonia* has an added benefit in that the heavy metals tend to accumulate on the surface of the cell. "If you let that happen for a period of time and then remove the bacteria from the soil, you can remove the heavy metal contaminants as well," Dunn explained.[136] Another potential application would be to link *Ralstonia*'s uptake of heavy metals to genes that cause bacteria to glow. Then the bugs could be used to indicate the presence of heavy metals in soil: the brighter the glow, the higher the metal concentration. It would be rather like one of those red pop-up thingies they stick in turkeys to let you know when they're fully cooked. "What we're doing is building on the diversity of biology," said Dunn. "Here's a bacterium that potentially could be used as a tool to help us clean up the environment and to monitor how well we're accomplishing that goal."[137]

Oil and Gas Journal put it quite simply in a 1999 article: "Bioremediation makes its alternatives look truly ridiculous."[138] Yet using biotech to clean and strengthen the environment has its opponents. With few exceptions, environmentalists seem to be waiting skeptically. The push toward bioremediation is almost entirely on the part of corporations and governments.[139] Organizations like Greenpeace are so anticapitalist and so committed to seeing biotech as an enemy of humanity that they cannot imagine something good for the environment coming out of this devil's toolbox.

Thus you may find one Sierra Club representative giving two cheers for bioremediation, as when Robert Hastings, a spokesman for the group at Southeastern Louisiana University in Hammond, said to *Science News* in 1996, "We need to cut down on the production of contaminants up front. But where we do have cleanup problems, phytoremediation does have potential."[140] Then, four years later, another Sierra Club representative rejected the idea of bioremediation on the very ground that it *is* cost-effective. "What drives the move to using natural attenuation is a *pathological* need to cut clean up costs," said Ross Vincent, a chemical engineer and chairman of the club's Environment Quality Strategy Team (emphasis added).[141] So trying to save money is a disease?

After three days of discussions at a conference in Atlantic City, New Jersey, EPA officials came to the conclusion that, as *Hazardous Waste* stated, "Economic factors play a key role in whether [toxic waste sites] are cleaned."[142] That's why cheap clean is so much better for all of us than expensive clean. It not only helps a company's bottom line when it does clean up, but helps the company decide whether it *will* clean up.

At sites requiring urgent action, there will always be the need to "dig and dump," said Steven McGrath, senior scientist at the Institute of Arable Crops Research in Hartfordshire, England, to the *New York Times.*[143] But he added, "Bioremediation techniques offer the promise of cleaning up the legacy of pollution in a cost-effective way. With biological methods, prices come down by orders of magnitude into an area where as a society we can afford it."[144] So with companies not working under regulatory or court orders, bioremediation may not just be the cheapest and the most environmentally benign way to do the job, it may prove the *only* way to get them to do the job.

CONCLUSION

A Biotech Future

Think of how the world has changed in the past four decades. Consider the popular *Star Trek* series that aired from 1965 to 1969. It was supposed to depict life in the twenty-third century, yet as astoundingly imaginative as the series was, it's amazing how many of the devices from the show are already in common use. These include tiny wireless phones that flip open, hypodermic injections applied with compressed air, doors that slide open as you approach them, computers that respond to voice commands, and computer storage on a thin square that fits in your front shirt pocket. (True, *they* didn't have front shirt pockets; we haven't caught up to the future in everything.)

The future is coming to us faster than even science fiction can imagine. Several technologies will carry it in.

One is "nanotechnology" (miniaturization down to the atomic level). It already has certain commercial applications that only hint at the incredible changes it will bring in the future.[1] Three areas of technology have already caused tremendous change: computers, telecommunications and robotics. Biotechnology is just beginning to enter its era of transcendence; it now lags only behind computer technology in patent applications.[2]

There's been lots of hype about biotech, some of which I've written about in this book.[3] There will be a lot more as "gene" becomes an ever more magic word. But as a professional skeptic, someone who has made his reputation by raining on other people's papier-mâché parades, I believe that biotechnology is more promising than we can even imagine.

Biotechnology *will* revolutionize medicine, *will* extend lifespans, *will* feed the hungry and provide a cornucopia of better foods to everyone, and *will* make the environment cleaner. We know it will because the process has already begun. Some of the advances won't come as quickly as forecast; others will come faster. Some won't come from the companies that say they'll bring them, but from others that have not yet spoken. Moreover, there are probably fewer overstatements than *understatements*—predictions by nonexperts of biotech wonders "somewhere down the road" that, unbeknownst to them, are actually already in late stages of testing.

There are also plenty of things to worry about concerning the future.

For example, will we really be better off when we can receive 2,500 channels on TV? What will happen when the characters in children's video games—the ones they splatter all over the screen with their joystick-controlled weapons—no longer merely appear lifelike but indeed look *identical* to human beings?[4]

Then there's the computer conundrum: Faster and smarter may not always be better. Software genius Ray Kurzweil's book *The Age of Spiritual Machines: When Computers Exceed Human Intelligence* predicts that by 2019 a cheap personal computer will match the human brain's processing power.[5] What will happen when these computers use that intelligence to redesign themselves quickly, over and over? Will they, as some have predicted, view us as nothing more than carbon-based infestations? Bill Joy, cofounder and chief scientist of Sun Microsystems, wrote a much-discussed essay entitled, "Why the Future Doesn't Need Us," invoking a doomsday scenario of artificial intelligence run amok based on the speed of computer advancement.[6] Less melodramatically, Hans Moravec, director of the Robotics Institute at Carnegie Mellon University in Pennsylvania, nonchalantly claimed that "the robots will eventually succeed us" and that "humans clearly face extinction." Famed scientist Stephen Hawking told a German magazine that "the danger is real that [computers] could develop intelligence and take over the world."[7] Kurzweil gave us "a 50 percent chance of survival," before adding, "But then, I've always been accused of being an optimist."[8]

What about social issues? One hears disparaging talk of so-called "techno-utopians," people who believe that technology can solve social problems, maybe *all* social problems.[9] But there aren't many such people and I certainly am not among them. Most technology "gurus" are smart enough to realize that "utopianism"—be it religious, agrarian or technological—is a synonym for "ignorance" and "fanaticism." It's difficult to conceive of a technology that will reverse society's path toward rudeness and coarseness, reduce the divorce rate or end racism.[10]

Should we worry comparably about developments in biotechnology? In a broad sense of the term "worry," yes. It would be absurd to say there won't be future problems with biotech when there already have been problems. But even defenders of biotechnology get it wrong when they defensively insist, "No new technology is completely safe." The fact is that no *old* technology is completely safe. What's safer, a kerosene lamp or light bulbs? Well, shorts in light sockets can cause a fire, too. What's safer, sawing off a soldier's wounded limb because it might become gangrenous, or administering antibiotics? Well, antibiotics can provoke fatal allergic reactions. What's safer, the cars of a decade ago or the automobiles of today with seat belts, air bags, crumple zones, passenger cages, child safety-seats, better headlamps, and sensors that dial 911 when your air bag goes off? Well, people still die in car crashes, and for all the lives that seat belts and air bags have saved, they've taken some as well.

What *is* fair to say is that new technologies tend to get safer and better faster, while the old technologies become stagnant. Certainly more modifications could have been made to oil lamps to make them a bit safer, but no change would have given them either the safety or the convenience of the electric light bulb.

What *is* different about new technology is that it intrinsically brings both hope and fear, the fear of the unknown. After world chess champion Garry Kasparov resigned during the final game with the IBM computer Deep Blue in 1997, he explained why: "I'm a human being. When I see something that is well beyond my understanding, I'm afraid."[11] Mark Twain supposedly said, "I'm all for progress. It's change I don't like." Actually, there's no evidence he ever said that, but it certainly sums up the feelings of many of us. We usually prefer change a lot more in retrospect. Change—whether it's going off to college, getting married, having children or adopting a new technology—can be spooky.

Perhaps I worry less about biotechnology than some of the other developing technologies because I understand it better. Maybe I find it reassuring to know that no matter how many genes we modify, or swap between this and that organism, or turn off or on, biotech always uses the building blocks of life that God and nature have bestowed upon us.

Certainly, I have no problem with the concept that for the first time in history we will be able to ensure that every citizen of the world gets enough food and nutrients and does so with agricultural practices that are safer and will allow more land to be returned to nature. I have no more qualms over a new food item from Syngenta or Calgene or DuPont in my grocery store than I do over one from Sri Lanka or Costa Rica or Djibouti. The corn I eat was invented by ancient Mexicans, and if it's modified slightly by Monsanto so it feeds more people, that's fine by me. I like the idea of poorer countries having air and water as clean as these are in the wealthier nations, and of rich countries having even cleaner air and water than they do now. I think it's great that people will no longer have to watch their children die from horrifying diseases. I think it's fine that people are going to live a lot longer, though I despair to think they may spend that time plopped in front of a holograph-generating television eating cheezypoofs.

I know there will be ethical problems with biotechnology because there already are. But I also know there are far more serious ethical problems facing us right now that have nothing to do with biotech. Depending on whose poll you consult, over 100 million Americans think we're allowing the murder of more than a million unborn children a year.[12] Whether or not you agree with them, this is a moral dilemma that swamps any ethical problem I can even conceive of regarding biotech. I also see it as an ethical issue that while obesity has become epidemic in the industrialized world, malnutrition remains endemic in the underdeveloped world. Should it not be an ethical issue that when I get bitten by a mosquito, I get a slight welt, but when a child on another continent gets bitten by a mosquito, it often leads to fever, chills and death?

But I know this, too: The future of biotechnology—guided not just by scientists but also by our elected leaders and the bureaucracy, by our best thinkers, by philanthropies and profit-minded corporations, and even by skeptical advocacy groups—is bright indeed. The process has begun and it's accelerating. The ultimate benefits are unimaginable, while the near-term ones are incredible. Hold onto the bar in front of you and don't stand up. We're in for one heck of a thrill ride.

Acknowledgments

I must first recognize the tireless efforts of my wife, Mary, who did so much of the grunt work on this project even as she developed both the general website (www.fumento.com) and the dedicated site for this book (www.bioevolution.org) and acted as webmistress for them. It's terribly convenient having one woman as both wife and mistress.

Several interns contributed their efforts, including Michael Chiswick-Patterson, Sarum Charumilind and Jasmeet Ahuja. Jesse Malkin, Matthew Kaufman and Shelley Padilla also read the manuscript in its entirety. I appreciate their comments, corrections and wisecracks scribbled in the margins. Adam Bellow read the manuscript early on and made many valuable suggestions. Too many scientists vetted various sections for me to list them, although I would like to single out the efforts of Bruce Ames, professor of biochemistry and molecular biology at Berkeley, and William Li, president of the Angiogenesis Foundation in Cambridge, Massachusetts. I also constantly drew on the expertise of such persons as: Greg Conko, director of food safety policy at the Competitive Enterprise Institute; C. S. Prakash, director of the Center for Plant Biotechnology Research at Tuskegee University; Dr. Henry Miller of the Hoover Institution; and Cindy Richard of the Council for Agricultural Science and Technology.

This book would not have been possible but for my employer, the Hudson Institute, especially President Herb London, D.C. Vice President Ken Weinstein, and the support staff. Generous financial support came from the Donner Foundation, Billy Richards and others who wish to remain anonymous.

Thanks also to Jay Ambrose, who asked me to become a weekly health and science columnist for the Scripps Howard News Service, thereby allowing me to reach out regularly to readers throughout North America not only on biotechnology but on other vital subjects as well.

Peter Collier at Encounter Books has been my publisher, editor and friend all in one. Matthew Guma, my agent at Arthur Pine Associates in New York, showed great faith in me by taking on an author who has been merely successful as opposed to the typical Pine client who has been incredibly successful.

I conclude with the disclaimer that in a book of this size and scope, despite tremendous efforts to prevent them, "Mistakes have been made," as a good politician would put it. May they be few in number, but they are mine alone.

APPENDIX

Biotech Companies Mentioned

Abbott Laboratories—
http://www.abbott.com
Acambis Inc.—
http://www.acambis.com
Advanced Cell Technology Inc.—
http://www.advancedcell.com
A/F Protein Inc.—
http://www.afprotein.com
Affymetrix Inc.—
http://www.affymetrix.com/index.affx
AgResearch—
http://www.agresearch.cri.nz
Amersham Biosciences—
http://www1.amershambiosciences.com/aptrix/upp01077.nsf/content/na_homepage
Amgen Inc.—
http://www.amgen.com
Artecel Sciences Inc.—
http://www.artecel.com
Atherysys Inc.—
http://www.athersys.com
AstraZeneca—
http://www.astrazeneca.com
Aventis Pharmaceutical Products Inc.—
http://www.aventis.com
Avesthagen—
http://www.avesthagen.com
AviGenics Inc.
Baxter—
http://www.baxter.com
Bayer AG—
http://www.bayer.com/en/index.php
Bayer CropScience—
http://www.bayercropscience.com
Biogen—
http://www.biogen.com/site/content/index.asp
BioHybrid Technologies
Biolex Inc.—
http://www.gamil.com/IDbiolex.html
Bionor—
http://www.bionor.no
Biosite Inc.—
http://www.biosite.com
BioTransplant Inc.—
http://www.corporate-ir.net/media_files/nsd/btrn/btrn_splash.html
Bristol-Meyers Squibb Company—
http://www.bms.com/landing/data/index.html

Calgene
Cambridge Antibody Technology—
http://www.cambridgeantibody.co.uk
CancerVax Corporation—
http://www.cancervax.com
Celera Genomics—
http://www.celera.com
CellExSys—
http://www.targetedgenetics.com/cellexsys
Cell Genesys Inc.—
http://www.cellgenesys.com/intro.html
Celltrans Inc.
Celmed BioSciences Inc.—
http://www.theratech.com
Centocor Inc.—
http://www.centocor.com/cgi-bin/site/index.cgi
Clontech—
http://www.clontech.com/index.shtml
Cook Biotech Inc.—
http://www.cooksis.com
Corixa Corporation—
http://www.corixa.com
CropTech Corporation—
http://www.croptech.com
Curis Inc.—
http://www.curis.com
deCODE genetics—
http://www.decode.com
Dekalb Genetics Corporation—
http://www.dekalb.com/index.html
Delta & Pine Land Company—
http://www.deltapineseed.com
DNA Sciences—
http://www.DNA.com
Dobelle Institute Lda—
http://www.artificialvision.com
Dow AgroSciences LLC—
http://www.dowagro.com/homepage/index.htm
Dow Chemical Company—
http://www.dow.com/Homepage/index.html
Dupont—
http://www.dupont.com
Dynavax Technologies—
http://www.Dynavax.com
EdenSpace Systems Corporation—
http://www.edenspace.com
Elan Corporation—
http://www.elan.com
Eli Lilly & Company—
http://www.lilly.com
Emisphere Technologies Inc.—
http://www.emisphere.com
EnSolve Biosystems—
http://www.ensolve.com
EPIcyte Pharmaceutical Inc.—
http://www.epicyte.com/home.html
Exelixis Inc.—
http://www.exelixis.com
Genentech—
http://www.gene.com/gene/index.jsp
Generics Group—
http://www.generics.co.uk
GeneTests—
http://www.genetests.org
Gene Works
Genmab—
http://www.genmab.com
Genset—
http://www.genxy.com
Genta Inc.—
http://www.genta.com

GenVec Inc.—
http://www.genvec.com
Genzyme—
http://www.genzyme.com
Genzyme Transgenics Corporation—
http://www.transgenics.com
Geron Corporation—
http://www.geron.com
GlaxoSmithKline—
http://www.twinrix.com
GlycoSciences—
http://www.ogs.com
Hoffman-LaRoche—
http://www.roche.com/home.html
Human Genome Sciences (HGS)—
http://www.hgsi.com
IBM—
http://www.ibm.com/us
IDEC Pharmaceuticals Corporation—
http://www.idecpharm.com
ImClone Systems—
http://www.imclone.com
Immerge BioTherapeutics Inc. (a joint venture of Novartis Pharma AG and BioTransplant Incorporated)
Imutran Ltd. (a British subsidiary of Novartis)—
http://www.pharmiweb.com/novartis
Incyte Genomics Inc.—
http://www.incyte.com
Integrated Coffee Technologies Inc.—
http://www.integratedcoffee.com
Integrated Plant Genetics—
http://www.ipgenetics.com
Integrated Protein Technologies (unit of Monsanto)
Johnson & Johnson—
http://www.jnj.com
Layton Biosciences—
http://www.laytonbio.com
Maltagen Forschung GmbH
Medicago Inc.—
http://www.medicago.com
MedImmune Inc.—
http://www.medimmune.com
Mendel Biotechnology Inc.—
http://www.mendelbio.com
Merck & Company Inc.—
http://www.merck.com
Méristem Therapeutics—
http://www.meristem-therapeutics.com
Millennium Pharmaceuticals Inc.—
http://www.mlnm.com
Modex Therapeutics Inc.—
http://www.modextherapeutics.ch
Monsanto—
http://www.monsanto.com/monsanto/layout/default.asp
Monsanto Far East Ltd.—
http://www.monsantopakistan.com/elsewhere/asia.html
Motorola Inc.—
http://www.motorola.com
MPB Cologne GmbH
Nexia Biotechnology—
http://www.nexiabiotech.com
Nextran (subsidiary of Baxter)—
http://www.baxter.com/investors/r_d/nextran/index.html

Novartis—
http://www.novartis.com
Novartis Pharmaceuticals Corporation—
http://www.pharma.us.novartis.com
Omnitech Robotics—
http://www.omnitech.com
Orchid BioSciences Inc.—
http://www.orchid.com
Osiris Therapeutics Inc.—
http://www.osiristx.com
Oxford Biomedica—
http://www.OxfordBiomedica.co.uk
OXiGENE Inc.—
http://www.oxigene.com
Pfizer Inc.—
http://www.pfizer.com/main.html
Pharming (subsidiary of Genzyme)—
http://www.pharming.com
Phytokinetics Inc.—
http://www.phytokinetics.com
Phytomedics Inc.—
http://www.phytomedics.com
Pioneer Hi-Bred International Inc.—
http://www.pioneer.com
Planet Biotechnology Inc.—
http://www.planetbiotechnology.com
Plasma Proteome Institute—
http://www.plasmaproteome.org
PPL Therapeutics Ltd.—
http://www.ppl-therapeutics.com
Proagro—
http://www.certiseurope.com
ProdiGene Inc.—
http://www.prodigene.com
Proteome Sciences—
http://www.proteome.co.uk
QinetiQ—
http://www.qinetiq.com/homepage.html
Qualigen Inc.—
http://www.qualigeninc.com/main.htm
ReProtect—
http://www.reprotect.com/index.shtml
Response Genetics Inc.—
http://www.responsegenetics.com
Scotts Company—
http://www.scotts.com
Selective Genetics Inc.—
http://www.selectivegenetics.com
SemBioSys Genetics Inc.—
http://www.sembiosys.ca
Senomyx Inc.—
http://www.senomyx.com
Serono International—
http://www.serono.com/index.jsp
StaufferBiotech
Stauffer Seeds—
http://www.staufferseeds.com
Stryker Biotech—
http://www.op1.com/index.cfm
SunGene GmbH—
http://www.sungene.de
Syngenta Seeds AG—
http://www.syngenta.com/en/index.aspx
Therics Inc.—
http://www.therics.com

TranXenoGen—
http://www.tranxenogen.com
ToBio LLC
Quest Diagnostics—
http://www.questdiagnostics.com
ValiGen—
http://www.valigen.net
Ventria Bioscience—
http://www.ventriabio.com
Vertex Pharmaceuticals Inc.—
http://www.vrtx.com
ViaCell—
http://www.viacellinc.com
Viacord—
http://www.viacord.com/home/home.asp
Viragen—
http://www.viragen.com
Visible Genetics—
http://www.visgen.com/
XTL Biopharmaceuticals—
http://www.xtlbio.com/home.html
Zyomyx Inc.—
http://www.Zyomyx.com

NOTES

Please note: At the time the manuscript was completed, all hyperlinks worked. But inevitably some will have since been redirected and others simply pulled from the Web. If the page has merely moved, you might be able to find it by truncating the address; that is, cutting off the last section after the last backslash (/) and then searching from that page. Repeat as necessary.

Introduction

1. Jeremy Rifkin, *The Biotech Century* (Los Angeles: J. P. Tarcher, 1999); as quoted in Sihoban Gorman, "Future Pharmers of America," *National Journal* 31:6 (6 February 1999): 355.
2. Jeremy Rifkin, *Algeny* (New York: Viking Press, 1983), p. 231.
3. As quoted in Mac-Wan Ho, "The Unholy Alliance," *Ecologist* 27:4 (July/August 1997), at: http://www.mtholyoke.edu/courses/jgrossho/archives/unholy-alliance.html
4. See Indur M. Goklany, "From Precautionary Principle to Risk-Risk Analysis," *Nature Biotechnology* 20:11 (November 2002): 1075.
5. Bill Gates, "Like Software, Biotechnology Will Change the World," 16 June 1996, at: http://www.microsoft.com/BillGates/columns/1996essay/ESSAY960618.asp
6. Wylie Wong, "Ellison Says Computing to Become 'Boring,'" *CNET News.com,* 21 February 2001, at: http://news.cnet.com/news/0-1007-200-4892011.html
7. As quoted in Art Hovey, "Genetically Altered Food Has Arrived," *Lincoln (Nebraska) Journal-Star,* 11 June 1999; the website for the National Association of Science Writers is: http://www.nasw.org/
8. Andrew Pollack with Lawrence K. Altman, "Large Trial Finds AIDS Vaccine Fails to Stop Infection," *New York Times,* 24 February 2003, p. A1; Steve Sternberg "Vaccine for AIDS Appears to Work," *USA Today,* 24 February 2003, p. A1.
9. "Europeans Suspicious of Biotech, Says Survey," *Agence France-Presse,* 27 April 2000. Residents of all fifteen member nations of the European Union were polled. Of these, only 38 percent said biotech was acceptable even to make crops more nutritious or tastier.

10. Ibid. Only 41 percent surveyed thought biotechnology will improve their lives over the next twenty years, compared with 47 percent three years earlier, while 23 percent thought biotech will have a negative effect on their lives, up from 19 percent previously.
11. Colored etching, "The Cow-Pock—the Wonderful Effects of the New Inoculation," by James Gillray, 1802, as appears in Kenneth F. Kiple, *Plague, Pox and Pestilence* (New York: Barnes & Noble, 1997), p. 78.
12. "Death and Money: The Electric Chair," at: http://inventors.about.com/science/inventors/library/weekly/aa102497.htm

Part One: Miracles in Medicine
The New Biotech Paradigm

1. Eli Lilly & Co., "Recombinant Enzymes," at: http://www.lilly.com/about/partnering/worldtrade/biotech/enzymes.html
2. A. M. Shapiro et al., "Islet Transplantation in Seven Patients with Type I Diabetes Mellitus Using a Glucocorticoid-free Immunosuppressive Regimen," *New England Journal of Medicine* 343:4 (27 July 2000): 230–38.
3. Drug costs as a percentage of overall healthcare spending are not increasing, as is widely believed. They remain at slightly below 10 percent. But since healthcare cost increases regularly outstrip the Consumer Price Index, then so do drug costs. See Pharmaceutical Research and Manufacturers of America, "The Best Value in Medicine Today: How Prescription Drugs Account for a Fraction of Health Cost Increases while Helping to Offset Other Health Costs," at: http://www.phrma.org/publications/publications//2002-11-11.615.pdf
4. Terence Chea, "No Mere Pest: Biotech Firm Uses Caterpillars to Produce Proteins That Aid in Diagnostic Tests, Medical Treatments," *Washington Post,* 30 July 2001, p. E1.
5. BIO, "Approved Biotechnology Therapeutics, 2002," at: http://www.bio.org/news/approved2002.asp
6. Biotechnology Industrial Organization, "Biotechnology Industry Statistics," at: http://www.bio.org/er/statistic.asp
7. Ibid.
8. Gary Walsh, "Biopharmaceutical Benchmarks," *Nature Biotechnology* 18:8 (August 2000): 831; Pharmaceutical Research and Manufacturers of America, "371 Biotechnology Medicines in Testing Promise to Bolster the Arsenal against Disease."
9. Walsh, "Biopharmaceutical Benchmarks."

10. C. B. Burge, R. A. Padgett and P. A. Sharp, "Evolutionary Fates and Origins of U12-type Introns," *Molecular Cell* 2:6 (December 1998):773–85; J. Hanke et al., "Alternative Splicing of Human Genes: More the Rule than the Exception?" *Trends in Genetics* 15:10 (October 1999): 389–90; D. Brett et al., "EST Comparison Indicates 38% of Human mRNAs Contain Possible Alternative Splice Forms," *FEBS Letters* 474:1 (26 May 2000): 83–86.
11. Stevens Kastrup Rehen et al., "Chromosomal Variation in Neurons of the Developing and Adult Mammalian Nervous System," *Proceedings of the National Academy of Sciences USA* 98:23 (6 November 2001): 13,361–66; Helen Pearson, "Brains May Be Genetic Mosaics," *Nature ScienceUpdate,* 16 December 2002, at: http://www.nature.com/nsu/021216/021216-2.html
12. Nicholas Wade, "Bioengineers Turn to Hens' Teeth," *New York Times,* 22 August 2000, p. F5; Jun Zou et al., "Microarray Profile of Differentially Expressed Genes in a Monkey Model of Allergic Asthma," *Genome Biology* 3:5 (11 April 2002), at: http://www.genomebiology.com/2002/3/5/research/0020/?mail=0000127
13. "Speaking the Language of Recombinant DNA," from "What Is Biotechnology?" Washington, D.C.: Biotechnology Industry Organization, 1989, at: http://www.accessexcellence.com/AB/IE/Speaking_Language_rDNA.html
14. David J. Lockhart and Elizabeth Winzeler, "Genomics, Gene Expression and DNA Arrays," *Nature* 405:6788 (June 2000): 827–36.
15. T. J. Sweeney et al, "Visualizing the Kinetics of Tumor-Cell Clearance in Living Animals," *Proceedings of the National Academy of Sciences USA* 96:21 (12 October 1999): 12,044–49.
16. Matt Ridley, *Genome: The Autobiography of a Species in 23 Chapters* (New York: HarperCollins, 1999), pp. 21–22.

Vanquishing Vaccines

1. "Types of Vaccines," *WebHealthCentre.Com,* at: http://www.webhealthcentre.com/general/im_types.htm
2. Centers for Disease Control and Prevention, "Vaccine Information," at: http://www.cdc.gov/ncidod/diseases/flu/fluvac.htm
3. Gary Walsh, "Biopharmaceutical Benchmarks," *Nature Biotechnology* 18:8 (August 2000): 831.
4. Rodney Hoff and James McNamara, "Therapeutic Vaccines for Preventing AIDS," *Lancet* 353:9166 (22 May 1999): 1723–24; "AIDS Failure 'No Need for Gloom,'" *Herald-Sun* (Victoria, Aus-

tralia), 25 February 2003, at: http://www.heraldsun.news.com.au/common/story_page/0,5478,6038889%255E421,00.html; C. O. Tacket et al., "Human Immune Responses to a Novel Norwalk Virus Vaccine Delivered in Transgenic Potatoes," *Journal of Infectious Diseases* 182:1 (July 2000): 302–5.

5. http://www.merck.com/; "Millions Will Die from a Preventable Liver Disease, Doctors Told," *Universal News Services,* 29 May 1987; http://www.whale.to/m/recombivax_hb.html
6. http://www.twinrix.com/
7. Beverly D. Lucas, "Now You Can Prevent Lyme Disease," *Patient Care* 33:11 (15 June 1999): 180; Lu-Ann Murdoch, "A Vaccine for Lyme Disease," *Drug News,* April 1999.
8. "Staph Vaccine Appears to Be Effective in Stopping Infections," *Associated Press,* 20 September 2000.
9. Laura Meckler, "New Anthrax Vaccine Takes Fewer Shots," *Associated Press,* 3 October 2002; Renae Merle, "Firm to Do Human Tests of New Anthrax Vaccine," *Washington Post,* 16 October, p. E5.
10. See Michael Fumento, "Ebola Fever Sweeps the West," *National Post* (Canada), 31 January 2001, at: http://www.nationalpost.com/search/story.html?f=/stories/20010131/458571.html; N. J. Sullivan et al., "Development of a Protective Vaccine for Ebola Virus Infection in Primates," *Nature* 408:6812 (30 November 2000): 605–9; Press Release, "Novel Vaccine Protects Monkeys from Ebola Infection," National Institute of Allergy and Infectious Diseases, 29 November 2000, at: http://www.niaid.nih.gov/newsroom/ebolavacc.htm
11. M. Houghton, "Strategies and Prospects for Vaccination against the Hepatitis C Viruses," *Current Topics in Microbiological Immunology* 242 (July 2000): 327–29; "*Yersinia Pestis* Defense Evaluation Research Agency Plans Clinical Trials," *R &D Focus Drug News;* Roger Highfield, "Bubonic Plague Vaccine Ready to Test on Humans," *Daily Telegraph* (London), 27 August 1999, p. 6; John Wagner, "Hope for a Vaccine against Cervical Cancer," *Scripps Howard News Service,* 1 June 2000; "Recombinant Human Papillomavirus Vaccine Decreases Tumor Size," *Vaccine Weekly,* 29 March 2000.
12. National Cancer Institute, "Do We Know What Causes Cervical Cancer?" at: http://www.cancer.org/docroot/CRI/content/CRI_2_4_2X_Do_we_know_what_causes_cervical_cancer_8.asp?sitearea=&level=

13. National Cancer Institute SEER data for cervical cancer, at: http://canques.seer.cancer.gov/cgi-bin/cq_submit?dir=seer1999&db=7&rpt=LINE&sel=1^0^43^^^2^0&x=Year%20of%20death^5,6,7,8,9,10,11,12,13,14,15,16,17,18,19,20,21,22,23,24,25,26,27,28,29,30,31,32,33,34,35&y=Race^0,1,2&dec=4
14. David Brown, "Vaccine May Be First for Cancer," *Washington Post,* 21 November 2002, p. A1.
15. Program for Appropriate Technology in Health (PATH), "Natural History of Cervical Cancer: Even Infrequent Screening of Older Women Saves Lives: Cervical Cancer Prevention Fact Sheet," November 2000, at: http://www.path.org/files/cxca-factsheet-natural-history.pdf
16. Personal telephone communication with Kathrin Jansen, 26 November 2002.
17. Laura A. Koutsky et al., "A Controlled Trial of a Human Papillomavirus Type 16 Vaccine," *New England Journal of Medicine* 347:21 (21 November 2002): 1645–51.
18. Denise Grady, "Vaccine Appears to Prevent Cervical Cancer," *New York Times,* 21 November 2002, p. A1.
19. Personal telephone communication with Kathrin Jansen, 26 November 2002..
20. Carina Krüger et al., "In Situ Delivery of Passive Immunity by Lactobacilli Producing Single-Chain Antibodies," *Nature Biotechnology* 20:7 (July 2002): 702–6; Rick Weiss, "Protecting Teeth with Bacteria That Bite Back," *Washington Post,* 22 July 2002, p. A7.
21. American Social Health Organization, "Finding Answers and Support for Herpes," at: http://www.ashastd.org/herpes/hrc/educate.html
22. Lawrence R. Stanberry et al., "Glycoprotein-D–Adjuvant Vaccine to Prevent Genital Herpes," *New England Journal of Medicine* 347:21 (21 November 2002): 1652–61; Ronald Kotulak, "Vaccines Promising in Preventing Herpes and Cervical Cancer," *Chicago Tribune,* 21 November 2002, p. 1.
23. Press Release, "New Survey Reveals Allergies Nearly Twice as Common as Believed—Afflicting More than One-Third of Americans," American College of Allergy, Asthma and Immunology, 29 July 1999, at: http://allergy.mcg.edu/news/survey.html
24. "Doctor, Explain Allergies and Hay Fever," American Academy of Otolaryngology–Head and Neck Surgery Inc., at: http://www.entnet.org/allergies_hayfever.html; "Genetic Vaccine from Dust

Mite Developed for Asthma Treatment," *Deutsche Presse-Agentur,* 7 July 2001.

25. http://www.hopkins-allergy.org/; http://www.dynavax.com/
26. As quoted in Press Release, "Johns Hopkins Researchers Say Human Trials of Vaccine Technique Prove Promising for Allergy Sufferers," *AScribe Newswire,* 22 March 2001.
27. Multilateral Initiative on Malaria, *The Intolerable Burden of Malaria: A New Look at the Numbers* (Bethesda, Maryland: MIM, 2002); to request a copy e-mail MIM Secretariat Coordinator Andréa Egan, Ph.D., at mim@nih.gov.
28. "Massive New Campaign against Malaria," *Inter Press Service,* 29 October 1998.
29. Stephanie James and Louis Miller, "Malaria Vaccine Development: Status Report," National Institute of Allergy and Infectious Diseases, Division of Microbiology and Infectious Diseases, at: http://www.niaid.nih.gov/dmid/malaria/malariavac.htm#c
30. "Trials of New Malaria Vaccine Start in Gambia," *Reuters,* 18 September 2000.
31. As quoted in Press Release, "Live Recombinant Vaccine Protects against Fungal Disease," NIH/National Institute of Allergy and Infectious Diseases, 29 November 2000, at: http://www.niaid.nih.gov/newsroom/blastomyces.htm
32. M. Wüthrich, H. I. Filutowicz and B. S. Klein, "Mutation of the WI-1 Gene Yields an Attenuated Blastomyces Dermatiditis Strain That Induces Host Resistance," *Journal of Clinical Investigation* 106:11 (November 2000): 1381–89.
33. Gary Walsh, "Biopharmaceutical Benchmarks," *Nature Biotechnology* 18:8 (August 2000): 831.
34. "Vaccine Shows Promise against Leukemia Relapse," *Medical Industry Today,* 18 May 2000; "Genetic Approach Targets Breast and Ovarian Cancers," *Cancer Weekly,* 27 June 2000; "Aphtons Therapy to Be Studied in Phase III Pancreatic Cancer Trials," *Medical Industry Today,* 12 April 2000; Elizabeth M. Jaffee et al., "Novel Allogenic Granulocyte-Macrophage Colony-Stimulating Factor-Secreting Tumor Vaccine for Pancreatic Cancer," *Journal of Clinical Oncology* 19:1 (January 2001): 145; "Combination Therapy to Be Tested for Non-Hodgkin's Lymphoma," *Medical Industry Today,* 27 April 2000; "Testing Begins for New Vaccine Designed to Prevent Recurrence of Brain Tumors," *Cancer Weekly,* 28 March 2000; "Allogeneic Tumor Cell Vaccine Attacks Colon Carcinoma," *Cancer Weekly,* 4 July 2000.

35. http://www.ncbi.nlm.nih.gov/entrez/query.fcgi?CMD=search&DB=PubMed
36. Janice M. Reichert and Cherie Paquette, "Therapeutic Cancer Vaccines on Trial," *Nature Biotechnology* 20:7 (July 2002): 659; Delthia Ricks, "Scientists: New Era in Cancer Care / Innovative Approaches, Treatments," *New York Newsday,* 14 May 2000, p. A7.
37. http://www.oxfordbiomedica.co.uk/
38. As quoted in "Cancer Vaccine 'Close,'" *BBC News,* 27 August 2000, at: http://news.bbc.co.uk/hi/english/health/newsid_896000/896438.stm; see also J. G. Sinkovics and J. C. Horvath, "Vaccination against Human Cancers," *International Journal of Oncology* 16:1 (January 2000): 81–96.
39. CancerVax, "Canvaxin," at: http://www.cancervax.com/vaccine/canvaxin.htm
40. CancerVax, "Clinical Data," at: http://www.cancervax.com/vaccine/immune_data.htm
41. Michael D. Lemonick and Alice Park, "Vaccines Stage a Comeback," *Time,* 21 January 2002, p. 70.
42. WaterFrog Productions, at: http://www.waterfrog.com/cancer_glossary.htm
43. See University of Victoria, Department of Biochemistry and Microbiology, "Declaring War on Cancer," at: http://web.uvic.ca/~pjr/teaching/BIOC102/War/L01Notes/L01P01.html
44. http://www.cellgenesys.com/products-cancer-vaccines.shtml
45. Smita K. Nair et al., "Induction of Cytotoxic T Cell Responses and Tumor Immunity against Unrelated Tumors Using Telomerase Reverse Transcriptase RNA Transfected Dendritic Cells," *Nature Medicine* 6:9 (September 2000): 1011–17.
46. Jan Karlseder, Agata Smogorzewska and Titia de Lange, "Senescence Induced by Altered Telomere State, Not Telomere Loss," *Science* 295:5564 (29 March 2002): 2446–49; University of Texas Southwestern Medical Center at Dallas, Shay/Wright Laboratory, "Basic Introduction to Telomeres," at: http://www.utsouthwestern.edu/home_pages/cellbio/shay-wright/intro/sw_intro.html
47. http://www.mc.duke.edu/; http://www.geron.com/
48. As quoted in Press Release, "First Potential 'Universal' Cancer Vaccine Shows Promise in Lab," 29 August 2000, at: http://www.dukenews.duke.edu/Med/vaccine1.htm; http://www.mc.duke.edu/

49. Jane A. Plumb et al., "Telomerase-Specific Suicide Gene Therapy Vectors Expressing Bacterial Nitroreductase Sensitize Human Cancer Cells to the Pro-Drug CB1954," *Oncogene* 20:53 (22 November 2001): 7797–7803; Marie Boyd et al., "Expression in UVW Glioma Cells of the Noradrenaline Transporter Gene, Driven by the Telomerase RNA Promoter, Induces Active Uptake of [131I]MIBG and Clonogenic Cell Kill," *Oncogene* 20:53 (22 November 2001): 7804–08; Smita K. Nair et al., "Induction of Cytotoxic T Cell Responses and Tumor Immunity Against Unrelated Tumors Using Telomerase Reverse Transcriptase RNA Transfected Dendritic Cells," *Nature Medicine* 6:9 (September 2000): 1011–17.
50. As quoted in Press Release, "First Potential 'Universal' Cancer Vaccine Shows Promise in Lab."
51. H. Szutorisz et al., "A Chromosome 3-Encoded Repressor of the Human Telomerase Reverse Transcriptase (hTERT) Gene Controls the State of hTERT Chromatin," *Cancer Research* 63:3 (1 February 2003): 689–95; Y. Mo et al., "Simultaneous Targeting of Telomeres and Telomerase as a Cancer Therapeutic Approach," *Cancer Research* 63:3 (1 February 2003): 579–85; "Blocking Genetic Switch Could Nip Cancer in the Bud," *Reuters*, 3 February 2003.
52. See National Institute of Allergies and Infectious Diseases, "Emerging Infectious Diseases Research—Introduction," at: http://www.niaid.nih.gov/eidr/intro.htm
53. See Mirko D. Grmek, *History of AIDS* (Princeton, New Jersey: Princeton University Press, 1997).
54. R. S. Lanciotti et al., "Origin of the West Nile Virus Responsible for an Outbreak of Encephalitis in the Northeastern United States," *Science* 286:5448 (17 December 1999): 2333–37.
55. http://www.acambis.com/
56. Michael E. Ruane, "NIH Funds Search for West Nile Vaccine," *Washington Post*, 2 August 2000, p. B1.
57. As quoted in ibid.
58. David Shook, "A New Breed of Microbe Hunters," *Business Week Online*, 9 August 2002.
59. A. G. Pletnev et al., "West Nile Virus/Dengue Type 4 Virus Chimeras That Are Reduced in Neurovirulence and Peripheral Virulence without Loss of Immunogenicity or Protective Efficacy," *Proceedings of the National Academy of Sciences* 99:10 (May 2002): 3036–41.

60. Michael Barbaro, "Spray Vaccine for Flu Wins FDA Clearance," *Washington Post,* 18 June 2003, p. A1.
61. As quoted in Laura Ruth, "Targeting Vaccines at Host of Human Diseases," *Genetic Engineering News* 21:3 (1 February 2001): 19.
62. See Terrance Chea, "MedImmunes Pain-Free Ambitions; If Approved by the FDA, FluMist Would Become First Vaccine Delivered as a Nasal Spray," *Washington Post,* 18 March 2002, p. E1.
63. Hongran Fan et al., "Immunization via Hair Follicles by Topical Application of Naked DNA to Normal Skin," *Nature Biotechnology* 17:9 (September 1999): 870–72.
64. As quoted in Anne Eisenberg, "Vaccines May Soon Have Punch but No Ouch," *New York Times,* 7 September 1999, p. F7.
65. http://cbi.swmed.edu/
66. As quoted in Eisenberg, "Vaccines May Soon Have Punch but No Ouch."
67. As quoted in ibid.
68. As quoted in ibid.

Miniature Pharmaceutical Factories and Medicinal Milk

1. See "Cloning Genes," *The MIT Biology Hypertextbook,* at: http://esg-www.mit.edu:8001/esgbio/rdna/cloning.html, or for an animated demonstration of gene splicing see: http://www.nytimes.com/library/national/science/splicing/index.html
2. "Protein Expression Systems: Expression of Recombinant Proteins," at: http://www.clontech.com/proteomics/pdf/ProteinExpression.pdf
3. "Non-Animal Trypsin: ProdiGene Begins Commercial Scale-Up Recombinant Protein from Plants," *Gene Therapy Weekly,* 2 May 2002, p. 10.
4. "Deep Vein Thrombosis and Pulmonary Embolism," *MedicineNet.com,* at: http://www.medicinenet.com/Script/Main/Art.asp?li=MNI&ArticleKey=12597; "Thrombocytopenia," *adam.com,* at: http://www.adam.com/ency/article/000586.htm
5. Philip Leder and Patricia Thomas, "Making Drugs from DNA," *Harvard Health Letter* 19:9 (July 1994): 9.
6. http://www.gene.com/gene/products/information/cardiovascular/tnkase/index.jsp; "FDA Approves Single-Injection Heart Clot Drug," *Washington Post,* 4 June 2000, p. A10.
7. "NINDS Gaucher's Disease Information Page," National Institute of Neurological Disorders and Stroke, at: http://www.ninds.nih.gov/health_and_medical/disorders/gauchers_doc.htm

8. C. L. Cramer, J. G. Boothe and K. K. Oishi, "Transgenic Plants for Therapeutic Proteins: Linking Upstream and Downstream Strategies," *Current Topics in Microbiology and Immunology* 240 (1999): 95–118.
9. "Cell Lines Instead of Placenta to Treat Gaucher Disease," European Federation of Biotechnology, at: http://www.efbweb.org/biotec/genetic/menu1_2.htm
10. http://www.genzymetherapeutics.com/cerezyme/
11. Matthew J. Herper, "Genomics Won't Mean Cheap Drugs," *Forbes.com,* 26 February 2000, at: http://www.forbes.com/2001/02/26/0226drugs.html
12. Lauran Neergaard, "FDA Approves Genetically Engineered Drug for Gaucher's Disease," *Associated Press,* 24 May 1994; Gary Walsh, "Biopharmaceutical Benchmarks," *Nature Biotechnology* 18:8 (August 2000): 831.
13. "When Beginning Growth Hormone," The MAGIC Foundation for Children's Growth, at: http://www.magicfoundation.org/ghdfaq.html
14. K. Mulligan, V. W. Tai and M. Schambelan, "Use of Growth Hormone and Other Anabolic Agents in AIDS Wasting," *Journal of Parenteral and Enteral Nutrition* 23, Sup. 6 (November/December 1999): S202–9.
15. National Institute of Neurological Disorders and Stroke, "NINDS Creutzfeldt-Jakob Disease Information Page," at: http://www.ninds.nih.gov/health_and_medical/disorders/cjd.htm; E. A. Croes et al., "Creutzfeldt-Jakob Disease 38 Years after Diagnostic Use of Human Growth Hormone," *Journal of Neurology, Neurosurgery, and Psychiatry* 72:6 (June 2002): 92–93.
16. Gary Walsh, "Biopharmaceutical Benchmarks," *Nature Biotechnology* 18:8 (August 2000): 831, table 1.
17. http://www.gene.com
18. Philip Leder and Patricia Thomas, "Making Drugs from DNA," *Harvard Health Letter* 19:9 (July 1994): 9.
19. http://www.serono.com/; http://www.pfizer.com/
20. http://www.rebif.com/user_site/glossaryPopUp.cfm?pop=1&gloss_id=26&item_site=2
21. http://www.serono.com/products/sclerosis.jsp?major=1&minor=1
22. http://www.rebif.com/user_site/user_page1.cfm?item_id=6#dosing

23. See American Association of Blood Banks, "All About Blood," at: http://www.aabb.org/All_About_Blood/FAQs/aabb_faqs.htm
24. "American Red Cross Addresses the Human Form of Mad Cow Disease," at: http://www.redcross.org/services/biomed/blood/supply/tse.html; Rachel K. Sobel, "A Bloody Mess," *U.S. News & World Report,* 3 September 2001, p. 37.
25. Barnaby J. Feder, "Scalpal! Clamp!... Hemopure?" *New York Times,* 4 March 2001, sec. 3, p. 1.
26. Ibid.; also see K. C. Lowe, "Blood Substitutes: Present and Future," *British Journal of Hospital Medicine* 34:4 (October 1985): 195.
27. Toshiaki Sato Yomiuri, "Japanese Scientists Develop Effective Red Blood Cell Substitute," *Daily Yomiuri* (Tokyo), 24 April 2001, p. 12.
28. See Charles Bickers, "A New Pulse Prepares to Beat," *Far Eastern Review* 164:19 (17 May 2001): 34, at: http://www.feer.com/2001/0105_17/p034innov.html
29. http://www.croptech.com
30. Andrew Pollack, "New Ventures Aim to Put Farms in Vanguard of Drug Production," *New York Times,* 14 May 2000, p. 1.
31. A. Menter, "Pathogenesis and Genetics of Psoriasis," *Cutis* 61, Sup. 2 (February 1988): 8–10.
32. National Psoriasis Foundation website at: http://www.psoriasis.org/
33. "The Research Pipeline," *National Psoriasis Foundation Bulletin* 30:4 (July/August 1999): 1.
34. Personal telephone communication of 20 September 2000.
35. Andrew Pollack, "F.D.A. Approves Biotechnology Drug for Psoriasis," *New York Times,* 31 January 2003, p. C1.
36. http://www.biogen.com/site/content/products_pipeline.asp?product=AMEVIVE#content
37. Personal telephone communication of 29 October 2002.
38. "Early Results Suggest Alefacept and Light Therapy May Clear Psoriasis," *Genetic Engineering News* 22:21 (December 2002): 49; Press Release, "Early Results Suggest Alefacept plus Light Therapy May Clear Psoriasis," 22 November 2002, at: http://www.prnewswire.com/cgi-bin/stories.pl?ACCT=105&STORY=/www/story/11-22-2002/0001846747
39. http://www.enbrel.com/
40. See "Aspirin Benefits," Columbia University's Health Education Program, at: http://www.goaskalice.columbia.edu/0802.html
41. See Product Information, "Enbrel," at: http://www.enbrel.com/patient/html/patpi.htm; Press Release, "Enbrel (Etanercept) Is

First Therapy Approved for Treatment of Psoriatic Arthritis," 16 January 2002, at: http://www.embrel.com/news/psoriatic.jsp?fvar=1

42. Justin Gillis, "FDA Approves Wider Use of Popular Arthritis Drug," *Washington Post,* 17 January 2002, p. E5.
43. Robert V. House, "Preclinical Immunotoxicology Assessment of Cytokine Therapeutics," in *Biotechnology and Safety Assessment,* ed. John A. Thomas and Roy L. Fuchs, 3rd ed. (San Diego: Academic Press, 2002), pp. 191–231; "Ask a Scientist: Cytokines," at: http://www.newton.dep.anl.gov/askasci/bio99/bio99849.htm
44. M. Feldman, F. M. Brennan and R. N. Maini, "The Role of Cytokines in Rheumatoid Arthritis," *Annual Review of Immunology* 14 (1996): 397–440; A. Grom et al., "Patterns of Expression of Tumor Necrosis Factor Alpha, Tumor Necrosis Factor Beta, and Their Receptors in Synovia of Patients with Juvenile Rheumatoid Arthritis and Juvenile Spondyloarthropathy," *Arthritis and Rheumatology* 39:10 (October 1996): 703–10; T. Saxne et al., "Detection of Tumor Necrosis Factor Alpha but Not Tumor Necrosis Factor Beta in Rheumatoid Arthritis Synovial Fluid and Serum," *Arthritis and Rheumatology* 31:8 (August 1988): 1041–45.
45. Dick Ahlstrom, "College Signs Deal with Biotech Firm," *Irish Times,* 23 March 2001, p. 59.
46. http://www.enbrel.com/jra/hew/hew01.jsp?fvar=0)
47. Ahlstrom, "College Signs Deal with Biotech Firm."
48. Personal telephone communication with Roy Fleischmann, University of Texas Southwestern Medical Center at Dallas, 7 May 2002.
49. As quoted in Robert Langreth, "Chemical Chameleon," *Forbes* 166:12 (30 October 2000): 368, at: http://www.forbes.com/forbes/2000/1030/6612368a.html
50. Ibid.
51. Naomi Aoki, "Clogged Pipeline Manufacturing Shortage Putting a Squeeze on Surging Success of Protein-Based Drugs," *Boston Globe,* 30 January 2001, p. D1.
52. http://www.remicade.com
53. http://www.kineretrx.com/
54. Lauran Neergaard, "FDA Approves New Class of Therapy for Rheumatoid Arthritis," *Associated Press,* 14 November 2001.
55. http://www.abbott.com/ (The drug was co-developed by Cambridge Antibody Technology, at: http://www.cambridgeantibody.com/; http://www.humira.com/).

56. "FDA Approves Rheumatoid Arthritis Medication," *Associated Press,* 2 January 2002; Press Release, "Abbott Laboratories Receives FDA Approval Earlier than Anticipated for Humira (Adalimumab) for the Treatment of Rheumatoid Arthritis (RA)," 31 December 2002, at: http://abbott.com/news/press_release.cfm?id=524
57. http://www.hgsi.com/
58. Michael Barbaro; "Hepatitis C Treatment May Halve Injections; HGS Sees Trials as Promising," *Washington Post,* 5 November 2002, p. E5.
59. Rutgers University, "Lupus Statistics," at: http://scils.rutgers.edu/~mmbuck/Statistics.htm; for more information on this and lupus in general, see the National Institute of Arthritis and Muskoskeletal Diseases, "HealthInfo," at: http://www.nih.gov/niams/healthinfo/lupusguide/chp1.htm#chp1_dd
60. Terence Chea, "Biotech Firm Links Protein to 2 Diseases; Finding May Lead to New Treatments," *Washington Post,* 30 October 2000, p. E1.
61. http://www.cambridgeantibody.co.uk/home.html
62. Press Release, "Human Genome Sciences to Initiate Human Clinical Trials of BLYS," Human Genome Sciences, 23 June 2000, at: http://www.hgsi.com/news/press/00-06-23_BLySIND.html; http://www.dow.com/
63. Chea, "Biotech Firm Links Protein to 2 Diseases"; Press Release, "Human Genome Sciences and Dow Agree to Develop HGS Radiolabeled B-Lymphocyte Stimulator," 30 October 2000, at: http://www.dow.com/dow_news/prodbus/20001030b_pb.html
64. As quoted in Calvin Coolidge, *The Autobiography of Calvin Coolidge* (New York: J. J. Little & Ives Co., 1929), p. 190.
65. Thomas J. Simpson, "Polyketide Biosynthesis," *Chemistry and Industry* 95:11 (5 June 1995): 4, at: http://biotec.mond.org/9511/951104.html
66. Blaine A. Pfeifer et al., "Biosynthesis of Complex Polyketides in a Metabolically Engineered Strain of *E. coli*," *Science* 291:5509 (2 March 2001): 1790–92.
67. As quoted in Tom Spears, "Killer *E. Coli* Has Its Good Side: Genetic Fiddling Produces Antibiotic That Kills Other Bacteria," *Ottawa Citizen,* 2 March 2001, p. A10.
68. Cathy Pollard-Colombo, "Emerging Multidrug-Resistant Organisms," *Physician Assistant* 23:2 (1 February 1999): 83.

69. Harold C. Neu, "The Crisis in Antibiotic Resistance," *Science* 257:5073 (21 August 1992): 1064; Stuart B. Levy, "Antimicrobial Resistance: Bacteria on the Defense—Resistance Stems from Misguided Efforts to Try to Sterilize Our Environment," *British Medical Journal* 317:7159 (5 September 1998): 612; Claude Carbon and Richard P. Bax, "Regulating the Use of Antibiotics in the Community," *British Medical Journal* 317:7159 (5 September 1998): 663.
70. World Health Organization, *Overcoming Antimicrobial Resistance* (Geneva, Switz.: World Health Organization, June 2000), p. 45; for example, see Laurie Garrett, *The Coming Plague: Newly Emerging Diseases in a World out of Balance* (New York: Penguin, 1995); Arno Karlen, *Man and Microbes: Disease and Plagues in History and Modern Times* (New York: Touchstone, 1995).
71. The comprehensive microbial resource content page with a full up-to-date listing is at: http://www.tigr.org/tigr-scripts/CMR2/CMR_DB_Check2.spl
72. Christian Drosten et al., "Identification of a Novel Coronavirus in Patients with Severe Acute Respiratory Syndrome," *New England Journal of Medicine* 348:20 (15 May 2003): 1967–76; "How Global Effort Found SARS Virus in a Matter of Weeks," *Wall Street Journal,* 16 April 2003, p. A1.
73. M. Barlow and Barry G. Hall, "Predicting Evolutionary Potential; In Vitro Evolution Accurately Reproduces Natural Evolution of the Tem Beta-lactamase," *Genetics* 160:3 (March 2002): 823–32; Alicia Chang, "Study Finds Laboratory Evolution Mimics Natural Evolution," *Associated Press,* 19 March 2002.
74. Matt Ridley, *Genome: The Autobiography of a Species in 23 Chapters* (New York: HarperCollins, 1999), pp. 46–47.
75. For more information, see Kathryn Brown, "Biotech Speeds Its Evolution," *Technology Review* 103:6 (November/December 2000): 84; Ellen Licking, "Evolution on Fast-Forward," *Business Week,* 27 September 1999, p. 140.
76. David N. Leff, "Teensiest Critters Dangle Greatest Boon: Bacteriophage Parasites Copy Penicillin's Bacteria Wipeout, Hint at New Antibiotics Class," *BioWorld Today* 12:122 (25 June 2001).
77. As quoted in Paroma Basu, "The Next Phage," *Technology Review Emerging Technologies,* 17 July 2001.
78. As quoted in Press Release, "Virus Found to Carry Antibiotic Against *E. Coli,*" Texas A&M University, 21 June 2001.
79. Ibid.

80. As quoted in Helen Pearson, "Anthrax Exposed and Killed: New Antibiotic Seeks and Destroys Bioterror Agent," *Nature ScienceUpdate*, 22 August 2002.
81. Ibid.; Nicole Johnston, "Enzyme Detects and Kills Anthrax," *Biomedcentral.com Scientist*, 22 August 2002, http://www.biomedcentral.com/news/20020822/04
82. John Heilprin, "Final Anthrax Cleanup Begins at Senate Office Building with Insertion of Poisonous Chlorine Dioxide Gas," *Associated Press*, 30 November 2001.
83. Pearson, "Anthrax Exposed and Killed: New Antibiotic Seeks and Destroys Bioterror Agent"; Johnston, "Enzyme Detects and Kills Anthrax."
84. For more on this condition, see "meningococcal septicemia," Meningitis Foundation of America," at: http://www.musa.org/meningococcal_septicemia.htm
85. http://www.txchildrens.org/
86. As quoted in Cindy Horswell, "Experimental Drug Credited with Saving Teen," *Houston Chronicle*, 23 January 2001, p. A13.
87. Gordon R. Bernard et al., "Efficacy and Safety of Recombinant Human Activated Protein C for Severe Sepsis," *New England Journal of Medicine* 344:10 (8 March 2001): 699–709.
88. Michael Matthay, "Severe Sepsis—A New Treatment with Both Anticoagulant and Antiinflammatory Properties," *New England Journal of Medicine* 344:10 (8 March 2001): 759–62.
89. Jeff Swiatek, "Lilly Expects Drug to Have Major Impact: Trials of Zovant Show That It Could Be a Landmark Treatment for a Killer Infection," *Indianapolis Star*, 10 February 2001, p. A1.
90. As quoted in Michael Lasalandra, "New Drug Better than Antibiotics in Treating Fatal Blood Infection" *Boston Herald*, 10 February 2001, p. 9; the website for Massachusetts General Hospital is: http://www.mgh.harvard.edu/
91. R. C. Bone, C. J. Grodzin and R. A. Balk, "Sepsis: A New Hypothesis for Pathogenesis of the Disease Process," *Chest* 112:1 (July 1997): 235–43.
92. D. C. Angus et al., "E5 Murine Monoclonal Antiendotoxin Antibody in Gram-Negative Sepsis: A Randomized Controlled Trial," *Journal of the American Medical Association* 283:13 (5 April 2000): 1723–30; M. S. Rangel-Frausto et al., "The Natural History of the Systemic Inflammatory Response Syndrome (SIRS): A Prospective Study," *Journal of the American Medical Association* 273:2 (11 January 1995): 117–23; D. Annane et al., "A 3-Level Prognostic

Classification in Septic Shock Based on Cortisol Levels and Cortisol Response to Corticotropin," *Journal of the American Medical Association* 283:8 (23 February 2000): 1038–45; "From the Centers for Disease Control. Increase in National Hospital Discharge Survey Rates for Septicemia—United States, 1979–1987," *Journal of the American Medical Association* 263:7 (16 February 1990): 937–38.

93. Walter T. Linde-Zwirble et al., "Epidemiology of Severe Sepsis in the United States: Analysis of Incidence, Outcome and Associated Costs of Care," *Critical Care Medicine* 29:7 (July 2001): 1303–10.
94. Jeff Swiatek, "Lilly Expects Drug to Have Major Impact: Trials of Zovant Show That It Could Be a Landmark Treatment for a Killer Infection," *Indianapolis Star,* 10 February 2001, p. A1.
95. For more on these products, see Lilly's page at: http://www.lillydiabetes.com/Products/default.cfm
96. As quoted in Michael Lasalandra, "New Drug Better than Antibiotics in Treating Fatal Blood Infection," *Boston Herald,* 10 February 2001, p. 9.
97. François Fourrier et al., "Septic Shock, Multiple Organ Failure, and Disseminated Intravascular Coagulation: Compared Patterns of Antithrombin III, Protein C, and Protein S Deficiencies," *Chest* 101:3 (March 1992): 816–23; J. A. Lorente et al., "Time Course of Hemostatic Abnormalities in Sepsis and Its Relation to Outcome," *Chest* 103:5 (May 1993): 1536–42; Joachim Boldt et al., "Changes of the Hemostatic Network in Critically Ill Patients—Is There a Difference between Sepsis, Trauma, and Neurosurgery Patients?" *Critical Care Medicine* 28:2 (February 2000): 445–50; D. Powars et al., "Epidemic Meningococcemia and Purpura Fulminans with Induced Protein C Deficiency," *Clinical Infectious Diseases* 17:2 (August 1993): 254–61.
98. For more on this cell line, see "Recombinant Protein Synthesis from HEK-293 Cells, Grown in Spinners at Small Scale upon Transient Transfection with Calcium-Phosphate DNA Coprecipitates," at: http://dcwww.epfl.ch/igc/cbue/page39.html
99. E-mail communication from Michael Bigelow, Eli Lilly & Co., 26 February 2001.
100. Tao Cheng et al., "Activated Protein C Blocks p53-mediated Apoptosis in Ischemic Human Brain Endothelium and Is Neuroprotective," *Nature Medicine* 9:3 (March 2003): 338–42; "Sepsis Drug May Also Help Prevent Stroke Damage, Researchers Report," *Associated Press,* 3 February 2002.
101. http://www.amgen.com/

102. Joseph L. Galloway, "An 'Insulin' for Anemia Sufferers," *U.S. News & World Report,* 12 June 1989, p. 13.
103. Paul Jacobs, "Cutting Edge / Frontiers: Four Fields That Have Been Shaped By, and Are Shaping, Southern California," *Los Angeles Times Magazine,* 25 July 1999, p. 28.
104. As quoted in Eric Johnson (*Yorba Linda Star*), "Man without Kidneys Is All Heart," *Orange County Register,* 24 June 1999, p. 1.
105. Ibid.
106. Jesse Hiestand, "Amgens Anti-Anemia Drug Gets Boost," *Los Angeles Daily News,* 20 March 2001, p. B1; Press Release, "Aranesp and Calcimimetics Clinical Data Presented at American Society of Nephrology Meeting," 15 October 2000, at: http://www.amgen.com/News/news00/pressRelease001015.html; "Weekly Dose of Amgens Aranesp Works for Anemia—Study," *Reuters,* 15 October 2000; Arlene Weintraub and David Shook, "A Blood Booster for Amgen?" *Business Week,* 4 December 2000, p. 104, at: http://www.businessweek.com/2000/00_49/b3710154.htm; Press Release, "First Efficacy Data from Cancer Trials of Amgen's Aranesp (Darbepoetin Alfa) Presented at ESMO," 16 October 2000, at: http://www.amgen.com/News/news00/pressRelease001016b.html
107. For the most detailed description of the late Miss Dolly's making and the ramifications thereof, see Ian Wilmut, Keith Campbell and Colin Tudge, *Dolly and the Age of Biological Control* (New York: Farrar, Straus & Giroux, 2000).
108. http://www.ppl-therapeutics.com/; http://www.ri.bbsrc.ac.uk/
109. For more information see Pseudomonas Genome Project, Cystic Fibrosis, at: http://www.pseudomonas.com/cystic_fibrosis.html; see National Heart, Lung and Blood Institute, NIH Publication No. 95-2020, 1995, at: http://www.nhlbi.nih.gov/health/public/lung/other/copd/copd_toc.htm; see Faculty of Medicine, Chiang Mai University, Chiang Mai, Thailand, "Congenital Lobar Emphysema," at: http://www.medicine.cmu.ac.th/dept/radiology/pedrad/lobarem.html
110. Marjorie Miller, "For Cloning Pioneer, Biotechnology Holds Promise of Medical Gains," *Los Angeles Times,* 9 April 2000, p. A8; Maria Burke, "Making Transgenic Animals for Human Proteins," *Bioventure View* 14:12 (1 December 1999): 5.
111. Burke, "Making Transgenic Animals for Human Proteins."

112. Genzyme Transgenics, "How Transgenics Works," at: http://www.transgenics.com/science/howitworks.html; Burke, "Making Transgenic Animals for Human Proteins."
113. Burke, "Making Transgenic Animals for Human Proteins."
114. http://www.agresearch.cri.nz/
115. Sarah L. Minden and Debra Frankel, "Plaintalk: A Booklet about MS for Families," National Multiple Sclerosis Society, at: http://www.nationalmssociety.org/Brochures-Plaintalk.asp
116. Kim Griggs, "'Human Cow Milk an MS Aid?" *Wired,* 2 August 2000, at: http://www.wired.com/news/print/0,1294,37921,00.html
117. http://www.pharming.com
118. Press Release, "Pharming Opens World's First Pharmaceutical Cattle Farm in Vienna, Wisconsin: State-of-the-Art Facility Will House Herd of Transgenic Female Calves," 25 October 1999, at: http://www.pharming.com/News/press_1999_25_10_TXT.html
119. XTL Biopharmaceuticals at: http://www.xtlbio.com/home.html; Press Release, "XTL Biopharmaceuticals and Pharming Group to Co-develop Lactoferrin to Treat Hepatitis C," at: http://www.pharming.com/News/press_2000_14_11_TXT.html
120. Burke, "Making Transgenic Animals for Human Proteins."
121. http://www.transgenics.com/science.html
122. "Datamonitor Inc. Says Therapeutic Protein Demand Likely to Outstrip Manufacturing Capacity," *Med Ad News* 20:10 (1 October 2001): 10.
123. "GTC Products in Development—Human Serum Albumin," Genzyme Transgenics, at: http://www.transgenics.com/products/humanserum.html
124. Burke, "Making Transgenic Animals for Human Proteins."
125. http://www.nexiabiotech.com/; Amanda Jelowicki, "Triplet Goats Cloned," *Montreal Gazette,* 27 April 1999, p. A1.
126. As quoted in Christopher Helman, "Charlotte's Goat," *Forbes* 167:4 (19 February 2001): 201, at: http://www.forbes.com/asap/2001/0219/101.html
127. Ibid.
128. E. W. Kieckhefer, "Goats with Spider Gene Produce Webs," *United Press International,* 18 May 2000.
129. David Brown, "U.S. Troops Injuries in Iraq Showed Body Armor's Value," *Washington Post,* 4 May 2003, p. A28.

130. "BioSteel Extreme Performance Fibers: Spider Silk Protein as a Biomaterial," at: http://www.nexiabiotech.com/biosteel.html
131. Anthoula Lazaris et al., "Spider Silk Fibers Spun from Soluble Recombinant Silk Produced in Mammalian Cells," *Science* 295:5554 (18 January 2002): 472–76; "Scientists Produce Spider Silk Protein Using Mammal Cells," *Associated Press,* 18 January 2002, p. A2; Press Release, "Nexia and U.S. Army Spin the World's First Man-Made Spider Silk Performance Fibers," at, http://www.eurekalert.org/pub_releases/2002-01/nbi-nau011102.php; see also Lawrence Osborne, "Got Silk," *New York Times Magazine,* 16 June 2002.
132. Michael K. Dyck et al., "Seminal Vesicle Production and Secretion of Growth Hormone into Seminal Fluid," *Nature Biotechnology* 17:11 (November 1999): 1087–90.
133. "Chickens Become Feathered Pharmaceutical Producers," *United Press International,* 3 December 2000, at: http://www.enn.com/news/wire-stories/2000/12/12032000/upi_chickens_40559.asp
134. Personal e-mail communication of 4 March 2001; Andrew Beaven, "Britney, the Genetically Modified Chicken," *Mail on Sunday* (London), 3 December 2000, p. 15.
135. http://www.viragen.com/; Auslan Cramb, "Altered Eggs Grow Cancer-Fighting Protein," *Chicago Sun-Times,* 4 December 2000, p. 27.
136. Andy Coghlan, "Big Breakfast," *New Scientist* 164:2212 (13 November 1999): 25, at: http://www.newscientist.com/ns/19991113/itnstory19991135.html
137. http://www.avigenics.com/
138. See AviGenics, "A Superior Production System," at: http://avigenics.com/protprod.html
139. As quoted in "Chickens Become Feathered Pharmaceutical Producers"; see also Alex J. Harvey et al., "Expression of Exogenous Protein in the Egg White of Transgenic Chickens," *Nature Biotechnology* 19:5 (April 2002): 396–99.
140. John Ingam, "Now for the GM Chicken Egg That Will Fight Cancer," *Express* (London), 11 November 1999.
141. Andrew Clark, "Focus: TranXenoGen," *Guardian* (London), 20 June 2000, p. 26; "TranXenoGen Puts Its Eggs in One Basket," *Boston Business Journal* 20:12 (28 April 2000): 1.
142. http://www.ciwf.co.uk/

143. As quoted in Ingam, "Now for the GM Chicken Egg That Will Fight Cancer."
144. Ibid.
145. See "The Flu and Flu Vaccine" at: http://www.909shot.com/flu-fax.htm

"Plantibodies"

1. Paul Raeburn, "Sabin and Salk Were Linked by Achievements—and a Bitter Feud," *Associated Press,* 3 March 1993.
2. Glynis Giddings et al., "Transgenic Plants as Factories for Biopharmaceuticals," *Nature Biotechnology* 18:11 (November 2000): 1151.
3. Ibid.
4. His personal website is: http://lsvl.la.asu.edu/plantbiology/faculty/arntzen.htm
5. B. Aylward et al., "Reducing the Risk of Unsafe Injections in Immunization Programmes: Financial and Operational Implications of Various Injection Technologies," *Bulletin of the World Health Organization* 73:4 (July 1995): 531.
6. John Donnelly, "Immunizations Plummet in Poorest Nations; Wars, Funding Cuts Blamed for Decline," *Boston Globe,* 13 November 2000, p. A1.
7. A. Kane et al., "Transmission of Hepatitis B, Hepatitis C and Human Immunodeficiency Viruses through Unsafe Injections in the Developing World: Model-Based Regional Estimates," *Bulletin of the World Health Organization* 77:10 (1 October 1999): 801; see also Aylward et al., "Reducing the Risk of Unsafe Injections in Immunization Programmes."
8. William H. R. Langridge, "Edible Vaccines," *Scientific American* 283:3 (September 2000): 66.
9. As quoted in Robert Lee Hotz, "Vaccines in Your Vegetables; Genetic Gardeners Are Trying to Grow Crops That Could Save Millions of Lives," *Los Angeles Times,* 23 November 1993, p. A1.
10. T. A. Haq et al., "Oral Immunization with a Recombinant Bacterial Antigen Produced in Transgenic Plants," *Science* 268:5211 (5 May 1995): 714–16.
11. C. O Tacket et al., "Immunogenicity in Humans of a Recombinant Bacterial Antigen Delivered in a Transgenic Potato," *Nature Medicine* 4:5 (May 1998): 607–9.
12. As quoted in M. B., "Edible Vaccines, in Raw Potatoes, Pass First Test in Humans," *Biotechnology Newswatch,* 4 May 1998, p. 1.

13. Tacket et al., "Immunogenicity in Humans of a Recombinant Bacterial Antigen Delivered in a Transgenic Potato."
14. For more, see American Chemical Society, Chemical Abstracts Service, "Mucous Membrane," at: http://www.cas.org/vocabulary/07899.html
15. Carol Featherstone, "M Cells: Portals to the Mucosal Immune System," *Lancet* 350:9086 (25 October 1997): 1230.
16. A. M. Walmsley and C. J. Arntzen, "Plants for Delivery of Edible Vaccines," *Current Opinion in Biotechnology* 11:2 (April 2000): 126–29.
17. Featherstone, "M Cells: Portals to the Mucosal Immune System."
18. http://www.llu.edu/llumc/; T. Arakawa, D. K. X. Chong and W. H. R. Landgridge, "Efficacy of a Food Plant-Based Oral Cholera Toxin B Subunit Vaccine," *Nature Biotechnology* 16:3 (March 1998): 292–97.
19. Paul Brown, "Vaccine in GM Fruit Could Wipe out Hepatitis B," *Guardian* (London), 8 September 2000.
20. As quoted in ibid.
21. Personal e-mail communication from David Wheat, the Bowditch Group, 7 February 2001. The companies are: AgriVax Inc., Los Angeles, California; Applied Phytologics Inc. (now Ventria Bioscience), Sacramento, California; Biolex Inc., Pittsboro, North Carolina; CropTech Development Corporation, Blacksburg, Virginia; EPIcyte Pharmaceutical Inc., San Diego, California; Integrated Protein Technologies (unit of Monsanto), St. Louis, Missouri; Large Scale Biology Corporation, Vacaville, California; Maltagen Forschung GmbH, Andernach, Germany; Medicago Inc., Sante-Foy, Quebec, Canada; Méristem Therapeutics, Clermont-Ferrand, France; MPB Cologne GmbH, Cologne, Germany; Phytomedics Inc., Dayton, New Jersey; Planet Biotechnology Inc., Mountain View, California; ProdiGene Inc., College Station, Texas; SemBioSys Genetics Inc., Calgary, Alberta, Canada; StaufferBiotech, Aurora, Nebraska; SunGene GmbH & Co. KgaA, Getersleben, Germany; ToBio LLC, Blacksburg, Virginia.
22. Hannah Cleaver, "Carrots Modified to Contain Hepatitis B Vaccine," *Reuters Health*, 10 May 2002.
23. David McNamee, "Transgenic Potatoes Produce Oral HBV Vaccine," *Lancet* 354:9191 (13 November 1999): 1707; Takeshi Arakawa et al., "A Plant-Based Cholera Toxin B Subunit-Insulin Fusion Protein Protects against the Development of Autoimmune Diabetes," *Nature Biotechnology* 16:10 (October 1998): 934–38; James E. Carter and William H. R. Langridge, "Plant-Based Vaccines for

Protection against Infectious and Autoimmune Diseases," *Critical Reviews in Plant Sciences* 21:2 (2002): 93–109; William H. R. Langridge, "Edible Vaccines," *Scientific American* 283:3 (September 2000): 66; T. Arakawa, D. K. Chong and W. H. Langridge, "Efficacy of a Food Plant-Based Oral Cholera Toxin B Subunit Vaccine," *Nature Biotechnology* 16:3 (March 1998): 292–97.

24. C. O. Tacket et al., "Human Immune Responses to a Novel Norwalk Virus Vaccine Delivered in Transgenic Potatoes," *Journal of Infectious Diseases* 182:1 (July 2000): 302–5.
25. Paul S. Mead et al., "Food-Related Illness and Death in the United States," *Emerging Infectious Diseases* 5:5 (September/October 1999): 607–25, table 2.
26. Kevin Bonham, "Biotechnology May Turn Farms into 'Farmaceutical' Delivery Systems," *Agweek*, 12 April 1999.
27. Robert Lee Hotz; "Vaccines in Your Vegetables—Genetic Gardeners Are Trying to Grow Crops That Could Save Millions of Lives," *Los Angeles Times*, 23 November 1993, p. A1.
28. "Hepatitis Antibody Made from GM Rice," *Daily Yomiuri* (Tokyo), 11 November 2000.
29. T. S. Mor, M. A. Gómez-Lim and K. E. Palmer, "Perspective: Edible Vaccines—A Concept Coming of Age," *Trends in Microbiology* 6:11 (November 1988): 449–53; A. Modelska, "Immunization against Rabies with Plant-Derived Antigen," *Proceedings of the National Academy of Sciences USA* 95:5 (3 March 1998): 2481–85; Bonham, "Biotechnology May Turn Farms into 'Farmaceutical' Delivery Systems."
30. "Bangalore Scientists Make Fruit-Based Rabies Vaccine," *Times of India Online*, 13 November 2000, at: http://www.timesofindia.com/today/13hlth1.hSm
31. J. S. Sandhu et al., "Oral Immunization of Mice with Transgenic Tomato Fruit Expressing Respiratory Syncytial Virus-f Protein Induces a Systemic Immune Response," *Transgenic Research* 9:2 (April 2000): 127–35.
32. Larry Zeitlin et al., "A Humanized Monoclonal Antibody Produced in Transgenic Plants for Immunoprotection of the Vagina against Genital Herpes," *Nature Biotechnology* 16:13 (December 1998): 1361–64; Paul Ellas, " 'Pharming' Patent Granted, Boosting Herpes-Fighting Corn Prospects," *Associated Press*, 9 July 2002.
33. D. E. Webster et al., "Successful Boosting of a DNA Measles Immunization with an Oral Plant-Derived Measles Virus Vaccine," *Journal of Virology* 76:15 (August 2002): 7910–12; Kendal Powell,

"Eat Up Your Vaccine," *Nature ScienceUpdate*, 22 July 2002, at; http://www.nature.com/nsu/020715/020715-16.html; Oscar J. M. Goddinj and Jan Pen, "Plants as Bioreactors," *Trends in Biotechnology* 13:9 (September 1995): 379–87.

34. http://www.prodigene.com/; http://www.tamu.edu/
35. Kristen Philipkoski, "Scientists Stalk AIDS Vaccine," *Wired News*, 30 November 2000, at: http://www.wired.com/news/print/0,1294,40416,00.html
36. See Michael Fumento, "Protein Power," *American Outlook* 5:4 (Fall 2002): 58.
37. Eli Lilly & Co., "Recombinant Enzymes," at: http://www.lilly.com/about/partnering/worldtrade/biotech/enzymes.html
38. Press Release, "Prodigene Launches First Large Scale-up Manufacturing of Recombinant Protein from Plant System," 13 February 2002, at http://www.prodigene.com/news_releases/02-02-24_trypsin.html
39. Novo Nordisk, "Novo History: Not Just Insulin," at: http://www.enzkey.com/press/history/7.asp
40. Press Release, "Prodigene Launches First Large Scale-up Manufacturing of Recombinant Protein from Plant System."
41. "Prodigene Begins Scale-up of Aprotinin from Plants," *Biotech Week*, 28 August 2002, p. 7.
42. Ilya Raskin et al., "Production of Recombinant Proteins in Tobacco Guttation Fluid," *Plant Physiology* 124:3 (November 2000): 927–34; "Recombinant Protein Secretion Technology (REPOST)" at: http://www.phytomedics.com/repo.html; Nell Boyce, "Drugs on Tap from Morning Dew," *New Scientist* 168:2266 (25 November 2000): 22, at: http://www.newscientist.com/news/news.jsp?id=ns226634/
43. As quoted in Boyce, "Drugs on Tap from Morning Dew."
44. Liz Fletcher, "Power to the Pea Plant," *New Scientist* 167:2245 (1 July 2000): 26.
45. http://www.phytomedics.com/
46. As quoted in Fletcher, "Power to the Pea Plant."
47. http://www.epicyte.com/home.html
48. Press Release, "Epicyte Receives Two SBIR Grants to Fund HIV and HSV Antibody Research," *BW HealthWire*, 2 October 2000, at: http://ipn.intelihealth.com/IPN/ihtIPN/WSIHW000/23883/7194/299234.html; personal telephone communication with Kris Briggs, Epicyte Pharmaceutical Inc., 29 August 2000.
49. Personal telephone communication of 29 August 2000.

50. Institute of Medicine, "Prospects for Immunizing against Respiratory Syncytial Virus" (Appendix N), in *Institute of Medicine: New Vaccine Development: Establishing Priorities,* vol. 2, *Disease Importance in Developing Countries* (Washington, D.C.: National Academy Press, 1986): 299–307.
51. "Respiratory Syncytial Virus," in *Red Book: Report of the Committee on Infectious Diseases,* ed. Georges Peter, 24th ed. (Elk Grove Village, Ill.: American Academy of Pediatrics, 1997): 443-44.
52. H. C. Meissner et al. "Immunoprophylaxis with Palivizumab, a Humanized Respiratory Syncytial Virus Monoclonal Antibody, for Prevention of Respiratory Syncytial Virus Infection in High Risk Infants: A Consensus Opinion," *Pediatric Infectious Disease Journal* 18:3 (March 1999): 223–31.
53. CDC, "Update: Respiratory Syncytial Virus Activity—United States, 1997–98 Season," *Morbidity and Mortality Weekly Report* 46:49 (12 December 1997): 1163–65.
54. D. N. Gerding et al., "Clostridium Difficile–Associated Diarrhea and Colitis," *Infection Control and Hospital Epidemiology* 16:8 (August 1995): 456–77; H. S. Gold and R. C. Moellering, "Antimicrobial-Drug Resistance," *New England Journal of Medicine* 335:19 (7 November 1996): 1445–53.
55. Personal telephone communication with Kris Briggs, 29 August 2000.
56. Press Release, "Ultra-High-Value BioPharming Is Becoming a Commercial Reality," at: http://www.staufferseeds.com/0703ultra.htm; Joan Olson, "Complete Feed from the Field," *Farm Industry News,* April 1999.
57. As quoted in Olson, "Complete Feed from the Field."
58. Press Release, "Ultra-High-Value BioPharming Is Becoming a Commercial Reality"; Olson, "Complete Feed from the Field."
59. http://www.dekalb.com/index.html
60. Press Release, "Dekalb Genetics/Monsanto Acquires Worldwide License for University of Connecticuts Interferon Technology," *PRNewswire,* 18 January 2000.
61. As quoted in Olson, "Complete Feed from the Field."
62. Janice M. Reichert, "Monoclonal Antibodies in the Clinic," *Nature Biotechnology* 19:9 (September 2001): 819.
63. Pharmaceutical Research and Manufacturers of America, "371 Biotechnology Medicines in Testing Promise to Bolster the Arsenal against Disease," at: http://www.phrma.org/newmedicines/resources/2002-10-21.93.pdf

64. http://www.elan.com; Peter Mitchell, "First Biotech Drug to Treat Psoriasis," *Nature Biotechnology* 20:7 (July 2002): 640; http://www.biogen.com/; "FDA Approves Treatment for Crohn's," *Biotechnology Newswatch,* 7 September 1998, p. 2.
65. See MedNets at: http://www.mednets.com/crohns.htm and the Crohn's and Colitis Foundation of America Questions and Answers page at: http://www.ccfa.org/Physician/crohnsb.html
66. David H. Miller et al., "A Controlled Trial of Natalizumab for Relapsing Multiple Sclerosis," *New England Journal of Medicine* 348:1 (2 January 2003): 15–23; Rob Stein, "New Drug Reduces Effects of MS, Crohn's: Immune System Targeting Improved," *Washington Post,* 2 January 2003, p. A1.
67. Robert Steyer, "Magic Bullet: Slow Development of a Versatile Medical Weapon," *St. Louis Post-Dispatch,* 20 February 1989, p. 12.
68. Harold M. Schmeck Jr., "Three Immunology Investigators Win Nobel Prize in Medicine," *New York Times,* 16 October 1984, p. A1; John Noble Wilford, "Nobel Winners: Three Quiet Men Conquering Hidden Mysteries," *New York Times,* 16 October 1984, p. C2.
69. For an extremely detailed description of the process, and some nifty drawings of mice, see John W. Kimball, "Monoclonal Antibodies," Kimball's Biology Pages, at: http://www.ultranet.com/~jkimball/BiologyPages/M/Monoclonals.html
70. Laura Beil, "Scientists Proposing New Paths to Disease Protection," *Dallas Morning News,* 3 January 1999.
71. Larry Zeitlin et al., "A Humanized Monoclonal Antibody Produced in Transgenic Plants for Immunoprotection of the Vagina against Genital Herpes," *Nature Biotechnology* 16:13 (December 1998): 1361–64; "Antibodies: Part One," *Medical & Healthcare Marketplace Guide* 1 (Philadelphia: Dorland Healthcare Information, 1998), sec. I-234.
72. http://www.biosite.com/discovery/tptfactsheet.html; http://www.genmab.com/html/intro_to_abs.html; Gary Stix, "The Mice That Warred," *Scientific American* 284:6 (June 2001): 34; Lisa M. Krieger, "Mice Injected with Genes Become Tiny Pharmaceutical Factories," *San Jose Mercury News,* 17 August 2000.
73. "Antibodies: Part One," *Medical & Healthcare Marketplace Guide* 1, sec. I-234.
74. "Antibodies: Part One," *Medical & Healthcare Marketplace Guide* 1, sec. I-234; Eugene L. Coodley, "Coronary Heart Disease: Diagnostic Techniques for Evaluating Myocardial Infarction Risk," *Consultant* 36:12 (December 1996): 2618; Susan H. Steiner, "Pharmacists

Continuing Education: Pregnancy and Ovulation: Home Diagnosis," *Drug Store News,* 23 November 1998, p. CP12; T. J. Pelkey, H. F. Frierson Jr. and D. E. Bruns, "Molecular and Immunological Detection of Circulating Tumor Cells and Micrometastases from Solid Tumors," *Clinical Chemistry* 42:9 (September 1996): 1369–81; Annika Lindblom and Annelie Liljegren, "Tumor Markers in Malignancies; Clinical Review," *British Medical Journal* 7232:320 (12 February 2000): 424.

75. Lawrence K. Altman, "A Discovery and Its Impact: Nine Years of Excitement," *New York Times,* 16 October 1984, p. C3.
76. http://www.jnj.com/who_is_jnj/products/orthoclone.htm; Christine Russell, "Revolutionary Biotech Therapy Gets Approval," *Washington Post,* 20 June 1986, p. A1.
77. U.S. Department of the Census, *Statistical Abstract of the United States 2001,* p. 99, table 105, at: http://www.census.gov/prod/2002pubs/01statab/vitstat.pdf
78. J. C. Seidell, "Obesity, Insulin Resistance and Diabetes—a Worldwide Epidemic," *British Journal of Nutrition,* 83, Sup. 1 (March 2000): S5–S8; H. E. Lebovitz, "Type 2 Diabetes: An Overview," *Clinical Chemistry* 45:8, part 2 (August 1999): 1339–45.
79. See: http://www.diabetes.org/ada/lifestyle.asp
80. American Diabetes Association, "The Dangerous Toll of Diabetes," at: http://www.diabetes.org/ada/facts.asp#toll
81. E. Larger and D. Dubois Laforgue, "Type 1 Diabetes Mellitus," *Annales De Medecine Interne* 150:3 (April 1999): 254–63; a vast amount of information that's user-friendly to diabetics and their loved ones can be found at the American Diabetes Association website at: http://www.diabetes.org
82. See O. J. Hines and H. A. Reber, "Median Pancreatectomy: Do the Risks Justify the Effort?" *Journal of the American College of Surgery* 190:6 (9 June 2000): 715–16.
83. B. J. Hering et al., eds., *International Islet Transplant Registry Newsletter Number* 7 (Giessen, Germany: Justus-Liebig University of Giessen, December 1996); Elena Portyansky, "New Monoclonal Antibody Aims to Thwart Organ Rejection; Hoffmann–La Roches Zenapax, Daclizumab," *Drug Topics* 2:142 (19 January 1998): 43.
84. A. M. Shapiro et al., "Islet Transplantation in Seven Patients with Type 1 Diabetes Mellitus Using a Glucocorticoid-Free Immunosuppressive Regimen," *New England Journal of Medicine* 343:4 (27 July 2000): 230–38; see also R. Paul Robertson, "Successful Islet

Transplantation for Patients with Diabetes—Fact or Fantasy?" *New England Journal of Medicine* 343:4 (27 July 2000): 289.

85. As quoted in Susan Okie, "Trying Life without Insulin Shots," *Washington Post,* 31 July 2000, p. A1.
86. Shapiro et al., "Islet Transplantation in Seven Patients with Type I Diabetes Mellitus Using a Glucocorticoid-free Immunosuppressive Regimen."
87. As quoted in Susan Okie, "Trying Life without Insulin Shots."
88. V. K. Ramiya, "Reversal of Insulin-Dependent Diabetes Using Islets Generated in Vitro from Pancreatic Stem Cells," *Nature Medicine* 6:3 (March 2000): 278–82; for an update see, Ammon B. Peck et al., "Use of in Vitro–Generated, Stem Cell–Derived Islets to Cure Type I Diabetes: How Close Are We?" *Annals of the New York Academy of Sciences* 958 (May 2002): 59–68.
89. As quoted in Ronald Kotulak, "Cell Transplants Give Diabetics Shot of Hope for Cure; All Type I Patients in Canadian Study Stop Taking Insulin," *Chicago Tribune,* 2 July 2000, p. 12.
90. K. C. Herold et al., "Anti-CD3 Monoclonal Antibody in New-Onset Type I Diabetes Mellitus," *New England Journal of Medicine* 346:22 (30 May 2002): 1740–42; Marilyn Chase, "Drug Holds Hope for Diabetics—Experiment Uses Antibody to Disarm Cells That Assault Pancreas," *Wall Street Journal,* 30 May 2002, p. D5.
91. Ed Edelson, "Genetic Engineering Shows Promise against Type I Diabetes," *HealthScoutNews,* 29 May 2002.
92. "Antibodies: Part One," *Medical & Healthcare Marketplace Guide* I (Philadelphia: Dorland Healthcare Information, 1998), sec. I-234.
93. Koon K. Teo, "Cardiology: Recent Advances; Clinical Review," *British Medical Journal* 316:7135 (21 March 1998): 911; A. A. Adgey, "An Overview of the Results of Clinical Trials with Glycoprotein IIb/IIIa Inhibitors," *European Heart Journal* 19, Sup. D9 (April 1998): D10–21.
94. http://www.centocor.com
95. EPILOG Investigators, "Platelet Glycoprotein IIb/IIIa Receptor Blockade and Low Dose Heparin during Percutaneous Coronary Revascularization," *New England Journal of Medicine* 336:24 (12 June 1997): 1689–96; http://www.centocor.com/images/product/prod_products_header.gif; http://www.reopro.com/us/index.cfm?CC=us
96. Press Release, "Results of New Study Show Efficacy Trends for ReoPro in Ischemic Stroke," 15 February 2003, at: http://newsroom.lilly.com/news/story.cfm?id=1161; see also, The Abcix-

imab in Ischemic Stroke Investigators, "Abciximab in Acute Ischemic Stroke: A Randomized, Double-Blind, Placebo-Controlled, Dose-Escalation Study," *Stroke* 31:3 (March 2000): 601, at: http://stroke.ahajournals.org/cgi/content/full/31/3/601

97. American Heart Association, "A Statement for Healthcare Professionals from a Special Writing Group of the Stroke Council, American Heart Association," http://www.americanheart.org/presenter.jhtml?identifier=1227
98. "FDA Approves Another Rheumatoid Arthritis Treatment," *Associated Press,* 10 November 1999.
99. Press Release, "FDA Approves Remicade for Treatment of Crohn's Disease," August 1998, at: http://www.prometheus-labs.com/patient-info/news_1998/08_remicade.htm
100. C. A. Mikula, "Anti-TNF Alpha: New Therapy for Crohn's Disease," *Gastroenterological Nursing* 22:6 (November/December 1999): 245–48.
101. D. H. Present et al., "Infliximab for the Treatment of Fistulas in Patients with Crohn's Disease," *New England Journal of Medicine* 340:18 (6 May 1999): 1398–405.
102. See New York Online Access to Health (NOAH) at: http://www.noah.cuny.edu/arthritis/arthritis.html
103. R. Maini et al., "Infliximab Chimeric Anti-Tumor Necrosis Factor: A Monoclonal Antibody versus Placebo in Rheumatoid Arthritis Patients Receiving Concomitant Methotrexate: A Randomised Phase III Trial," *Lancet* 354:9194 (4 December 1999): 1932–39.
104. As quoted in Press Release, "FDA Approves Remicade for Rheumatoid Arthritis Combination Treatment," 11 November 1999, at: http://www.pslgroup.com/dg/145046.htm
105. Peter E. Lipsky et al., "Infliximab and Methotrexate in the Treatment of Rheumatoid Arthritis," *New England Journal of Medicine* 343:22 (30 November 2000): 1594–602; J. Rautenstrauch, "Neue Therapien bei Rheumatoider Arthritis. So Stoppen Sie den Gelenkfrass" [New Therapies in Rheumatoid Arthritis: Stopping Joint Deterioration], *MMW-Fortschritte der Medizin* 142:3 (20 January 2000): 4–8.
106. As quoted in Laura Beil, "Scientists Proposing New Paths to Disease Protection," *Dallas Morning News,* 3 January 1999.
107. Ibid.
108. Larry Zeitlin, Richard A. Cone and Kevin J. Whaley, "Using Monoclonal Antibodies to Prevent Mucosal Transmission of Epidemic

Infectious Diseases," *Emerging Infectious Disease* 15:1 (January/February 1999): 54–64; at: http://www.cdc.gov/ncidod/eid/vol5no1/zeitlin.htm

109. http://www.idecpharm.com/
110. See Leukemia & Lymphoma Society, "Facts and Statistics about Leukemia, Lymphoma, Hodgkin's Disease and Myeloma" (chart), at: http://www.leukemia-lymphoma.org/CMS/q?action=static&v=PF&pageID=18
111. CancerNet, "Types of Non-Hodgkin's Lymphoma," at: http://cancernet.nci.nih.gov/wyntk_pubs/non-hodgkins.htm#5
112. See "Study: 28 Percent of Rituxan Responders in Remission," *Medical Industry Today,* 9 December 1998.
113. Caroline Helwick, "Antibody Shows Promise in Lymphocytic Leukemia [Rituximab]," *Medical Post* 36(4 January 2000]): 32.
114. "Study: 28 Percent of Rituxan Responders in Remission."
115. Helwick, "Antibody Shows Promise in Lymphocytic Leukemia [Rituximab]."
116. Ibid.
117. http://www.corixa.com; http://us.gsk.com/
118. "Studies Evaluate New Therapies for Non-Hodgkin's Lymphoma," *Medical Industry Today,* 5 December 2000; "Drug Combination Improves Lymphoma Survival Rates," *Reuters,* 4 December 2000), at: http://www.cnn.com/2000/HEALTH/cancer/12/04/health.lymphoma.reut/
119. Jane Salodof MacNeil, "Bexxar Response Is Durable for Advanced NHL," *Medscape Medical News,* 12 December 2002, at: http://www.medscape.com/viewarticle/446273
120. "FDA Approves Bexxar for Non-Hodgkin's Lymphoma," *Associated Press,* 1 July 2003.
121. Justin Stebbing, E. Copson and S. O'Reilly, "Herceptin (Trastuzamab) in Advanced Breast Cancer," *Cancer Treatment Reviews* 26:4 (August 2000):287–90.
122. "Herceptin: First of Its Kind, Genentech's Breast Cancer Drug Slated for FDA Panel Review," *BioWorld Today* 9:169 (2 September 1998).
123. Andrew Pollack, "In the Works: Drugs Tailored to Individual Patients," *New York Times,* 20 December 1999, p. C8.
124. James Hawkins, as quoted in Cynthia Star, "Biomolecular 'Secret Agents' Try to Baffle Bad Genes," *Drug Topics* 137:6 (22 March 1993): 21.

125. For an attractive slide-show presentation of how antisense works, see "Antisense Explained," on the Vitravene web page, at: http://www.vitravene.com/UNITED_STATES/html/ANTISENSEEXP/anti_frset.html; for an entire book on the subject, see Stanley T. Crooke, ed., *Antisense Drug Technology: Principles, Strategies, and Technologies* (New York: Marcel Dekker Inc., 2001).
126. http://www.vitravene.com/
127. National Center for Infectious Diseases, "Cytomegalovirus (CMV) Infection," at: http://www.cdc.gov/ncidod/diseases/cmv.htm
128. http://www.genta.com/; Press Release, "Investigator Reports Major Anti-tumor Activity with Gentas Antisense Product G3139 in Malignant Menanoma,"4 April 2000, at: http://www.genta.com/pres_2000_0404.html
129. American Cancer Society, "Melanoma Skin Cancer Overview," at: http://www3.cancer.org/cancerinfo/load_cont.asp?ct=50&doc=77&Language=English
130. Daniel Q. Haney, "Genetic Approach to Cancer Promising," *Associated Press,* 4 April 2000.
131. http://www.genta.com/programs/ctrials.htm
132. Alan Dove, "Antisense and Sensibility," *Nature Biotechnology* 20:2 (February 2002): 122, table 1.

The Book of Angiogenesis

1. As quoted in John Crewdson and Judy Peres, " 'Smart-Bomb' Drugs Transform Cancer War; Hopeful Researchers Predict a Treatment Explosion," *Chicago Tribune,* 27 February 2000, p. 1.
2. http://www.imclone.com
3. As quoted in Crewdson and Peres, " 'Smart-Bomb' Drugs Transform Cancer War."
4. Judah Folkman, "Tumor Angiogenesis: Therapeutic Implications," *New England Journal of Medicine* 18:285 (18 November 1971): 1182–86; Chuan-Yuan Li et al., "Initial Stages of Tumor Cell-Induced Angiogenesis: Evaluation via Skin Window Chambers in Rodent Models," *Journal of the National Cancer Institute* 92:2 (19 January 2000): 143–47.
5. Judah Folkman, "The Vascularization of Tumors," *Scientific American* 234:5 (May 1976): 58–64, 70–73; Sharon Begley, Claudia Kalb and Theodore Gideonse, "One Man's Quest to Cure Cancer," *Newsweek* 131:20 (18 May 1998): 54. (For ten years, according to *Newsweek,* "whenever Folkman got up to speak at a scientific meeting, he would 'hear people laughing in the corner,' he says. Or the

room would empty out. 'Everybody had to go to the bathroom at once,' he says.")

6. Personal telephone communication of 18 July 2000; Kaitlin Gurney, "Two Chosen to Receive City of Medicine Community Service Awards," *Raleigh News & Observer* (N.C.), 31 August 1999, p. B3; Robert Cooke, *Dr. Folkman's War: Angiogenesis and the Struggle to Defeat Cancer* (New York: Random House, 2001).
7. C. von Bulow et al., "Endothelial Capillaries Chemotactically Attract Tumor Cells," *Journal of Pathology* 193:3 (March 2001): 367–76.
8. Nancy J. Nelson, "Inhibitors of Angiogenesis Enter Phase III Testing," *Journal of the National Cancer Institute* 90:13 (1 July 1998): 960–63.
9. Stephen B. Fox, "Tumor Angiogenesis and Prognosis," *Histopathology* 30:3 (March 1997): 294–301; Stephen B. Fox, K. C. Gatter and A. L. Harris, "Tumor Angiogenesis," *Journal of Pathology* 179:3 (July 1996): 232–37; L. M Ellis and I. J. Fidler, "Angiogenesis and Metastasis," *European Journal of Cancer* 32A:14 (December 1996): 2451–60.
10. "Metastasis," National Institute of Medicine MedLine Plus, at: http://www.nlm.nih.gov/medlineplus/ency/article/002260.htm
11. Personal telephone communication of 18 July 2000.
12. Günter Gastl et al., "Angiogenesis as a Target for Tumor Treatment," *Oncology* (Switzerland) 54:3 (May/June 1997): 177–84.
13. http://www.angio.org/
14. Personal telephone communication of 18 July 2000; see also I. Dissing, "Metastasis from an Unknown Tumor," *Acta Radiologica: Therapy, Physics, Biology* 15:2 (April 1976): 117–28.
15. Patricia A. D'Amore, "Antiangiogenesis as a Strategy for Antimetastasis," *Seminars in Thrombosis and Hemostasis* 14:1 (January 1988): 73–78.
16. For a readily understandable explanation, see "Chemotherapy and You: A Guide to Self-Help during Cancer Treatement," NIH Publication #99-1136, revised June 1999, at: http://cancernet.nci.nih.gov/peb/chemo_you/index.html#1_how_does_it_work
17. P. Boffetta and J. M. Kaldor, "Secondary Malignancies Following Cancer Chemotherapy," *Acta Oncology* 33:6 (1994): 591–98.
18. Benjamin Bonavida et al., "Cross-Resistance of Tumor Cells to Chemotherapy and Immunotherapy: Approaches to Reverse Resistance and Implications in Gene Therapy," *Oncology Reports,* 4, Sup. 1 (January/February 1997): 201–5; C. H. Graham et al., "Rapid

Acquisition of Multicellular Drug Resistance after a Single Exposure of Mammary Tumor Cells to Antitumor Alkylating Agents," *Journal of the National Cancer Institute* 86:13 (6 July 1994): 975–82.
19. Thomas Boehm et al., "Antiangiogenic Therapy of Experimental Cancer Does Not Induce Acquired Drug Resistance," *Nature* 390:6658 (27 November 1997): 404–7; Robert S. Kerbel, "A Cancer Therapy Resistant to Resistance," *Nature* 390:6658 (27 November 1997): 335–36; Robert S. Kerbel, "Inhibition of Tumor Angiogenesis as a Strategy to Circumvent Acquired Resistance to Anti-cancer Therapeutic Agents," *Bioessays* 13:1 (January 1991): 31–36.
20. Peter F. Carmeliet and Rakesh K. Jain, "Angiogenesis in Cancer and Other Diseases," *Nature* 407:6801 (14 September 2000): 249–57; Rakesh K. Jain and Peter Carmeliet, "Vessels of Death or Life," *Scientific American* 285:6 (December 2001): 38.
21. Personal telephone communication with William Li, 18 July 2000.
22. Ibid.; see also C. Kuhnen et al., "Patterns of Expression and Secretion of Vascular Endothelial Growth Factor in Malignant Soft-Tissue Tumors," *Journal of Cancer Research and Clinical Oncology* 126:4 (April 2000): 219–25.
23. As quoted in Gina Kolata, "A Cautious Awe Greets Drugs That Eradicate Tumors in Mice," *New York Times*, 3 May 1998, p. A1, with correction, 8 May 1998.
24. "Antiangiogenic Therapy," The Angiogenesis Foundation, at: http://www.angio.org/providers/oncology/oncology.html
25. Pharmaceutical Research and Manufacturers of America, "371 Biotechnology Medicines in Testing Promise to Bolster the Arsenal against Disease," at: http://www.phrma.org/newmedicines/resources/2002-10-21.93.pdf
26. http://www.oxigene.com/; http://www.bms.com/
27. "Combretastatin: An Anti-Tumor Vascular Targeting Agent," OxiGene at: http://www.oxigene.com/prCombfr.html
28. Press Release, "OXiGENE Announces Promising Clinical Results for Lead Tumor Vascular Targeting Agent, Combretastatin A4," 5 April 2000, at: http://www.oxigene.com/press/4-5-00.htm
29. See Nancy J. Nelson, "Angiogenesis Research Is on Fast Forward," *Journal of the National Cancer Institute* 91:10 (19 May 1999): 820–22.
30. "FDA Approves Celecoxib," NCI's CancerTrials, at: http://search.nci.nih.gov/search97cgi/s97_cgi
31. As quoted in John Crewdson and Judy Peres, "Cancer-Drug Treatment: Less Might Prove More," *Chicago Tribune*, 2 April 2000, p. 3.

32. J. L. Masferrer et al., "Antiangiogenic and Antitumor Activities of Cyclooxygenase-2 Inhibitors," *Cancer Research* 60:5 (1 March 2000): 1306–11.
33. K. M. Leahy, A. T. Koki and J. L. Masferrer, "Role of Cyclooxygenases in Angiogenesis," at: http://www.bentham.org/cmc-sample/masferrer/masferrer.htm
34. M. K. Urban, "Cox-2 Specific Inhibitors Offer Improved Advantages over Traditional NSAIDS," *Orthopedics* 23, Sup. 7 (July 2000): S761–64.
35. R. M. Tamimi et al., "Prospects for Chemoprevention of Cancer," *Journal of Internal Medicine* 251:4 (April 2002): 286–30.
36. Lauran Neergaard, "FDA Approves Infamous Thalidomide," *Associated Press,* 16 July 1998.
37. Robert J. D'Amato et al., "Thalidomide Is an Inhibitor of Angiogenesis," *Journal of the Proceedings of the National Academy of Sciences USA* 91:9 (26 April 1994): 4082–85.
38. "EntreMed on Antiangiogenesis," at: http://www.slip.net/~mcdavis/database/angio156.htm; F. E. Davies et al., "Thalidomide and Immunomodulatory Derivatives Augment Natural Killer Cell Cytotoxicity in Multiple Myeloma," *Blood* 98:1 (July 2001): 210–16; "Thalidomide Wreaks Havoc on Multiple Myeloma via Natural Killer Cells," *Cancer Weekly,* 24 July 2001.
39. For example, "Antiangiogenesis Combinations May Be Beneficial," *Cancer Weekly Plus,* 25 January 1999, discussing M. Lingen, P. J. Polverini, and N. P. Bouck, "Retinoic Acid and Interferon alpha Act Synergistically as Antiangiogenic and Antitumor Agents against Human Head and Neck Squamous Cell Carcinoma," *Cancer Research* 58:23 (1 December 1998): 5551–58; Helena J. Mauceri, "Combined Effects of Angiostatin and Ionizing Radiation in Antitumor Therapy," *Nature* 394:6690 (16 July 1998): 287–91.
40. Peter F. Carmeliet and Rakesh K. Jain, "Angiogenesis in Cancer and Other Diseases," *Nature* 407:6801 (14 September 2000): 249–57; Rakesh K. Jain and Peter Carmeliet, "Vessels of Death or Life," *Scientific American* 285:6 (December 2001): 38.
41. Personal telephone communication of 18 July 2000.
42. Carmeliet and Jain, "Angiogenesis in Cancer and Other Diseases"; Jain and Carmeliet, "Vessels of Death or Life," citing the work of Pietro M. Gullino of the National Cancer Institute in 1976.

43. As quoted in Gina Kolata, "A Cautious Awe Greets Drugs That Eradicate Tumors in Mice," *New York Times,* 3 May 1998, p. A1; see also "Correction," *New York Times,* 8 May 1998.
44. http://www.genvec.com/products_gene_therapy_cad.asp#cad
45. Press Release, "BIOBYPASS(R) Shows Statistically Significant Positive Results in Patients with Severe Coronary Artery Disease," at: http://www.corporate-ir.net/ireye/ir_site.zhtml?ticker=GNVC&script=410&layout=-6&item_id=358269; see also, Steven Sternberg, "Doctors: Gene Therapy Boosts Hearts Oxygen Supply, *USA Today,* 21 November 2002, p. 9D.
46. Kate Murphy, "A Kinder, Gentler Heart Bypass," *Business Week,* 27 October 1997, at: http://www.businessweek.com/1997/43/b3550167.htm

Medicinal Matchmaking

1. Nicholas Wade, "Scientist Reveals Secret of Genome: It's His," *New York Times,* 27 April 2002, p. A1; J. Craig Venter et al., "The Sequence of the Human Genome," *Science* 291:5507 (16 February 2001): 1304–51, at: http://www.sciencemag.org/cgi/content/full/291/5507/1304
2. Human Genome Project Information, "Facts about Genome Sequencing," at: http://www.ornl.gov/hgmis/faq/seqfacts.html#whose
3. See William E. Evans and Mary V. Relling, "Pharmacogenomics: Translating Functional Genomics into Rational Therapeutics," *Science* 286:5439 (15 October 2000): 487–91.
4. In reality, SNP is often widely, if wrongly, used as a catch-all label for many different types of subtle sequence variation. See Anthony J. Brookes, "The Essence of SNPs," *Gene* 234:2 (8 July 1999): 177–86, at: http://www.cgr.ki.se/cgb/groups/brookes/Articles/essence_of_snps_article.pdf
5. Brookes, "The Essence of SNPs." (Incidentally, SNPs that occur with less than 1% frequency are called "idiomorphisms.")
6. See Karen Schmidt, "Just for You," *New Scientist* 160:2160 (14 November 1998): 32; Andrew Pollack, "Company Seeking Donors of DNA for a 'Gene Trust,' " *New York Times,* 1 August 2000, p. A1.
7. "Breast Cancer," NCI's CancerNet, at: http://cancernet.nci.nih.gov/wyntk_pubs/breast.htm#25
8. Natalie Angier, "Vexing Pursuit of Breast Cancer Gene," *New York Times,* 12 July 1994, p. C1.

9. R. Wooster et al., "Localization of a Breast Cancer Susceptibility Gene, BRCA2, to Chromosome 13q12-13," *Science* 265:5181 (30 September 1994): 2088–90; Y. Miki et al. "A Strong Candidate for the Breast and Ovarian Cancer Susceptibility Gene BRCA 1," *Science* 266:5182 (7 October 1994): 66–71; R. Wooster et al., "Identification of the Breast Cancer Susceptibility Gene BRCA2," *Nature* 378:6559 (21–28 December 1995): 789–92; Jeffrey P. Struewing et al., "The Carrier Frequency of the BRCA1 185delAG Mutation Is Approximately 1 Percent in Ashkenazi Jewish Individuals," *Nature Genetics* 11:2 (October 1995): 198–200; Robert Cooke, "Tracing Roots; Genetic Link to Breast Cancer for Jews," *New York Newsday*, 30 August 1995, p. A7.
10. Brad Keoun, "Ashkenazim Not Alone: Other Ethnic Groups Have Breast Cancer Gene Mutations, Too," *Journal of the National Cancer Institute* 89:1 (1 January 1997): 8–9.
11. Robert Cooke, "Cancer Test's Benefits/Genetic Risks Seen Early Can Aid Survival," *New York Newsday*, 8 December 1996, p. A6.
12. C. P. Archambeault, "Mass Antimalarial Therapy in Veterans Returning from Korea," *Journal of the American Medical Association* 154 (1954): 1411–15; See "An Introduction to G6PD Deficiency: The Most Common Human Enzyme Deficiency in the World," at: http://www.rialto.com/g6pd/index.htm
13. L. Bertilsson, M. L. Dahl and G. Tybring, "Pharmacogenetics of Antidepressants: Clinical Aspects," *Acta Psychiatrica Scandinavica Supplementum* 391 (1997): 14–21.
14. F. R. Sallee, C. L. DeVane, and R. E. Ferrell, "Fluoxetine-Related Death in a Child with Cytochrome P-450 2D6 Genetic Deficiency," *Journal of Child and Adolescent Psychopharmacology* 10:1 (Spring 2000): 27–34.
15. Kathryn A. Phillips et al., "Potential Role of Pharmacogenomics in Reducing Adverse Drug Reactions," *Journal of the American Medical Association* 286:18 (14 November 2001): 2270–79.
16. http://www.orchid.com/
17. Gautam Naik, "Custom-Tailored Medicine—Rx4U: Drug Firms Get Personal," *Wall Street Journal*, 25 March 2002, p. B1.
18. http://www.mayo.edu/mcr/
19. Penni Crabtree, "The Quest for 'Snips;' Biotechs Seek the Genomic Keys to an Era of Personalized Medicine," *San Diego Union-Tribune*, 14 January 2001, p. H1.
20. http://www.pharma.us.novartis.com/

21. As quoted in Andrew Pollack, "Company Seeking Donors of DNA for a 'Gene Trust,'" *New York Times,* 1 August 2000, p. A1.
22. Andrew Pollack, "E-Finance: Gene Hunters Say Patients Are a Bankable Asset," *Guardian* (London), 2 August 2000, p. 20; Tom Hollon, "Gene Pool Expeditions: Estonians or Subjects of the Crown of Tonga: Whose Gene Pool Hides Gold?" *Scientist* 15:4 (19 February 2001): 1, at: http://www.thescientist.com/yr2001/feb/hollon_p1_010219.html
23. http://www.decode.com; http://sunsite.berkeley.edu/biotech/iceland/index.html
24. Nicholas Wade, "A Genomic Treasure Hunt May Be Striking Gold," *New York Times,* 18 June 2002, p. F1.
25. V. B., "DeCode CEO Defends Peeking into Iceland's Genetic Records," *Biotechnology Newswatch,* 20 November 2000, p. 4; Ruth Chadwick, "The Icelandic Database—Do Modern Times Need Modern Sagas," *British Medical Journal* 319:7207 (14 August 1999): 441.
26. Einar Årnason, "Genetic Heterogeneity of Icelanders," *Annals of Human Genetics* 67: Pt. 1 (January 2003): 5–6.
27. Wade, "A Genomic Treasure Hunt May Be Striking Gold."
28. Sigurlaug Sveinbjornsdottir et al., "Familial Aggregation of Parkinson's Disease in Iceland," *New England Journal of Medicine* 343:24 (14 December 2000): 1765–70.
29. http://www.roche.com/; Matt Crenson, "Icelanders' DNA Offers Promise," *Associated Press,* 14 February 2001; Press Release, "Roche and DeCode Genetics Announce Major Progress in Turning Their Genomic Discoveries in Schizophrenia and Paod into Novel Drugs and Diagnostics," at: http://www3.roche.com/med-corp-detail-2001?id=585
30. Andrew Pollack, "Company Seeking Donors of DNA for a 'Gene Trust,'" *New York Times,* 1 August 2000, p. A1.
31. Randall Osborne, "DNA Sciences Inc. Seeks Genetic Data from Public," *BioWorld Today* 11:148 (2 August 2000); "The DNA Sciences Gene Trust Project," at: http://www.dna.com/sectionHome/0,1687,Jm5hdj10aGVfZ2VuZV90cnVzdA==,00.html
32. As quoted in David Stipp, "A DNA Tragedy," *Fortune* 142:10 (30 October 2000): 170.
33. http://corp.gsk.com/; Adrian Michaels and David Pilling, "The Gene Rush," *Financial Times* (London), 18 March 2000, p. 12.

34. http://www.genxy.com/; "Takeover Proves Key to Growth," *Financial Times* (London), 28 October 1999, p. 3.
35. Their website is http://snp.cshl.org/
36. These include APBiotech (now Amersham Biosciences), AstraZeneca, Bayer, Bristol-Myers Squibb, F. Hoffmann–La Roche, Glaxo Wellcome, Aventis/Hoechst Marion Rouse, Novartis, Pfizer, Searle (now part of Pfizer), and SmithKline Beecham (now part of GlaxoSmithKline), the Wellcome Trust, IBM, and Motorola.
37. Clive Cookson, "Markers on the Road to Avoiding Illness," *Financial Times,* 3 March 1998, p. 18; "DNA Microarray (Genome Chip)," at: http://www.gene-chips.com/
38. http://www.affx.com/
39. Ben Hirschler, " 'Gene Chips Come of Age as Costs Plummet," *Reuters,* 24 May 2001.
40. Ibid.
41. Press Release, "IBM and Mayo Clinic Collaborate on System to Advance Medical Science," Mayo Clinic Rochester, 25 March 2002, at: http://www.mayo.edu/comm/mcr/news_2038.html; David P. Hamilton, "Mayo Clinic Plans Database of Every Patient's History," *Wall Street Journal,* 25 March 2002, p. B1.
42. As quoted in Hamilton, "Mayo Clinic Plans Database of Every Patient's History."
43. As quoted in Hirschler, " 'Gene Chips' Come of Age as Costs Plummet."
44. Erika Jonietz, "Personal Genomes," *Technology Review* 104:8 (October 2001): 30. Weinstock's personal website is: http://public.bcm.tmc.edu/pa/weinstock2.htm; see also: http://public.bcm.tmc.edu/pa/index.html
45. Press Release, "Survey Finds 103 New AIDS Medicines in the Pipeline," Pharmaceutical Research and Manufacturers Association, 29 November 2000, at: http://www.phrma.org/press/newsreleases//2000-11-29.185.phtml
46. David Brown, "Study Finds Drug-Resistant HIV in Half of Infected Patients," *Washington Post,* 19 December 2001, p. A2.
47. As quoted in Andrew Pollack, "When Gene Sequencing Becomes a Fact of Life," *New York Times,* 17 January 2001, p. C1; http://www.visgen.com/
48. Sarah Staples, "FDA Okays Visible Genetics HIV-Resistance Test," *Genetic Engineering News* 21:18 (15 October 2001): 42; Press Release, "FDA Grants Market Clearance of Visible Genetics HIV Genotyping System," 26 September 2001, at: http://www.visgen.com/

Investor_Relations/Press_Releases/press_release.shtml?press103.shtml

49. Lauran Neergaard, "Gene-Based AIDS Test Hits Market," *Associated Press,* 27 September 2001.
50. Gary R. Cohan, physician and managing director of the Pacific Oaks Medical Group in Beverly Hills, California, as quoted in Pollack, "When Gene Sequencing Becomes a Fact of Life."
51. Pollack, "When Gene Sequencing Becomes a Fact of Life."
52. http://www.questdiagnostics.com/
53. FDA Center for Drug Evaluation and Research, "Lotronex Information," at: http://www.fda.gov/cder/drug/infopage/lotronex/lotronex.htm
54. Andrea Knox, "Safety Withdrawal of 11 Drugs in 4 Years Appears Unprecedented," *Philadelphia Inquirer,* 7 January 2001.
55. Denise Grady, "Anesthesia Drug Is Removed from Market after the Deaths of 5 Patients," *New York Times,* 31 March 2001, p. A13.
56. Denise Grady, "FDA Pulls a Drug, and Patients Despair," *New York Times,* 30 January 2001, p. F1.
57. Allen D. Roses, "Pharmacogenetics and the Practice of Medicine," *Nature* 405:6788 (15 June 2000): 857–65.
58. Chris Sander, "Genomic Medicine and the Future of Health Care," *Science* 287:5460 (17 March 2000): 1977–78, at: http://www.biotech-info.net/genomic_medicine.html
59. See Gautam Naik, "Custom-Tailored Medicine—Rx4U: Drug Firms Get Personal," *Wall Street Journal,* 25 March 2002, p. B1.
60. http://www.med.uc.edu/
61. As quoted in Gina Kolata, "Using Gene Tests to Decide a Patient's Treatment," *New York Times,* 20 December 1999, p. A1.
62. Stephen B. Liggett et al., "The Ile164 Beta2-adrenergic Receptor Polymorphism Adversely Affects the Outcome of Congestive Heart Failure," *Journal of Clinical Investigation* 102:8 (15 October): 1534–39.
63. As quoted in Kolata, "Using Gene Tests to Decide a Patient's Treatment."
64. John Weinstein, "Pharmacogenomics—Teaching Old Drugs New Tricks," *New England Journal of Medicine* 343:19 (9 November 2000): 1408.
65. http://www.responsegenetics.com/
66. Personal telephone communication with David M. Smith, Vice Chairman of the Board, Response Genetics Inc., 10 April 2001.

Ingenious Gene Therapy

1. http://www.usc.edu/schools/medicine/
2. As quoted in Leon Jaroff and Alice Park, "The Future of Medicine—Fixing the Genes—Gene Therapy, Heralded in the Early 1990s, Then Stalled by One Setback after Another, Is Finally Starting to Live Up to Its Promise," *Time,* 11 January 1999, p. 68, at: http://www.time.com/time/magazine/article/0,9171,17692,00.html
3. See "Gene Therapy Clinical Trials," at: http://www.wiley.co.uk/genetherapy/clinical/
4. J. M. Wilson, "Adenoviruses as Gene-Delivery Vehicles," *New England Journal of Medicine* 334:18 (2 May 1996): 1185–87.
5. For more information, see "The SCID Homepage" at: http://www.scid.net/
6. Rick Weiss, "Genetic Therapy Apparently Cures 2," *Washington Post,* 28 April 2000, p. A1.
7. Salima Hacein-Bey-Abina et al., "Sustained Correction of X-Linked Severe Combined Immunodeficiency by ex Vivo Gene Therapy," *New England Journal of Medicine* 346:16 (18 April 2002): 1185–93; see also the accompanying editorial, F. S. Rosen, "Successful Gene Therapy for Severe Combined Immunodeficiency," *New England Journal of Medicine* 346:16 (18 April 2002): 1241–43; Linda A. Johnson, "Gene Therapy Breakthrough in France," *Associated Press,* 18 April 2002.
8. A. Aiuti et al., "Correction of ADA-SCID by Stem Cell Gene Therapy Combined with Nonmyeloablative Conditioning," *Science* 296:5577 (28 June 2002): 2410–13.
9. As quoted in Rick Weiss, "New Genes Cure Rare Disorder in Two Children; Procedure Boosts Hope for Therapy," *Washington Post,* 28 June 2002, p. A3.
10. See "Genetics Heralds a Medical Revolution, but at What Cost?" *Inter Press Service,* 3 February 1999.
11. Laura Johannes, "Second Chance: Gene Therapy, Much Maligned, Is Promising in Some Cancer Trials," *Wall Street Journal,* 4 May 2000, p. A1.
12. http://www.wiley.co.uk/genetherapy/clinical/
13. National Cancer Institute, "Brain Tumors," NIH Publication No. 95-1558, 28 August 1998, at: http://cancernet.nci.nih.gov/wyntk_pubs/brain.htm; American Brain Tumor Association, "Facts and Statistics," at: http://www.abta.org/primer/facts.htm; as

quoted in Press Release, "Genetically Engineered Poliovirus Fights Brain Tumors," American Society for Microbiology, 21 May 2001.

14. Press Release, "Genetically Engineered Poliovirus Fights Brain Tumors."
15. Ibid.
16. As quoted in E. J. Mundell, "Polio Virus Targets, Kills Brain Tumors: Study," *Reuters Health*, 21 May 2001; see also, Matthias Gromeier et al., "Intergeneric Poliovirus Recombinants for the Treatment of Malignant Glioma," *Proceedings of the National Academy of Sciences USA* 97:12 (6 June 2000): 6803–8.
17. Gromeier et al., "Intergeneric Poliovirus Recombinants for the Treatment of Malignant Glioma"; see also David J. Solecki et al., "Expression of the Human Poliovirus Receptor/CD155 Gene Is Activated by Sonic Hedgehog," *Journal of Biological Chemistry* 277:28 (12 July 2002): 25,697–702.
18. E. J. Mundell, "Polio Virus Targets, Kills Brain Tumors: Study," *Reuters Health*, 21 May 2001.
19. http://www.salk.edu/; as quoted in Leon Jaroff and Alice Park, "The Future of Medicine—Fixing the Genes—Gene Therapy, Heralded in the Early 1990s, Then Stalled by One Setback after Another, Is Finally Starting to Live Up to Its Promise," *Time*, 11 January 1999, p. 68, at: http://www.time.com/time/magazine/article/0,9171,17692,00.html
20. As quoted in ibid.
21. http://www.pitt.edu/~rsup/mgb/
22. Muscular Dystrophy Association, "Facts about Muscular Dystrophy (MD)," at: http://www.mdausa.org/publications/fa-md-9.html; Nathan Seppa, "New Tools for Muscular Dystrophy Research," *Science News* 152:10 (6 September 1997): 151.
23. Bing Wang, Juan Li and Xiao Xiao, "Adeno-Associated Virus Vector Carrying Human Minidystrophin Genes Effectively Ameliorates Muscular Dystrophy in *mdx* Mouse Model," *Proceedings of the National Academy of Sciences USA* 97:25 (5 December 2000): 13,714.
24. http://health.ucsd.edu/
25. Virginia L. Waters, "Conjugation between Bacterial and Mammalian Cells," *Nature Genetics* 29:4 (1 December 2001): 375–76.
26. Emma Hitt, "Promiscuous Bacteria May Improve Gene Therapy," *Reuters Health*, 19 November 2001.
27. http://www.med.upenn.edu/~ihgt/

28. Sheryl Gay Stolberg, "The Biotech Death of Jesse Gelsinger," *New York Times Magazine,* 28 November 1999, pp. 137–40, 149–50.
29. Sheryl Gay Stolberg, "Institute Restricted after Gene Therapy Death," *New York Times,* 25 May 2000, p. A20.
30. Huntly Collins, "Penn Team Finds Clue to Gene-Drug Death," *Philadelphia Inquirer,* 26 January 2001, p. A1.
31. Sheryl Gay Stolberg, "Trials Are Halted on a Gene Therapy," *New York Times,* 4 October 2002, p. A1; Andrew Pollack, "Second Cancer Is Attributed to Gene Used in FDA Test," *New York Times,* 17 January 2003, p. A24.
32. http://www.uni-jena.de/
33. Uday K. Tirlapur and Karsten Konig, "Targeted Transfection by Femtosecond Laser," *Nature* 418:6895 (2002): 290–91; Philip Ball, "Laser Delivers DNA," *Nature ScienceUpdate,* 18 July 2002, at: http://www.nature.com/nsu/020715/020715-7.html
34. Q. L. Lu et al., "Microbubble Ultrasound Improves the Efficiency of Gene Transduction in Skeletal Muscle *in Vivo* with Reduced Tissue Damage," *Gene Therapy* 10:5 (March 2003): 396–405; "Breakthrough in Gene Therapy Delivery System," *BBC News,* 24 February 2003, at: http://news.bbc.co.uk/go/pr/fr/-/1/hi/health/2780325.stm; Stephen Pincock, "Microbubble Delivery May Aid Gene Therapy Effect," *Reuters Health,* 24 February 2003.
35. http://www.ox.ac.uk/; M. G. Davies et al., "Vascular Tissue Alteration of Arterial Vasomotor Function *in Vitro* by Gene Transfer with a Replication-Deficient Adenovirus," *Vascular Surgery* 31:2 (March/April 1997): 131–36.
36. As quoted in "Genetically-Modified Veins Could Save Lives," *BBC News,* 7 October 1999, at: http://news2.thls.bbc.co.uk/hi/english/health/newsid_467000/467355.stm
37. See "Angioplasty," at: http://www.rad.bgsm.edu/patienteduc/Angioplasty.htm
38. Eric Schmitt, "Cheney Complains of Pains in Chest; Artery Is Cleared," *New York Times,* 6 March 2001, p. A1.
39. http://www.chop.edu/index.html
40. Bruce D. Klugherz et al., "Gene Delivery from a DNA Controlled-Release Stent in Porcine Coronary Arteries," *Nature Biotechnology* 18:11 (November 2000): 1181–84; see also Press Release, "Gene-Coated Device for Blocked Arteries?" *Reuters Health,* 31 October 2000. Gene therapy is also being experimented with to halt damage to arteries that stents may cause, which itself may lead to reclogging. See Mikko O. Laukkanen, "Adenovirus-Mediated Extracellu-

lar Superoxide Dismutase Gene Therapy Reduces Neointima Formation in Balloon-Denuded Rabbit Aorta," *Circulation* 106:15 (8 October 2002): 1999–2003.

41. Lebers Congenital Amaurosis, "The Low Vision Gateway," at: http://www.lowvision.org/lebers_congenital_amaurosis.htm
42. "Retinitis Pigmentosa," StLukesEye.com, at: http://www.stlukes-eye.com/retinitis_pigmentosa.htm
43. http://www.upenn.edu/; G. M. Acland et al., "Gene Therapy Restores Vision in a Canine Model of Childhood Blindness," *Nature Genetics* 28:1 (May 2001): 92–95.
44. As quoted in Keith Mulvihill, "Gene Therapy Helps Blind Dogs See," *Reuters Health,* 27 April 2000, at: http://www.novaob.com/viewArticle?ID=74453
45. G. M. Acland et al., "Gene Therapy Restores Vision in a Canine Model of Childhood Blindness," *Nature Genetics* 28:1 (May 2001): 92–95.
46. http://www.med.upenn.edu/ophth/
47. As quoted in Rick Weiss, "Gene Treatment Restores Vision in Blind Dogs; Study Offers Humans Hope," *Washington Post,* 28 April 2001, p. A1.
48. Ibid.
49. As quoted in Malcolm Ritter, "Gene Therapy Restores Vision in Dogs with Genetic Disease," *Associated Press,* 27 April 2001; http://www.blindness.org/
50. http://www.chop.edu/index.html; Mark A. Kay et al., "Evidence for Gene Transfer and Expression of Factor IX in Hemophilia B Patients Treated with an AAV Vector," *Nature Genetics* 24:3 (March 2000): 257–61.
51. See P. M. Mannucci and E. G. Tuddenbam, "The Hemophilias: Progress and Problems," *Seminal Hematology* 36:4, Sup. 7 (October 1999): 104–17; National Institutes of Health, "Hemophilia Update: 1997," at: http://www.nhlbi.nih.gov/health/public/blood/other/hemo_97.htm; National Heart, Lung, and Blood Institute, "Hemophilia," at: http://www.nhlbi.nih.gov/health/public/blood/other/hemophel.htm
52. http://genome-www.stanford.edu/genetics/
53. As quoted in Joseph B. Verrengia, "Gene Experiment Boosts Blood Clotting in Hemophiliacs," *Associated Press,* 2 March 2000.
54. http://www.chmkids.org/chm/; http://www.hemophilia.org/home.htm

55. As quoted in Huntly Collins, "Early Trial of Gene Therapy for Hemophilia Goes Well," *Pittsburgh Post-Gazette,* 2 March 2000, p. A3, at: http://www.phillynews.com/2000/Mar/01/blood02.htm
56. See the National Parkinson Disease Inc. website at: http://www.parkinson.org/pdedu.htm
57. "Actor Michael J. Fox Has Parkinson's Disease," *Washington Post,* 26 November 1998, p. C6.
58. See the National Parkinson Disease Inc. website at: http://www.parkinson.org/pdedu.htm
59. http://www.parkinson.org/texthtms/ttreamen.htm
60. Jeffrey H. Kordower et al., "Neurodegeneration Prevented by Lentiviral Vector Delivery of GDNF in Primate Models of Parkinson's Disease," *Science* 290:5492 (27 October 2000): 767–73; as quoted in Robert Cooke, "Gene-Therapy New Hope for Parkinson's: Study on Monkeys Finds Brain Damage Reversed," *New York Newsday,* 27 October 2000, p. A6.
61. http://www.uphs.upenn.edu/ihgt/

The Genomic Generation

1. From "An Aspirin a Day Keeps the Doctor at Bay," at: http://www.almaz.com/nobel/medicine/aspirin.html
2. See for example, Diana Taylor, "Great Bio-Treasure Hunt in Australia's Barrier Reef," *Reuters,* at: http://news.excite.com/news/r/010131/20/science-australia-marine-dc
3. Source: Pharmaceutical Research and Manufacturers of America, see "Why Do Medicines Cost So Much?" at: http://www.phrma.org/publications/publications/brochure/questions/whycostmuch.phtml.
4. Tim Chapman, "Lab Automation and Robotics," *Nature* 421:6923 (6 February 2003): 661–66, at: http://www.nature.com/cgi-taf/DynaPage.taf?file=/nature/journal/v421/n6923/full/421661a_fs.html; source: Pharmaceutical Research and Manufacturers of America, see "Why Do Medicines Cost So Much?"
5. As quoted in Michael D. Lemonick et al., "Brave New Pharmacy," *Time,* 15 January 2001, p. 58, at: http://www.time.com/time/health/article/0,8599,93963,00.html; estimates of the ultimate number of targets vary greatly. See also Ronald Rosenberg, "Bioinformatics: Putting It All Together," *Boston Globe,* 23 October 2000, p. C2, quoting Mark Schwartz of Trega Biosciences in San Diego saying, "Until the human genome project, there have been 500 disease targets that drug makers have focused on, but now there are

predictions of 10,000 disease targets, and that creates a lot more work in chemistry and information processing."

6. Sue Pearson, "Identifying Novel Targets in the Post-Genomic Era," *Genetic Engineering News* 22:20 (15 November 2002): 62.
7. As quoted in Michael D. Lemonick et al., "Brave New Pharmacy."
8. http://public.bcm.tmc.edu/departments/breastcenter.html
9. For more, see "Drug Design" at: http://www.ipc.pku.edu.cn/drug_design/Drug_Design.html
10. As quoted in Shari Roan, "Weapon in the War on Cancer," *Los Angeles Times,* 26 October 1998, p. S1.
11. As quoted in "Herceptin: First of Its Kind Genentech's Breast Cancer Drug Slated for FDA Panel Review," *BioWorld Today* 9:169 (2 September 1998).
12. http://www.pharma.novartis.com/
13. For more information see "Chronic Myelogenous Leukemia," *CancerNet,* at: http://cancernet.nci.nih.gov/cgi-bin/srchcgi.exe?DBID=pdq&TYPE=search&SFMT=pdq_statement/1/0/0&Z208=208_01031P
14. See A. M. Carella, "New Insights in Biology and Current Therapeutic Options for Patients with Chronic Myelogenous Leukemia," *Haematologica* 82:4 (July/August 1997): 478–95.
15. Janice Billingsly, "Leukemia Pill Shows Enormous Promise," 13 July 2000, at: http://www.newscout.com; Stephen D. Moore, "Blood Test: News about Leukemia Unexpectedly Puts Novartis on the Spot," *Wall Street Journal,* 6 June 2000, p. A1.
16. As quoted in Billingsly, "Leukemia Pill Shows Enormous Promise."
17. Brian J. Druker et al., "Efficacy and Safety of a Specific Inhibitor of the BCR-ABL Tyrosine Kinase in Chronic Myeloid Leukemia," *New England Journal of Medicine* 344:14 (5 April 2001):1031–37.
18. As quoted in Susan Okie, "A New Drug Shows Promise in Battling Form of Leukemia," *Washington Post,* 5 April 2001, p. A2.
19. As quoted in Nicholas Wade, "Swift Approval for a New Kind of Cancer Drug," *New York Times,* 11 May 2001.
20. U.S. Food and Drug Administration, "FDA Approves Gleevec to Treat Gastrointestinal Stromal Cancer," FDA Talk Paper, 1 February 2002, at: http://www.fda.gov/bbs/topics/ANSWERS/2002/ANS01134.html
21. For more information about the drug and clinical trials, see the manufacturers website, at: http://www.novartisoncology.com, or call 1-800-340-6843. A particularly inspiring story about one trial

involving liver cancer is Terry Jennings, "Trial by Gleevec," *Washington Post,* 2 July 2002, p. H1.

22. As quoted in John Crewdson and Judy Peres, " 'Smart-Bomb' Drugs Transform Cancer War; Hopeful Researchers Predict a Treatment Explosion," *Chicago Tribune,* 27 February 2000, p. 1.
23. "Creating New Kinase Inhibitors," *Drug Discovery/Technology News* 4:2 (February 2001).
24. As quoted in Ken Garber, "Tyrosine Kinase Inhibitor Research Presses On Despite Halted Clinical Trial," *Journal of the National Cancer Institute* 92:12 (21 June 2000): 967–69; http://www.icr.ac.uk/cctherap/
25. http://www.astrazeneca.com
26. F. Ciardiello "Epidermal Growth Factor Receptor Tyrosine Kinase Inhibitors as Anticancer Agents," *Drugs* 60, Sup. 1 (2000): 25–32; Garber, "Tyrosine Kinase Inhibitor Research Presses On Despite Halted Clinical Trial."
27. "AstraZeneca Reveals New Oncology Data," *Marketletter,* 20 November 2000; F. M. Sirotnak et al., "Efficacy of Cytotoxic Agents against Human Tumor Xenografts Is Markedly Enhanced by Coadministration of ZD1839 (Iressa), an Inhibitor of EGFR Tyrosine Kinase," *Clinical Cancer Research* 6:12 (December 2000): 4885–92.
28. F. Ciardiello et al., "Anti-tumor Effect and Potentiation of Cytotoxic Drugs Activity in Human Cancer Cells by ZD-1839 (Iressa), an Epidermal Growth Factor Receptor-Selective Tyrosine Kinase Inhibitor," *Clinical Cancer Research* 6:5 (May 2000): 2053–63.
29. "FDA Approves New Type of Drug for Lung Cancer," *FDA News,* 5 May 2003, at: http://www.fda.gov/bbs/topics/NEWS/2003/NEW00901.html
30. Nicholas Wade, "Reading the Book of Life," *New York Times,* 13 February 2001, p. F1; Sue Goetinck Ambrose, "Recipe of Life Revealed," *Dallas Morning News,* 11 February 2001, p. 1A; Todd Ackerman, "Road Map to the Core of Mankind," *Houston Chronicle,* 13 February 2001, p. A1.
31. Nicholas Wade, "Once Again, Scientists Say Human Genome Is Complete," *New York Times,* 15 April 2003, p. F1; CNN, "Human Genetic Code 'Essentially Complete,' " at: http://www.cnn.com/2003/HEALTH/04/14/human.genome.ap/index.html
32. Seth Borenstein, "Scientists Crack Code of Life: The Human Genome Project Breakthrough to Aid Battle against Almost All Diseases, May Cut Cancer Deaths to Zero," *Detroit Free Press,* 27 June

2000, p. 1A. (Note that this was also written over half a year before even the "draft" was announced!)

33. Lee Chyen Yee, "Scientists Expect to Complete Human Genome in 2003," *Reuters*, 14 April 2002; Michael Fumento, "I Know Why the Caged Genome Sings," *American Outlook* 4:12 (January/February 2001): 13, at: http://www.fumento.com/genome.html; for a brief explanation of the project, its purpose, and its importance, see Leslie Roberts, "The Gene Hunters," *U.S. News & World Report*, 3 January 2000, p. 34; for a more detailed explanation of how incomplete the allegedly completed project still is, see Richard C. Strohman, "Five Stages of the Human Genome Project," *Nature Biotechnology* 17:2 (February 1999): 112.
34. See Andrew Pollack, "Scientist Quits the Company He Led in Quest for Genome," *New York Times*, 23 January 2002, p. C1.
35. http://www.incyte.com
36. See David Smoller, "Is Gene Sequencing Over?" *Genomics & Proteomics*, at: http://www.genpromag.com/feats/0109guest.asp
37. Personal telephone communication of 6 March 2001.
38. Matt Ridley, *Genome: The Autobiography of a Species in 23 Chapters* (New York: HarperCollins, 1999), p. 54.
39. "Mutation Protects against HIV, Ups Hepatitis C Risk," *Reuters Health*, 7 February 2001.
40. Harold Chen,"Wolf-Hirschhorn Syndrome," at: http://emedicine.com/cgi-bin/foxweb.exe/showsection@d:/em/ga?book=ped&topicid=2446
41. Betsy A. Hosler et al., "Linkage of Familial Amyotrophic Lateral Sclerosis with Frontotemporal Dementia to Chromosome 9q21-q22," *Journal of the American Medical Association* 284:13 (4 October 2000): 1664–69.
42. As quoted in Edward Edelson, "Genetic Clue Found in Lou Gehrig's Disease," *HealthSCOUT*, 5 October 2000, at: http://www.healthscout.com/cgi-bin/WebObjects/Af.woa/6/wo/Uy6000ic800QO700DH/13.0.13.5.64.3.3.7.3.3.1
43. Institute Curie, "P53 Story: Everything You Want to Know about p53 but You Never Bother to Ask," at: http://perso.curie.fr/Thierry.Soussi/p53_story.html
44. Matt Ridley, *Genome* (New York: HarperCollins, 1999), p. 236; C. C. Harris and M. Hollstein, "Clinical Implications of the P53 Tumor-Suppressor Gene," *New England Journal of Medicine* 329:18 (28 October 1993): 18–27.

45. See William G. Kaelin Jr., "The Emerging P53 Gene Family," *Journal of the National Cancer Institute* 1999:91 (7 April 1999): 594–98.
46. A. Hartmann et al, "Overexpression and Mutations of P53 in Metastatic Malignant Melanomas," *International Journal of Cancer* 67:3 (1996): 313–17; M. C. Saenz-Santamaria et al, "P53 Expression Is Rare in Cutaneous Melanomas," *American Journal of Dermatopathology* 17:4 (August 1995): 344–49.
47. D. Schadendorf et al, "Chemosensitivity Testing of Human Malignant Melanoma, A Retrospective Analysis of Clinical Response and *in Vitro* Drug Sensitivity," *Cancer* 73:1 (1 January 1994): 103–8.
48. María S. Soengas et al., "Inactivation of the Apoptosis Effector Apaf-1 in Malignant Melanoma," letter, *Nature* 409:6817 (11 January 2001): 207–11; see also, Lidia Wasowicz, "Clues to Treating Deadly Melanoma," *United Press International,* 10 January 2001, at: http://www.applesforhealth.com/cluemelan2.html
49. Rickey W. Johnstone, Astrid A. Ruefli and Scott W. Lowe, "Apoptosis: A Link between Cancer Genetics and Chemotherapy," *Cell* 108:2 (25 January 2002):153–64.
50. E-mail communication from María S. Soengas, 11 March 2003.
51. http://www.med.umich.edu/medschool/
52. Peter A. Jones, "Cancer: Death and Methylation," *Nature* 409:6817 (11 January 2001): 141–44.
53. American Cancer Society, "Melanoma Skin Cancer Overview," at: http://www.nlm.nih.gov/medlineplus/melanoma.html
54. Martin Widschwendter et al., "Methylation and Silencing of the Retinoic Acid Receptor-[beta]2 Gene in Breast Cancer," *Journal of the National Cancer Institute* 2000:92 (17 May 2000): 826–32; Alice B. Glover et al., "Azacitidine: 10 Years Later," *Cancer Treatment Reports* 71:10 (July/August 1987): 737–46.
55. Soengas et al., "Inactivation of the Apoptosis Effector Apaf-1 in Malignant Melanoma."
56. As quoted in Patricia Reaney, "Scientists Find New Clue to Tackle Skin Cancer," *Reuters,* 10 January 2001.
57. "Drug Delivery," Emisphere Technologies Inc., at: http://www.emisphere.com/new/drugdelivery.htm; A. Leone-Bay, D. R. Paton, and J. J. Weidner, "The Development of Delivery Agents That Facilitate the Oral Absorption of Macromolecular Drugs," *Medical Research Reviews* 20:2 (March 2000): 169–86; Alex Barnum, "Biotech Companies Shift Focus to Drugs from Small Molecules," *Toronto Star,* 13 May 1991, p. B5.

58. David Pilling, "Genentech to Develop Small-Molecule Drugs," *Financial Times* (London), 28 August 2000, p. 20.
59. http://www.emisphere.com/html/pr_oralinsulin_061603.html
60. http://www.emisphere.com/html/pc_oi.html
61. As quoted in David Pilling, "Genentech to Develop Small-Molecule Drugs," *Financial Times* (London), 28 August 2000, p. 20.
62. "Proteomics, Transcriptomics: What's in a Name?" *Nature* 402:6763 (16 December 1999): 715; see also Kate Fodor, "After Genome, Scientists Tackle Even Bigger Challenge," *Reuters Health,* 1 February 2001.
63. Oliver Morton, "Gene Machine," *Wired,* July 2001, at: http://www.wired.com/wired/archive/9.07/blue_pr.html
64. http://www.ogs.com/
65. As quoted in David Pilling, "Real-Time Molecular Movie Producer Targets Proteins," *Financial Times* (London), 15 November 2000, p. 3.
66. As quoted in Jon Cohen, "The Proteomics Payoff," *Technology Review* 104:8 (October 2001): 54.
67. http://www.proteome.co.uk/
68. http://plasmaproteome.org/bio_leigh.htm
69. http://www.molsci.org/
70. As quoted in Cohen, "The Proteomics Payoff."
71. Anjana Ahuja, "Proteins: The Next Frontier," *Times* (London), 4 January 2001.
72. See Andrew Pollack, "Three Companies Will Try to Identify All Human Proteins," *New York Times,* 5 April 2001, p. C4.
73. As quoted in Victoria Griffith, "Proteins Added to the Medical Map," *Financial Times* (London), 2 October 1998, p. 10.
74. As quoted in Cohen, "The Proteomics Payoff."
75. Anjana Ahuja, "Proteins: The Next Frontier," *Times* (London), 4 January 2001.
76. As quoted in David Voss, "Protein Chips," *Technology Review* 104:4 (May 2001): 35; http://www.zyomyx.com/
77. See Peter Mitchell, "A Perspective on Protein Microarrays," *Nature Biotechnology* 3:20 (March 2002): 225; Voss, "Protein Chips"; Ellen Licking and John Carey, "First Silicon. Then DNA. And Soon, Protein Chips," *Business Week,* 10 April 2000, p. 142, at: http://www.businessweek.com/2000/00_15/b3676120.htm
78. http://www.vrtx.com/
79. Barry Werth, *The Billion-Dollar Molecule* (New York: Simon & Schuster, 1994).

80. "Chemogenomics Virtual Tour (Quicktime movie)," at: http://www.vrtx.com/Chemogenomics.html
81. Press Release, "Vertex Pharmaceuticals Reports Second Quarter and First Half 2002 Financial Results," 18 July 2002, at: http://www.vrtx.com/Pressreleases2002/pr071802.html; Biotechnology Investment Tracking Systems, "Autoimmune Disorders," at: http://www.bitsplace.com/bitsplace/Autoimmunediscomp.html#anchor9216723
82. Products/Pipeline Chart, at: http://www.vrtx.com/PipelineChart.html
83. http://www.exelixis.com/
84. Kirk McMillan and Adam Galan, "Tutorial: Exelixis Improves Screening Success," *Genetic Engineering News* 21:9 (1 May 2001): 34.

Gene Testing

1. Rick Weiss, "Ignorance Undercuts Gene Tests Potential," *Washington Post*, 2 December 2000, p. A1.
2. http://www.genetests.org/
3. Wendy Uhlmann, genetic counselor at the University of Michigan, as quoted in Weiss, "Ignorance Undercuts Gene Test's Potential." According to Weiss, testing is increasing by about 30% per year.
4. "NINDS Spina Bifida Information," National Institute of Neurological Disorders and Stroke, at: http://www.ninds.nih.gov/health_and_medical/disorders/spina_bifida.htm; "General Information about Spina Bifida: Fact Sheet Number 12," April 2000, *National Information Center for Children and Youth with Disabilities*, at: http://www.nichcy.org
5. "What Is Down Syndrome?" The March of Dimes, at: http://www.modimes.org/HealthLibrary2/FactSheets/Down_syndrome.htm
6. See "The Genetics of Complex Traits," Texas Tech University Principles of Genetics, at: http://www.pssc.ttu.edu/pss3421/genetics%20of%20complex%20traits.htm
7. See "Penetrance," at: http://www.genxy.com/Science/index.html#PENETRANCE
8. R. A. Eeles, "Screening for Hereditary Cancer and Genetic Testing, Epitomized by Breast Cancer," *European Journal of Cancer* 35:14 (December 1999): 1954–62.
9. Weiss, "Ignorance Undercuts Gene Tests Potential."

10. Linda S. Kinsinger et al., "Chemoprevention of Breast Cancer: A Summary of the Evidence for the U.S. Preventive Services Task Force," *Annals of Internal Medicine* 137:1 (2 July 2002): 59–69.
11. Liesbeth Bergman, "Risk and Prognosis of Endometrial Cancer after Tamoxifen for Breast Cancer," *Lancet* 356:9233 (9 September 2000): 881–87.
12. Bruce M. Psaty et al., "Diuretic Therapy, the Alpha-Adducin Gene Variant, and the Risk of Myocardial Infarction or Stroke in Persons with Treated Hypertension," *Journal of the American Medical Association* 287:13 (3 March 2002): 1680–89; Carol Smith, "Genes Play Growing Part in Therapy; Doctors Eventually May Be Able to Tailor Drugs to Individuals," *Seattle Post-Intelligencer,* 3 April 2002, p. B6.
13. W. Smalley et al., "Use of Nonsteroidal Anti-inflammatory Drugs and Incidence of Colorectal Cancer: A Population-Based Study," *Archives of Internal Medicine* 159:2 (25 January 1999): 161–66; Randall E. Harris, S. Kasbari and W. B. Farrar, "Prospective Study of Nonsteroidal Anti-inflammatory Drugs and Breast Cancer," *Oncology Reports* 6:1 (January/February 1999): 71–73; M. M. Taketo "Cyclooxygenase-2 Inhibitors in Tumorigenesis (Part II)," *Journal of the National Cancer Institute* 90:21 (4 November 1998): 1609–20; Daniel Hwang et al., "Expression of Cyclooxygenase-1 and Cyclooxygenase-2 in Human Breast Cancer," *Journal of the National Cancer Institute* 90:6 (18 March 1998): 455–60; R. E. Harris, K. K. Namboodiri and W. B. Farrar, "Nonsteroidal Anti-inflammatory Drugs and Breast Cancer," *Epidemiology* 7:2 (March 1996): 203–5.
14. U.S. Food and Drug Administration, "FDA Approves Celebrex for New Indication," FDA Talk Paper, 23 December 1999; see also J. K. Kelly et al., "The Pathogenesis of Inflammatory Polyps," *Diseases of the Colon and Rectum* 30:4 (April 1987): 251–54.
15. X. Z Ding, W. G. Tong and T. E. Adrian, "Blockade of Cyclooxygenase-2 Inhibits Proliferation and Induces Apoptosis in Human Pancreatic Cancer Cells," *Anticancer Research* 20:4 (July/August 2000): 2625–31; M. T. Yip-Schneider et al., "Cyclooxygenase-2 Expression in Human Pancreatic Adenocarcinomas," *Carcinogenesis* 21:2 (February 2000): 139–46.
16. Bas A. int Veld et al., "Nonsteroidal Anti-inflammatory Drugs and the Risk of Alzheimer's Disease," *New England Journal of Medicine* 345:21 (22 November 2001): 1515–21; Susan Okie, "Ibuprofen, Similar Medicines May Slash Alzheimer's Risk," *Washington Post,* 22 November 2001, p. A4; Food and Drug Administration, "New

Prescribed Uses of Aspirin: Questions and Answers," 20 October 1998, at: http://www.fda.gov/cder/news/aspirin/aspirin_QA.htm

17. "Breast Cancer Adjuvant Therapy Patients Benefit from Exercise," *ACS News*, 21 December 2000, at: http://www2.cancer.org/zine/index.cfm?fn=001_12212000_0; Keith Mulvihill, "Large Study Provides Clues to Inherited Breast Cancer," *Reuters Health*, 6 October 2000.
18. See the website of the National Cancer Institute Early Detection Network, at: http://cancer.gov/prevention/cbrg/edrn/
19. http://www.unimi.it/engl/
20. Ida Martinelli et al., "Mutations in Coagulation Factors in Women with Unexplained Late Fetal Loss," *New England Journal of Medicine* 343:14 (5 October 2000): 1015–18. (Note: The researchers also claimed a threefold increased risk for women with mutations for another clotting factor, factor V, but the increase was not statistically significant.)
21. "Mutations Linked to Miscarriages Late in Pregnancy," *Reuters Health*, 4 October 2000.
22. Helen Lippman, "Genetic Discrimination: Just Say No," *Medical Laboratory Observer* 32:4 (1 April 2000): 8.
23. Council for Responsible Genetics, "Genetic Discrimination Legislation in the United States, August 2000," at: http://www.gene-watch.org/programs/GD_onepage_00.html
24. Sarah Schafer, "EEOC Sues to Halt Worker Gene Tests," *Washington Post*, 10 February 2001, p. A1.
25. Thomas D. Bird, "Hereditary Neuropathy with Liability to Pressure Palsies," at: http://www.geneclinics.org/profiles/hnpp/details.html
26. Matt Curry, "Railroad Gene Test Baffles Ethicists," *Associated Press*, 18 February 2001.
27. Ibid.
28. Daniel Eisenberg, "Viaticals Let Sick People Sell Their Life Insurance, But Investing in Deaths May Not Always Be the Best," *Time*, 29 November 1999, p. 112(S). (Not available in all editions.)
29. Eric Pianin, "D.C. Yields on AIDS Insurance; Council Reluctantly Agrees to Change City Hiring Policy," *Washington Post*, 30 November 1988, p. A1.
30. Lisa S. Howard, "U.K. Companies Get OK to Use Specific Genetic Test," *National Underwriter*, 23 October 2000, p. 44.

31. "Huntington Disease," National Center for Biotechnology Information, at: http://www.ncbi.nlm.nih.gov/disease/Huntington.html
32. See Matt Ridley, *Genome: The Autobiography of a Species in 23 Chapters* (New York: HarperCollins, 1999), pp. 55–56.
33. Tod Zwillich, "Women's Insurance Not Affected by Cancer Gene Status," *Reuters Health,* 18 January 2000.
34. For a thoughtful and iconoclastic essay on ethics of the use of genetic testing, see Michael Kinsley, "Genetic Correctness," *Washington Post,* 18 April 2000, p. A29. (He had no easy answers either, but at least he asked some of the right questions.)
35. See Anastasia Toufexis, Thomas McCarroll and Raji Samghabadi, "Convicted by Their Genes: A New Forensic Test Is Revolutionizing Criminal Prosecutions," *Time,* 31 October 1988, p. 74.
36. Adam Cohen, "Innocent, After Proven Guilty; More Inmates Are Being Set Free Thanks to DNA Tests—and a Pioneering Law Clinic," *Time,* 13 September 1999, p. 26; Ronald Bailey, "Unlocking the Cells; Post-Conviction DNA Testing of Prisoners," *Reason* 31:8 (January 2000): 50.
37. Paul Duggan, "DNA Frees Man, Condemns Another," *Washington Post,* 16 August 2000, p. A2.
38. For example, see Adam Cohen, "Innocent, after Proven Guilty; More Inmates Are Being Set Free Thanks to DNA Tests."
39. W. Goodwin, A. Linacre and P. Vanezis, "The Use of Mitochondrial DNA and Short Tandem Repeat Typing in the Identification of Air Crash Victims," *Electrophoresis* 20:8 (June 1999): 1707–11; J. A. Thomson, "Validation of Short Tandem Repeat Analysis for the Investigation of Cases of Disputed Paternity," *Forensic Science International* 100:1-2 (15 March 1999): 1–16; "DNA Evidence Being Used to Prosecute Poachers," *Associated Press State & Local Wire,* 2 November 1998. (It's used to detect whether the meat that somebody's selling is really from a specific "road kill" or not. If it's not from the road kill, it means they shot the animal.) D. Corach et al., "Additional Approaches to DNA Typing of Skeletal Remains: The Search for 'Missing' Persons Killed During the Last Dictatorship in Argentina," *Electrophoresis* 18:9 (August 1997): 1608–12.
40. "Obituary of Prince Rostislav Romanoff, Prince Whose DNA Samples Helped to Disprove the Claim of Anna Anderson to Be the Last Tsar's Daughter Anastasia," *Daily Telegraph* (London), 14 January 1999, p. 27.

41. Tim O'Neil and Valerie Schremp, "St. Louisan's Remains Were in Tomb of Unknowns; Blassie's Family Had Sought Burial Here," *St. Louis Post-Dispatch,* 30 June 1998, p. A1.
42. Brooke A. Masters, "Empty Tomb Haunts Arlington; Finding Unknown Soldier from Vietnam Is Unlikely," *Washington Post,* 26 February 1999, p. B1.
43. Lawrence K. Altman, "Now, Doctors Must Identify the Dead," *New York Times,* 25 September 2001, p. D1; Sally Jenkins, "The Quest for 6,347 Identities," *Washington Post,* 27 September 2001, p. A1.

"Baby by Versace"?

1. As quoted in Colum Lynch, "U.S. Seeks to Extend Ban on Cloning," *Washington Post,* 27 February 2002, p. A8.
2. Fox Studios, *The Boys from Brazil,* Franklin J. Schaffner, director, 1978; 20th Century Fox, *Star Wars Episode II: Attack of the Clones,* George Lucas, director, 2002.
3. See William Saletan, "Fool Me Twice," *Slate,* 9 January 2003, at: http://www.slate.msn.com/id/2076561/
4. Jose B. Cibelli et al., "The First Human Cloned Embryo," *Scientific American* 286:1 (January 2002): 44–51; Rick Weiss, "First Human Embryos Are Cloned in U.S.," *Washington Post,* 26 November 2001, p. A1; a *Nature Biotechnology* editorial ripped ACT, not only for not doing what they had claimed but for contributing to what it said was eliciting "comical media coverage." See "Publish and Be Damned," *Nature Biotechnology* 20:1 (January 2002): 1.
5. Gina Kolata and Andrew Pollack, "A Breakthrough on Cloning? Perhaps, or Perhaps Not Yet," *New York Times,* 27 November 2001, p. A1.
6. Inder Verma, "Ban Reproductive but Not Therapeutic Cloning," *San Diego Union-Tribune,* 19 April 2002, p. B9.
7. Peter McGuffin and Irving I. Gottesman, "Risk Factors for Schizophrenia," *New England Journal of Medicine* 341:5 (29 July 1999): 370–71.
8. Alison Pike and Robert Plomin, "Importance of Nonshared Environmental Factors for Childhood and Adolescent Psychopathology," *Journal of the American Academy of Child and Adolescent Psychiatry* 35:5 (May 1996): 560–70; Malcolm Ritter, "Twins More Identical than Clones, Scientists Say," *Associated Press,* 30 December 2002.
9. Charles Arthur and Jeremy Laurance Health, "Cloning," *Independent on Sunday* (London), 11 January 1998, p. 8, at: http://

www.nepalonline.net/kwc/Science_Popularation/cloning_questions_answers.htm

10. Jeffrey Kluger and Sora Song, "Here, Kitty, Kitty!" *Time,* 25 February 2002): 58.
11. For a photo of both cats, see: http://images.google.com/imgres?imgurl=www.accessexcellence.org/WN/graphics/cat2_160.jpg&imgrefurl=http://www.accessexcellence.org/WN/SU/SU102001/copycat.html&h=192&w=200&prev=/images%3Fq%3D%2522carbon%2Bcopy%2522%2Bcat%26svnum%3D10%26hl%3Den%26lr%3D%26ie%3DUTF-8%26oe%3DUTF-8%26safe%3Doff%26sa%3DG
12. Claire Ainsworth, "You Are What You Eat: Embryos That Get Off to a Poor Start Are Marked for Life," *New Scientist* 167:2257 (23 September 2000): 18.
13. See Gina Kolata, "Researchers Find Big Risk of Defect in Cloning Animals," *New York Times,* 25 March 2001, p. A1.
14. Kevin Eggan et al., "Hybrid Vigor, Fetal Overgrowth, and Viability of Mice Derived by Nuclear Cloning and Tetraploid Embryo Complementation," *Proceedings of the National Academy of Sciences USA* 98:11 (22 May 2001): 6209–14; A. Dinnyes et al., "Somatic Cell Nuclear Transfer: Recent Progress and Challenges," *Cloning Stem Cells* 4:1 (April 2002): 81–90; Ian Wilmut, "Cloning and Stem Cells," *Cloning Stem Cells* 4:1 (April 2002): 1; Maggie Fox, "Clones Flawed: Study Determines All Clones Are Genetically Abnormal," *Associated Press,* 11 September 2002.
15. Calvin R. Simerly et al., "Molecular Correlates of Primate Nuclear Transfer Failures," *Science* 300:5617 (11 April 2003): 297; Rick Weiss, "Study Shows Problems in Cloning People; Researchers Find Replicating Primates Is Harder than for Other Mammals," *Washington Post,* 11 April 2003, p. A12.
16. See Ronald Bailey, "The Twin Paradox: What Exactly Is Wrong with Cloning People?" *Reason* 29:1 (May 1997): 52, at: http://www.reason.com/9705/col.bailey.html
17. See Laura Ciampa, "Infanticide: Children as Chattel," *American Feminist* 6:4 (Winter 1999): 6, at: http://www.feministsforlife.org/taf/1999/winter/chldchat.htm; see John Pomfret, "In China's Countryside, 'It's a Boy!' Too Often," *Washington Post,* 29 May 2001, p. A1.
18. Rick Weiss, "Alzheimer's Gene Screened from Newborn," *Washington Post,* 27 February 2002, p. A1; Jill R. Murrell et al., "Early-Onset Alzheimer Disease Caused by a New Mutation (V717L) in the

Amyloid Precursor Protein Gene," *Archives of Neurology* 57:6 (June 2000): 885–87; Yury Verlinsky et al., "Preimplantation Diagnosis for Early-Onset Alzheimer Disease Caused by V717L Mutation," *Journal of the American Medical Association* 287:8 (27 February 2002): 1018–21.

19. Mark S. Frankel and Audrey R. Chapman, *Human Inheritable Genetic Modifications: Assessing Scientific, Ethical, Religious and Policy Issues* (Washington, D.C.: American Association for the Advancement of Science, 2000), p. 9; at: http://www.aaas.org/spp/dspp/sfrl/germline/report.pdf
20. See Clifton E. Anderson, "Genetic Engineering: Dangers and Opportunities," *Futurist* 34:2 (1 March 2000): 20; Gina Kolata, "Gene Technique Could Shape Future Generations," *New York Times*, 22 November 1994, p. A1.
21. Paul Rabinow, "Getting to Know You," *Nature Biotechnology* 20:6 (June 2002): 545.
22. Francis Fukuyama, *Our Posthuman Future: Consequences of the Biotechnology Revolution* (New York: Farrar, Strauss & Giroux, 2002), pp. 76–102.
23. Johan Wessberg et al., "Real-Time Prediction of Hand Trajectory by Ensembles of Cortical Neurons in Primates," letter, *Nature* 408:6810 (16 November 2000): 361–65; Sandra Blakeslee, "Brain Signals Shown to Move a Robot's Arm," *New York Times*, 16 November 2000, p. A20.
24. See Don Colburn, "Wired for Sound; Deaf since Birth, Jeffrey Stibick Decided to Try a Cochlear Implant at Age 59," *Washington Post*, 3 October 2000, p. Z12.
25. "Limbaugh Has Operation to Regain Hearing," *Philadelphia Inquirer*, 22 December 2001, p. C8.
26. http://www.cochlear.com/MissAmerica.asp
27. "Artificial Vision for the Blind," at: http://www.artificialvision.com/vision/index.html; see also John Elliott, "Bionic Eye Restores Partial Sight to Woman Blind for 40 Years," *Sunday Times* (London), 10 December 2000; "New Artificial Vision System May Open New Horizons for the Blind," *Medical Industry Today*, 18 January 2000.
28. Anne Eisenberg, "A Chip that Mimics Neurons, Firing Up the Memory," *New York Times*, 20 June 2002, p. G7.
29. G. Q. Maquire and Ellen McGee, "Implantable Brain Chips? Time for Debate," *Hastings Center Report* 29:1 (1 January 1999): 7; Eisenberg, "A Chip that Mimics Neurons, Firing Up the Memory."

30. Thomas E. Ricks, "U.S. Arms Unmanned Aircraft," *Washington Post,* 18 October 2001, p. A1.
31. Robert P. Hey, "New Push for Robotic Aircraft," *Christian Science Monitor,* 22 June 2001, p. 1.
32. "About Omnitech Robotics," at: http://www.omnitech.com/company.htm
33. Dinesh D'Souza, *The Virtue of Prosperity: Finding Values in an Age of Techno-Affluence* (New York: Free Press, 2000), p. 223.
34. Freeman J. Dyson, *The Sun, the Genome, and the Internet* (New York: Oxford University Press, 1999), p. xii.

Will Biotech Break the Bank?

1. Eric J. Topol et al., "Comparison of Two Platelet Glycoprotein IIb/IIIa Inhibitors, Tirofiban and Abciximab, for the Prevention of Ischemic Events with Percutaneous Coronary Revascularization," *New England Journal of Medicine* 344:25 (21 June 2001): 1888–94.
2. For example, Stuart O. Schweitzer, *Pharmaceutical Economics and Policy* (London and New York: Oxford University Press, 1997); Frank A. Sloan, ed., *Valuing Health Care: Costs, Benefits, and Effectiveness of Pharmaceuticals and Other Medical Technologies* (Cambridge, England: Cambridge University Press, 1996); W. Duncan Reekie and Michael H. Weber, *Profits, Politics, and Drugs* (New York, Holmes & Meier, 1979).
3. W. Michael Cox and Richard Alm, *Myths of Rich and Poor* (New York: Basic Books, 1999), pp. 41–44.
4. S. C. Fagan et al., "Cost-Effectiveness of Tissue Plasminogen Activator for Acute Ischemic Stroke," *Neurology* 50:4 (April 1998): 883–90.
5. R. F. Legg, "Cost Benefit of Sumatriptan to an Employer," *Journal of Occupational and Environmental Medicine* 39:7 (July 1997): 652–57. For more such comparisons, see Pharmaceutical Research and Manufacturers Association, "How Can We Calculate the Value of Pharmaceuticals?" at: http://www.phrma.org/publications/publications/brochure/questions/calculatevalue.phtml#fig3
6. The enzyme is Glucose-6-phosphate dehydrogenase, or G6PDH.
7. Organization for Economic Cooperation and Development, *Biotechnology for Clean Industrial Products and Processes* (Paris: OECD, 1998): p. 71, table 3.8, citing J. Wiesner, "Production-integrated Environmental Protection," *Ullmann's Encyclopedia of Industrial Chemistry,* vol. B8 (1995), pp. 213–309.
8. Huntly Collins, "Early Trial of Gene Therapy for Hemophilia Goes Well," *Pittsburgh Post-Gazette,* 2 March 2000, p. A3.

9. Alan Dove, "Cell-Based Therapies Go Live," *Nature Biotechnology* 20:4 (April 2002): 339.
10. For example, Layton Biosciences at: http://www.laytonbio.com/lead.htm; for example, CellExSys, at: http://www.targetedgenetics.com/cellexsys/; for example, Geron Corp. at: http://www.geron.com/merixbio/print.pr_082900.html
11. Andrew Pollack, "New Ventures Aim to Put Farms in Vanguard of Drug Production," *New York Times,* 14 May 2000, p. 1.
12. As quoted in ibid.
13. Personal telephone communication with Kris Briggs, 29 August 2000.
14. See Angelo DePalma, "Addressing Process Speed and Optimization," *Genetic Engineering News* 22:21 (December 2002): 44.
15. B. D. Purdy and K. I. Plaisance, "Infection with the Human Immunodeficiency Virus: Epidemiology, Pathogenesis, Transmission, Diagnosis, and Manifestations," *American Journal of Hospital Pharmacy* 46:6 (June 1989): 1185–209.
16. http://www.bionor.no/
17. For a greater description and photos, see Binor's website at: http://www.bionor.no/hiv_test.html
18. "HIV/AIDS Diagnostics: You Name It, It's There," *Genesis Report-Dx* 8:1 (July 1998): 21.
19. http://www.qualigeninc.com
20. John-Manual Andriote, "No-Wait Prostate Test Results," *Washington Post,* 6 February 2001, p. H6; "Blood Testing System Enters Point-of-Care Market in U.S.," *Medical Industry Today,* 12 July 2000.
21. "Blood Testing System Enters Point-of-Care Market in U.S." For an animated demonstration of how the system works, see Qualigen's website at: http://www.qualigeninc.com
22. Thomas Kupper, "Qualigens Blood-Test Device Has FDA's Nod; Small Carlsbad Biotech Working on Refinements," *San Diego Union-Tribune,* 11 July 2000, p. C1.
23. http://urology.ucdmc.ucdavis.edu/
24. As quoted in "Blood Testing System Enters Point-of-Care Market in U.S."

Part Two: The Fountain of Youth

Turning Back the Clock

1. See Michael Fumento, *The Fat of the Land: The Obesity Epidemic and How Overweight Americans Can Help Themselves* (New York: Viking, 1997), pp. 131–63.

2. See John R. Wilmoth et al., "Increase of Maximum Life-Span in Sweden, 1861–1999," *Science* 289:5488 (29 September 2000): 2366.
3. His homepage is: http://www.demog.berkeley.edu/~jrw/
4. John Wilmoth, as quoted in Roger Highfield, "We Can Live Beyond 120, Claims New Study," *Daily Telegraph* (London), 29 September 2000, p. 14.
5. "Researchers Find That Bowhead Whales Can Live 200 Years," *Associated Press,* 19 December 2000.
6. See Thomas B. Kirkwood and Steven N. Austad, "Why Do We Age?" *Nature* 408:6808 (2 November 2000): 233–38.
7. Stuart D. Tyner et al. "p53 Mutant Mice That Display Early Ageing-Associated Phenotypes," *Nature* 415:6867 (3 January 2002): 45–53.
8. See, for example, Justin Gillis, "A Critical Protein That Protects Animals from Cancer in Their Early Years Appears, in Later Life, to Cause Much of the Deterioration Associated with Aging, According to a Provocative New Study," *Washington Post,* 3 January 2002, p. A1.
9. As quoted in John Whitfield, "Cancer-Proof Mice Age Prematurely," *Nature Science Update,* 3 January 2002.
10. Stanley J. Colcombe et al., "Aerobic Fitness Reduces Brain Tissue Loss in Aging Humans," *Journals of Gerontology Series A: Biological Sciences and Medical Sciences* 58:2 (February 2003): 176–80; Arthur F. Kramer and Stanley J. Colcombe, "Fitness Effects on the Cognitive Function of Older Adults: A Meta-Analytic Study," *Psychological Science* 14:2 (March 2003): 125–30; Press Release, "Study Is First to Confirm Link between Exercise and Changes in Brain," 27 January 2003, at: http://www.news.uiuc.edu/scitips/03/0127exercise.html
11. Aubrey D. N. J. de Grey et al., "Time to Talk SENS: Critiquing the Immutability of Human Aging," *Annals of the New York Academy of Sciences* 959 (April 2002): 452–62.
12. Jan Karlseder, Agata Smogorzewska and Titia de Lange, "Senescence Induced by Altered Telomere State, not Telomere Loss," *Science* 295:5564 (29 March 2002): 2446–49; University of Texas Southwestern Medical Center at Dallas, Shay/Wright Laboratory, "Basic Introduction to Telomeres," at: http://www.utsouthwestern.edu/home_pages/cellbio/shay-wright/intro/sw_intro.html
13. For a longer discussion of telomeres in general and telomeres and aging, see generally Elizabeth H. Blackburn and Carol W. Greider,

eds., *Telomeres* (Plainview, New York: Cold Spring Harbor Laboratory Press, 1999).

14. His personal website is: http://www.genetics.utah.edu/faculty/rcawthon.html
15. Richard M. Cawthon et al., "Association between Telomere Length in Blood and Mortality in People Aged 60 Years or Older," *Lancet* 361:9355 (1 February 2003): 393–95; Celeste Biever, "Shorter Telomeres Mean Shorter Life," *New Scientis,* 31 January 2003, at: http://www.newscientist.com/news/news.jsp?id=ns99993337
16. http://www3.utsouthwestern.edu/
17. Andrea G. Bodnar et al., "Extension of Life-Span by Introduction of Telomerase into Normal Human Cells," *Science* 279:5349 (16 January 1998): 349–52; Terence Monmaney, "A Real Blast: Defusing 'Genetic Time Bomb,'" *Los Angeles Times,* 18 January 1998, p. A1.
18. As quoted in Rick Weiss, "Scientists Discover Way to Prolong Life of Cells; Enzyme Seems to Put Aging 'on Hold,'" *Washington Post,* 14 January 1998, p. A1.
19. Walter D. Funk et al., "Telomerase Expression Restores Dermal Integrity to *in Vitro*-Aged Fibroblasts in a Reconstituted Skin Model," *Experimental Cell Research* 258:2 (August 2000): 270–78.
20. Press Release, "Geron Publication Describes *in Vivo* Results of Telomerase Activation," *Business Wire,* 13 November 2000.
21. Michael Fossel, *Reversing Human Aging* (New York: William Morrow, 1996), pp. 158–61.
22. Ulf M. Lindström et al., "Artificial Human Telomeres from DNA Nanocircle Templates," *Proceedings of the National Academy of Sciences USA* 99:25 (10 December 2002): 15,953–58; "Enzyme Treatment Could Lengthen Cell Life," *United Press International,* 18 November 2002.
23. See Titia de Lange, "Telomeres and Senescence: Ending the Debate," *Science* 279:5349 (16 January 1998): 334–35.
24. Titia de Lange and Ronald DePinho, "Unlimited Mileage from Telomerase?" *Science* 283:5404 (12 February 1999): 947.
25. H. Matsunaga, "Beta-galactosidase Histochemistry and Telomere Loss in Senescent Retinal Pigment Epithelial Cells," *Investigative Ophthalmology and Visual Science* 40:1 (January1999): 197–202; Karl L. Rudolph et al., "Inhibition of Experimental Liver Cirrhosis in Mice by Telomerase Gene Delivery," *Science* 287:5456 (18 February 2000): 1253–58; Lorene Leiter, "Liver Cirrhosis in Mice Inhibited by Telomerase Gene Therapy," *Web Weekly* (News from the Harvard Medical Community), 21 February 2000, at: http://www.hms.

harvard.edu/webweekly/archive/2000/2_21/headlines.html; Edwin Chang and Calvin B. Harley, "Telomere Length and Replicative Aging in Human Vascular Tissues," *Proceedings of the National Academy of Sciences USA* 92:24 (21 November 1995): 11,190–94.

26. Leiter, "Liver Cirrhosis in Mice Inhibited by Telomerase Gene Therapy."
27. As quoted in Nicholas Wade, "Teaching the Body to Heal Itself; Work on Cells Signals Fosters Talk of a New Medicine," *New York Times,* 7 November 2000, p. F1.
28. http://www.advancedcell.com/
29. As quoted in Wade, "Teaching the Body to Heal Itself."
30. Roger Highfield, "'Methuselah Mice' Resist Aging with Gene Defect," *Daily Telegraph* (London), 18 November 1999, p. 17.
31. Enrica Migliaccio et al., "The P66shc Adaptor Protein Controls Oxidative Stress Response and Life Span in Mammals," *Nature* 402:6759 (1999): 309–13; David N. Leff, "Long-Lived Mice Stall Aging Process," *BioWorld Today* 10:221 (18 November 1999).
32. http://www.ieo.cilea.it/inglese/Welcome.html
33. As quoted in Patricia Reaney, "Scientists Find Protein to Control Lifespan in Mice," *Reuters,* 17 November 1999.
34. http://www.caltech.edu/; his personal website is: http://www.cco.caltech.edu/~biology/brochure/faculty/benzer.html; Yi-Jyun Lin, Laurent Seroude and Seymour Benzer, "Extended Life-Span and Stress Resistance in the Drosophila Mutant Methuselah," *Science* 282:5390 (30 October 1998): 943.
35. As quoted in Elizabeth Pennisi, "Single Gene Controls Fruit Fly Life-Span," *Science* 282:5390 (30 October 1998): 856.
36. http://www.agingresearch.org/; as quoted in Richard A. Knox, "Fruit Fly Study Ties Gene, Life Span; Implications Seen for Human Aging," *Boston Globe,* 30 October 1998, p. A12.
37. As quoted in Nicola Dixon, "Gene Tweaking Safely Doubles Life-span," *New Scientist,* 24 October 2002, at: http://www.newscientist.com/news/news.jsp?id=ns99992969; http://www.ucsf.edu/tetrad/
38. Andrew Dillin, Douglas K. Crawford and Cynthia Kenyon, "Timing Requirements for Insulin/IGF-1 Signaling in *C. elegans,*" *Science* 298:5594 (25 October 2002): 830–34; Dixon, "Gene Tweaking Safely Doubles Lifespan."
39. As quoted in Pennisi, "Single Gene Controls Fruit Fly Life-Span."

40. Martin Holzenberger, "IGF-1 Receptor Regulates Lifespan and Resistance to Oxidative Stress in Mice," *Nature* 42:6919 (9 January 2003): 182–87; Roger Highfield, "Mice Genes Raise Hopes of Longer Life," *Daily Telegraph* (London), 11 December 2002, p. 18.
41. Laura A. Herndon et al., "Stochastic and Genetic Factors Influence Tissue-Specific Decline in Aging *C. elegans*," *Nature* 419:6909 (24 October 2002): 808–14; see also accompanying commentary, Thomas B. L. Kirkwood and Caleb E. Finch, "Aging: The Old Worm Turns More Slowly," *Nature* 419 (2002): 794–95.
42. Monica Driscoll, as quoted in William McCall, "Study on Worms May Hold Clues to Aging," *Associated Press*, 23 October 2002.
43. As quoted in ibid.
44. For more on this program, see website at: http://www.healthandage.com/positivelifestyles/new_england/index.htm
45. As quoted in Jennifer Maddox, "The Constant Search for Good Health in Old Age," *Press Journal* (Vero Beach, Fl.), 2 June 1999, p. C1.
46. Ibid.
47. As quoted in Karyn Miller-Medzon, "Seniors: Healthful Habits Lead to Longevity," *Boston Herald*, 4 November 1999, p. B1.
48. As quoted in Dorothy Grant, "Methuselah's Secrets: If You Ask Centenarians the Secret to Their Long Lifespans, Be Prepared for Some Surprising Answers," *Medical Post* 36 (7 March 2000): 44.
49. See Gary Taubes, "The Famine of Youth," *Scientific American* 11:2 (Summer 2000): 44.
50. Richard Weindruch, "Caloric Restriction and Aging," *Scientific American* 274:1 (January 1996): 32–38.
51. Mark A. Lane, "Nonhuman Primate Models in Biogerontology," *Experimental Gerontology* 35:5 (1 August 2000): 533–41; Nell Boyce, "How to Defy Death," *New Scientist* 165:2231 (25 March 2000): 20.
52. http://www.medsch.ucla.edu/
53. As quoted in Taubes, "The Famine of Youth."
54. See Michael Fumento, *The Fat of the Land* (New York: Viking, 1997), pp. 33–35.
55. If you think I'm kidding, see the photo on page 44 in Taubes, "The Famine of Youth."
56. http://web.mit.edu/biology/
57. Su-Ju Lin, Pierre-Antoine Defossez and Leonard Guarente, "Requirement of NAD and *SIR2* for Life-Span Extension by Calorie Restriction in *Saccharomyces cerevisiae*," *Science* 289:5487 (22 Sep-

tember 2000): 2126–28; Leonard Guarente and Cynthia Kenyon, "Genetic Pathways That Regulate Aging in Model Organisms," *Nature* 408:4808 (9 November 2000): 255–62.

58. As quoted in Lidia Wasowicz, "Molecule Slows Aging by Silencing Genes," *United Press International*, 23 February 2000.
59. As quoted in ibid.
60. As quoted in Alan Macdermid, "Staying Young in Old Age; Scientists Aim to Develop Pill to Stop the Clock on Illnesses That Dog Twilight Years," *Herald* (Glasgow), 17 February 2000, p. 11.
61. Heidi A. Tissenbaum and Leonard Guarente, "Increased Dosage of a sir-2 Gene Extends Lifespan in Caenorhabditis Elegans," *Nature* 410:6825 (8 March 2001): 227–30.
62. As quoted in Patricia Reaney, "Genetic Tinkering Makes Roundworms Live Longer," *Reuters*, 7 March 2001.
63. http://genetics.uchc.edu/
64. Blanka Rogina, Stephen L. Helfand and Stewart Frankel, "Longevity Regulation by Drosophila Rpd3 Deacetylase and Caloric Restriction," letter, *Science* 298:5599 (29 November 2002): 1745.
65. As quoted in Lisa Richwine, "U.S. Study Explores Why Eating Less Extends Life," *Reuters*, 28 November 2002.
66. Karen T. Chang and Kyung-Tai Min, "Regulation of Lifespan by Histone Deacetylase," *Ageing Research Reviews* 1:3 (June 2002): 313–26.
67. As quoted in Nicholas Wade, "A Pill to Extend Life? Don't Dismiss the Notion Too Quickly," *New York Times*, 22 September 2000, p. A20; http://www.ssc.wisc.edu/aging/
68. Ibid.
69. As quoted in Nell Boyce, "How to Defy Death," *New Scientist* 165:2231 (25 March 2000): 20.
70. His personal website is: http://www.biochem.ucr.edu/faculty/spindler.html
71. Shelley X. Cao et al., "Genomic Profiling of Short- and Long-Term Caloric Restriction Effects in the Liver of Aging Mice," *Proceedings of the National Academy of Sciences USA* 98:10 (11 September 2001): 10,630–35; Susan Okie, "Low-Calorie Diet Slows Aging in Mice in Study," *Washington Post*, 4 September 2001, p. A10.
72. "Reversing Aging Rapidly with Short-Term Calorie Restriction," Life Extension Foundation–Funded Research Breakthrough Published in the *Proceedings of the National Academy of Sciences*, An Interview

with Stephen R. Spindler, Ph.D., at: http://www.lef.org/featured-articles/spindler_press_release01.html

73. Ibid.
74. Personal telephone communication of 9 October 2000.
75. His personal website is: http://www.bruceames.org; "Ames Awarded Medal of Honor," NIEHS News, *Environmental Health Perspectives* 107:5 (May 1999), at: http://ehpnet1.niehs.nih.gov/docs/1999/107-5/niehsnews.html#ames; see S. Robert Lichter and Stanley Rothman, *Environmental Cancer—A Political Disease?* (New Haven: Yale University Press, 1999), p. 162, table 5.2. (Ames was rated first by polled members of the American Association for Cancer Research for "level of confidence in expertise on environmental cancer.")
76. Personal telephone communication of 9 October 2000.
77. For an animation of ATP synthesis in mitochondria, see: http://www.sp.uconn.edu/~terry/images/anim/ATPmito.html
78. Arlan Richardson, director of the Longevity and Aging Center at the University of Texas, San Antonio, as quoted in David N. Leff, "Black-Hatted Oxygen, No Friend of the Aged: Probe in Rodents of Oxidative DNA Damage, Low-Calorie Diet Elucidates Aging Ills in Humans," *BioWorld Today* 12:163 (22 August 2001). The study is: Michelle L. Hamilton et al., "Does Oxidative Damage to DNA Increase with Age?" *Proceedings of the National Academy of Sciences USA* 98:18 (28 August 2001): 10,469–74.
79. As quoted in Andrew Hon et al., "Interview with Bruce Ames," *Berkeley Scientific Journal* 4:1 (Spring 2000): 13.
80. As quoted in Roger Highfield, "Mice Genes Raise Hopes of Longer Life," *Daily Telegraph* (London), 11 December 2002, p. 18.
81. http://www.hnrc.tufts.edu/scientists/people/bhale.php; Barbara Shukitt-Hale as quoted in Jane E. Brody, "Studies of the Infirmities of Aging Dogs Offer Insights for Humans," *New York Times,* 5 February 2002, p. D4.
82. Brody, "Studies of the Infirmities of Aging Dogs Offer Insights for Humans."
83. Mary Sano et al., "A Controlled Trial of Selegiline, Alpha-tocopherol, or Both as Treatment for Alzheimer's Disease: The Alzheimer's Disease Cooperative Study," *New England Journal of Medicine* 336:17 (24 April 1997): 1216–22; Danielle Laurin et al., "Vitamin E and C Supplements and Risk of Dementia," letter, *Journal of the American Medical Association* 288:68 (13 November 2002): 2266–68; Juhie Bhatia, "Taking Certain Antioxidant Vitamin

Supplements Will Not Prevent Dementia in Old Age, New Study Findings Suggest," *Reuters Health*, 2 December 2002.

84. Santy Daya, "Stress and the Brain," *Science in Africa*, June 2001, at: http://www.scienceinafrica.co.za/2001/june/stress.htm
85. Carol Jenkins et al., "Antioxidants: Their Role in Pregnancy and Miscarriage," *Antioxidants and Redox Signaling* 2:3 (Fall 2000): 623–28; Pervin Vural et al., "Antioxidant Defense in Recurrent Abortion," *Clinica Chimica Acta* 295:1-2 (May 2000): 169–77.
86. Jiankang Liu et al., "Memory Loss in Old Rats Is Associated with Brain Mitochondrial Decay and RNA/DNA Oxidation: Partial Reversal by Feeding Acetyl-L-Carnitine and/or R-Lipoic Acid," *Proceedings of the National Academy of Sciences USA* 99:4 (19 February 2002): 2356–61.
87. Jiankang Liu, David W. Killilea and Bruce N. Ames, "Age-Associated Mitochondrial Oxidative Decay: Improvement of Carnitine Acetyltransferase Substrate-Binding Affinity and Activity in Brain by Feeding Old Rats Acetyl-L-Carnitine and/or R-Lipoic Acid," *Proceedings of the National Academy of Sciences USA* 99:4 (19 February 2002): 1876–81, figure 3.
88. Personal telephone communication of 18 October 2002.
89. Liu, Killilea and Ames, "Age-Associated Mitochondrial Oxidative Decay."
90. For an illustration and description of the maze and how it works, see: http://www.wiley.co.uk/cp/cpns/fig852.htm
91. Jiankang Liu et al., "Memory Loss in Old Rats Is Associated with Brain Mitochondrial Decay and RNA/DNA Oxidation," figure 1.
92. Tory M. Hagen et al., "Feeding Acetyl-L-Carnitine and Lipoic Acid to Old Rats Significantly Improves Metabolic Function while Decreasing Oxidative Stress," *Proceedings of the National Academy of Sciences USA* 99:4 (19 February 2002): 1870–75.
93. Personal telephone communication of 9 October 2000.
94. S. Melov et al., "Extension of Life-Span with Superoxide Dismutase/catalase Mimetics" *Science* 189:5484 (1 September 2000): 1567–69.
95. Clifford W. Shults et al., "Effects of Coenzyme Q10 in Early Parkinson Disease," *Archives of Neurology* 59:10 (October 2002): 1541–50; Lindsey Tanner, "Dietary Supplement May Slow Parkinson's Disease," *Associated Press*, 14 October 2002.
96. For example, Gabriele Siciliano, "Coenzyme Q10, Exercise Lactate and CTG Trinucleotide Expansion in Myotonic Dystrophy," *Brain Research Bulletin* 56:3-4 (October/November 2001): 405–10.

97. Joel Garreau, "Forever Young: Suppose You Soon Can Live to Well over 100, As Vibrant and Energetic as You Are Now," *Washington Post*, 13 October 2002, p. F1; http://www.juvenon.com; personal telephone communication of 26 October 2002.
98. As quoted in Jennifer Maddox, "Scientists Search for Good Health in Old Age," *Cleveland Plain Dealer*, 31 May 1999, p. 10A.
99. See Yousry Naguib, "Antioxidants: A Technical Overview," *Nutraceuticals World* 2:2 (March/April 1999): 40–42.
100. Andrew Hon et al., "Interview with Bruce Ames," *Berkeley Scientific Journal* 4:1 (Spring 2000): 13.
101. Personal telephone communication of 9 October 2000.
102. Ibid.
103. For more on upper limits of antioxidants, see Kristine Napier, "Too Many Vitamins? Health Effects of Taking Megadoses of Vitamins," *Harvard Health Letter* 31:3 (January 1996): 1.
104. Institute of Medicine, *Dietary Reference Intakes for Vitamin A, Vitamin K, Arsenic, Boron, Chromium, Copper, Iodine, Iron, Manganese, Molybdenum, Nickel, Silicon, Vanadium, and Zinc* (Washington, D.C.: National Academy Press, 2001); Lauran Neergaard, "Too Much Vitamin A Dangerous; A Report Sets Nutrient Guidelines," *Associated Press*, 9 January 2001; "Easy on the A; Vitamin A Consumption Levels by Women Who Are Not Pregnant or Who Are Pregnant Should Not Be Excessive Due to Higher Risks of Birth Defects," *Tufts University Diet & Nutrition Letter* 13:11 (January 1996): 3.
105. Jonathan Swift, *Gulliver's Travels* (New York: Oxford University Press, 1977), pp. 204–12.
106. Kenneth G. Manton and XiLiang Gu, "Changes in the Prevalence of Chronic Disability in the United States Black and Nonblack Population above Age 65 from 1982 to 1999," *Proceedings of the National Academy of Sciences USA* 98:11 (22 May 2001): 6354–59; Ceci Connolly, "Aging Americans Are Staying Healthier, Study Finds," *Washington Post*, 8 May 2001, p. A10.
107. Elizabeth R. Barton-Davis et al., "Viral Mediated Expression of Insulin-like Growth Factor I Blocks the Aging-Related Loss of Skeletal Muscle Function," *Proceedings of the National Academy of Sciences USA* 95:26 (22 December 1998): 15,603–7.
108. As quoted in Janet Raloff, "New Gene Therapy Fights Frailty," *Science News* 154:25 & 26 (19 & 26 December 1998): 388; http://www.med.upenn.edu/
109. http://www.med.cornell.edu/
110. As quoted in Raloff, "New Gene Therapy Fights Frailty."

111. http://www.uams.edu/medcenter/; as quoted in Raloff, "New Gene Therapy Fights Frailty."
112. See generally, Michael Fumento, *The Fat of the Land* (New York: Viking, 1997).
113. "$50 Million Earmarked for Alzheimer's Study," *Medical Industry Today,* 18 July 2000.
114. Elena Portyansky Beyzarov, "Then There Were Three," *Drug Topics* 144:10 (15 March 2000): 19.
115. As quoted in Karen Springen, Adam Rogers and Thomas Hayden, "Alzheimer's: Losing More than Memory," *Newsweek* 131:24 (15 June 1998): 52; see also H. Aguero-Torres, L. Fratiglioni and B. Winblad, "The Natural History of Alzheimer's Disease and Other Dementias: Review of the Literature in the Light of the Findings from the Kungsholmen Project," *International Journal of Geriatric Psychiatry* 13:11 (November 1998): 755–66.
116. David E. Smith et al, "Age-Associated Neuronal Atrophy Occurs in the Primate Brain and Is Reversible by Growth Factor Gene Therapy," *Proceedings of the National Academy of Sciences USA* 96:19 (14 September 1999): 10,893–98.
117. As quoted in Paul Recer, "Withered Brain Cells Restored with Gene Therapy, Researchers Say," *Associated Press,* 14 September 1999.
118. Ibid.
119. For more on theories about the cause of Alzheimer's and experimental therapies see J. Madeleine Nash, "The New Science of Alzheimer's," *Time,* 17 July 2000, p. 50.
120. James M. Conner et al., "Nontropic Actions of Neurotrophins: Subcortical Nerve Growth Factor Gene Delivery Reverses Age-Related Degeneration of Primate Cortical Cholinergic Innervation," *Proceedings of the National Academy of Sciences USA* 98:4 (13 February 2001): 1941–46.
121. As quoted in Press Release, "UCSD Study Reports Vital Brain Cell Connections Restored with Gene Therapy in Aged Monkeys: Alzheimer's Patients Sought for Human Clinical Trials," at: http://health.ucsd.edu/news/2001/02_05_Tusz.html
122. "Osteoporosis," American Academy of Orthopaedic Surgeons, at: http://orthoinfo.aaos.org/brochure/thr_report.cfm?Thread_ID=13&topcategory=Osteoporosis&searentry=osteoporosis
123. Richard D. Wasnich, "What Is a Fragility Fracture," in *The Aging Skeleton,* ed. Clifford Rosen, Julie Glowacki and John P. Bilezikian (New York: Academy Press, 1999).

124. "Osteoporosis: Fast Facts," National Osteoporosis Foundation, at: http://www.nof.org/osteoporosis/stats.htm
125. "Osteoporosis," American Academy of Orthopaedic Surgeons.
126. Personal telephone communication of 22 March 2001.
127. For more on Evista, see Lilly's site at: http://www.better-bones.com/lilly.htm
128. Paul Recer, "FDA Approves Drug to Stimulate Bone Growth," *Associated Press,* 26 November 2002.
129. Personal telephone communication with the author, 22 March 2001.
130. http://www.osteo.org/
131. As quoted in David Eggert, "Lilly's New Bone Drug Lauded," *Indianapolis Star,* 24 June 2000, p. C1.
132. Personal telephone communication with the author, 22 March 2001.
133. Robert M. Neer et al., "Effect of Parathyroid Hormone (1-34) on Fractures and Bone Mineral Density in Postmenopausal Women with Osteoporosis," *New England Journal of Medicine* 344:10 (10 May 2001): 1434–41; "Osteoporosis," MayoClinic.com, at: http://www.mayoclinic.com/home?id=DS00128
134. Personal telephone communication of 22 March 2001.
135. Robert Marcus et al., "The Skeletal Response to Teriparatide Is Largely Independent of Age, Initial Bone Mineral Density, and Prevalent Vertebral Fractures in Postmenopausal Women with Osteoporosis," *Journal of Bone and Mineral Research* 18:1 (January 2003): 18–23; Press Release, "New Study Shows FORTEO Especially Effective in Older Patients," 30 December 2002, at: http://newsroom.lilly.com/news/story.cfm?ID=1152
136. Eggert, "Lilly's New Bone Drug Lauded."
137. Centers for Disease Control and Prevention, "Prevalence of Self-Reported Arthritis or Chronic Joint Symptoms among Adults—United States, 2001" *Morbidity and Mortality Weekly Report* 51:42 (25 October 2002): 948–50, at: http://www.cdc.gov/mmwr/preview/mmwrhtml/mm5142a2.htm; Daniel Yee, "Study Finds Joint Problems Increasingly Widespread," *Associated Press,* 24 October 2002.
138. Joseph DDagnese, "Brothers with Heart," *Discover* 22:7 (July 2001): 36, at: http://www.discover.com/july_01/featbros.html
139. Vladimir Martinek, M.D., Freddie H. Fu, M.D. and Johnny Huard, Ph.D., "Gene Therapy and Tissue Engineering in Sports Medicine," *Physician and Sportsmedicine* 28:2 (February 2000): 34–51, at:

http://www.physsportsmed.com/issues/2000/02_00/huard.htm; http://www.medsch.ucla.edu/; "Gene Therapy May Heal Some Injuries," *Gene Therapy Weekly,* 16 March 2000.
140. "Researchers Test New Way to Form Bones," *Medical Industry Today* 3:4 (May 1996).
141. http://www.op1.com/
142. Fran Simon, "Tulane Pioneers Implant to Induce Bone Healing," *Tulane University Magazine,* at: http://www2.tulane.edu/article_news_details.cfm?ArticleID=3422
143. "Australian Backing for Stryker's OP-1," *Marketletter,* 26 February 2001; Stryker Inc., at: http://www.op1.com
144. http://www.selectivegenetics.com/
145. As quoted in Randolph E. Schmid, "Gene Therapy Gives Dogs a New Bone," *Associated Press,* 28 June 1999.
146. Jeffrey Bonadio et al., "Localized, Direct Plasmid Gene Delivery *in Vivo:* Prolonged Therapy Results in Reproducible Tissue Regeneration," *Nature Medicine* 5:7 (July 1999): 753–59; see also, "Fracture Repair," at: http://www.selectivegenetics.com/selective5.html
147. "Osteoporosis," American Academy of Orthopaedic Surgeons, at: http://orthoinfo.aaos.org/brochure/thr_report.cfm?Thread_ID=13&topcategory=Osteoporosis&searentry=osteoporosis

Starfish for a Day

1. Joseph D'Dagnese, "Brothers with Heart," *Discover* 22:7 (July 2001): 36, at: http://www.discover.com/july_01/featbros.html
2. As quoted in Helen Pearson, "The Regeneration Gap," *Nature ScienceUpdate,* 22 November 2001: 388–90.
3. Ibid.
4. As quoted in Nicholas Wade, "Teaching the Body to Heal Itself; Work on Cells' Signals Fosters Talk of a New Medicine," *New York Times,* 7 November 2000, p. F1.
5. Catherine Arnst, "I Can See Clearly Now," *Business Week,* 31 July 2000, p. 40.
6. As quoted in ibid.; see also Ivan R. Schwab and R. Rivkah Isseroff, "Bioengineered Corneas—The Promise and the Challenge," *New England Journal of Medicine* 343:2 (13 July 2000): 136.
7. Arnst, "I Can See Clearly Now."
8. Philip Cohen, "Organs without Donors," *New Scientist* 159:2142 (11 July 1998): 4.
9. http://www.osiristx.com/; as quoted in Wade, "Teaching the Body to Heal Itself."

10. C. S. Young et al., "Tissue Engineering of Complex Tooth Structures on Biodegradable Polymer Scaffolds," *Journal of Dental Research* 81:10 (October 2002): 695–700.
11. http://www.childrenshospital.org/
12. Frank Oberpenning et al., "*De Novo* Reconstitution of a Functional Mammalian Urinary Bladder by Tissue Engineering," *Nature Biotechnology* 17:2 (February 1999): 133–34; "Doctors Herald Grow-Your-Own Organs," *BBC News,* at: http://news.bbc.co.uk/hi/english/sci/tech/newsid_265000/265713.stm
13. http://www.curis.com/; personal telephone communication with Doros Platika, president and CEO of Curis Inc., 8 August 2001.
14. Leigh Hopper, "Scientists Grow Complete Heart Valve in Lab for First Time," *Houston Chronicle,* 8 November 1999, p. A1.
15. http://www.med.umich.edu/intmed/
16. H. David Humes et al., "Bioartificial Kidney for Full Renal Replacement Therapy," *Seminars in Nephrology* 20:1 (January 2000): 71–82; David J. Mooney and Antonios G. Mikosm, "Growing New Organs," *Scientific American* 280:4 (April 1999): 60.
17. "Prosthesis 'Natural' Breast Implants Being Developed?" *Cancer Weekly Plus,* 19 January 1998; Lois Rogers, "Scientists Build Living Breasts," *Sunday Times* (London), 4 January 1998; D. Lazovich et al., "Breast Conservation Therapy in the United States Following the 1990 National Institutes of Health Consensus Development Conference on the Treatment of Patients with Early Stage Invasive Breast Carcinoma," *Cancer* 86:4 (15 August 1999): 629–37; Lynn Ries, "Use of Lumpectomy for Breast Cancer," *Journal of the National Cancer Institute* 87:3 (March 1995): 339.
18. Tae Gyun Kwon, James J. Yoo and Anthony Atala, "Autologous Penile Corpora Cavernosa Replacement Using Tissue Engineering Techniques," *Journal of Urology* 168:4 (October 2002): 1754–58.
19. As quoted in Sylvia Pagán Westphal, "Tissue Engineers Grow Penis in the Lab," *New Scientist* 175:2360 (4 September 2002): 14.
20. http://www.mgh.harvard.edu/
21. As quoted in Joseph D'Dagnese, "Brothers with Heart," *Discover* 22:7 (July 2001): 36, at: http://www.discover.com/july_01/featbros.html
22. Nenad Bursac et al., "Cardiac Muscle Tissue Engineering: Toward an *in Vitro* Model for Electrophysiological Studies," *American Journal of Physiology—Heart and Circulatory Physiology* 277:2 Pt. 2 (August 2001): H433–44; Press Release, "Advances in Engineering Heart Tissue Reported by MIT Scientists, Colleagues," MIT, 28 Septem-

ber 1999, at: http://web.mit.edu/newsoffice/nr/1999/heart.html

23. Doug Garr, "The Human Body Shop," *Technology Review* 104:3 (April 2001): 104.
24. http://www.therics.com/tissue.html
25. See Therics Inc., "Tissue Engineering," at: http://www.therics.com/tissue.html

Stupendous Stem Cells

1. For a readable discussion on stem cells, see National Institutes of Health, "Stem Cells: A Primer," at: http://www.nih.gov/news/stemcell/primer.htm
2. http://www.medsch.ucla.edu/
3. Personal telephone communication of 12 July 2001.
4. J. Gill et al., "Generation of a Complete Thymic Microenvironment by MTS24^{+} Thymic Epithelial Cells," *Nature Immunology* 3:7 (July 2002): 635–42.
5. http://www.med.monash.edu.au/; Michael Perry, "Australian Scientists Say They Can Rebuild Immune System," *Wired News,* 18 June 2002.
6. As quoted in Nicholas Wade, "Teaching the Body to Heal Itself; Work on Cells Signals Fosters Talk of a New Medicine," *New York Times,* 7 November 2000, p. F1.
7. Connie Mack, "I'm Pro-Life—and in Favor of Stem Cell Research," *Wall Street Journal,* 19 June 2001, p. A22.
8. William Safire, "Stem Cell Hard Sell," *New York Times,* 5 July 2001, p. A17.
9. http://www.marrow.org/NMDP/about_nmdp_idx.html
10. http://www.mdxn.ch/
11. http://www.epidex.com/index.html
12. Michael J. Young et al., "Neuronal Differentiation and Morphological Integration of Hippocampal Progenitor Cells Transplanted to the Retina of Immature and Mature Dystrophic Rats," *Molecular and Cellular Neuroscience* 16:3 (September 2000): 197–205.
13. http://www.eri.harvard.edu/; as quoted in Amy Norton, "Stem Cells May One Day Boost Fading Eyesight," *Reuters Health,* 27 September 2000.
14. As quoted in James Chapman, "Headline: Blindness Cure Is in Sight with Cell Transplant Breakthrough," *Daily Mail* (London), 28 September 2000.

15. Eriko Tateishi-Yuyama et al., "Therapeutic Angiogenesis for Patients with Limb Ischaemia by Autologous Transplantation of Bone-Marrow Cells: A Pilot Study and a Randomised Controlled Trial," *Lancet* 360:9331 (10 August 2002): 427–35; Emma Ross, "Study Finds Stem Cells May Save Limbs," *Associated Press,* 8 August 2002.
16. As quoted in Ross, "Study Finds Stem Cells May Save Limbs."
17. American Heart Association, "Heart Attack and Angina Statistics," at: http://www.americanheart.org/Heart_and_Stroke_A_Z_Guide/has.html
18. Gregory G. Schwartz, "Exploring New Strategies for the Management of Acute Coronary Syndromes," *American Journal of Cardiology* 19:86 (19 October 2000): 44J–49J; Ali H. Mokdad et al., "The Spread of the Obesity Epidemic in the United States, 1991–1998," *Journal of the American Medical Association* 282:16 (27 October 1999): 1519–22; CDC, "Obesity Epidemic Increases Dramatically in the United States; CDC Director Calls for National Prevention Effort," at: http://www.cdc.gov/nccdphp/dnpa/obesity-epidemic.htm
19. Federico Quaini et al., "Chimerism of the Transplanted Heart," *New England Journal of Medicine* 1:346 (3 January 2002): 5–15; David Brown, "Heart May Be Able to Repair Damage," *Washington Post,* 7 June 2001, p. A1.
20. "Cardiovascular System: Myocardial Infarction," University of Calgary, at: http://rehab.educ.ucalgary.ca/courses/edis/551.60/Cardiovascular/mi-overview.html
21. http://www.uni-rostock.de/
22. http://www.hku.hk/; Christof Stamm et al., "Autologous Bone-Marrow Stem-Cell Transplantation for Myocardial Regeneration," letter, *Lancet* 361:9351 (4 January 2003): 45–46; Hung-Fat Tse et al., "Angiogenesis in Ischaemic Myocardium by Intramyocardial Autologous Bone Marrow Mononuclear Cell Implantation," letter, *Lancet* 361:9351 (4 January 2003): 47–49; "Stem Cells Offer Hope to Heart Attack Victims," *Reuters,* 2 January 2002, at: http://abcnews.go.com/wire/SciTech/reuters20030102_482.html
23. Donald Orlic et al., "Bone Marrow Cells Regenerate Infarcted Myocardium," letter, *Nature* 410:6829 (5 April 2001): 701–4; see also, Joseph B. Verrengia, "Cell Injections May Be Able to Repair Heart Attack Damage; Also Moved in Previous Cycle," *Associated Press,* 31 March 2001.

24. Albert Hagège et al., *Journal of the American College of Cardiology* 41:7 (2 April 2003): 1078–83.
25. http://www.aphp.fr/hopitaux/hegp.htm; as quoted in Shaoni Bhattacharya, "Stem Cells Can Mend Human Hearts," *New Scientist*, 7 February 2003, at: http://www.newscientist.com/news/print.jsp?id=ns99993367
26. Arjun Deb et al., "Bone Marrow–Derived Cardiomyocytes Are Present in Adult Human Heart: A Study of Gender-Mismatched Bone Marrow Transplantation Patients," *Circulation* 107:9 (11 March 2003): 1247–49; "Mayo Clinic Proves New Heart Muscle Cells Can Come from Bone Marrow," *ScienceDaily*, 11 March 2003, at: http://www.sciencedaily.com/releases/2003/03/030311074254.htm
27. For example, see Richard K. Burt et al., "Induction of Tolerance in Autoimmune Diseases by Hematopoietic Stem Cell Transplantation: Getting Closer to a Cure?" *International Journal of Hematology* 76, Sup. 1 (2002 August): 226–47; Ann E. Traynor et al, "Hematopoietic Stem Cell Transplantation for Severe and Refractory Lupus. Analysis after Five Years and Fifteen Patients," *Arthritis and Rheumatism* 46:11 (2002 November): 2917–23; Richard K. Burt et al. "Induction of Tolerance in Autoimmune Diseases by Hematopoietic Stem Cell Transplantation: Getting Closer to a Cure?" *Blood* 99:3 (1 February 2002): 768–84; Elizabeth Raetz, P. G. Beatty and R. H. Adams, "Treatment of Severe Evans Syndrome with an Allogeneic Cord Blood Transplant," *Bone Marrow Transplant* 20:5 (September 1997): 427–29; T. Yamaoka, "Regeneration Therapy of Pancreatic Beta Cells: Towards a Cure for Diabetes?" *Biochemical and Biophysical Research Communications* 296:5 (6 September 2002): 1039–43; Athanasios Fassas et al., "Hematopoietic Stem Cell Transplantation for Multiple Sclerosis. A Retrospective Multicenter Study," *Journal of Neurology* 249:8 (August 2002): 1088–97.
28. National Institute of Neurological Disorders and Stroke, "NINDS Amyotrophic Lateral Sclerosis Information Page," at: http://www.ninds.nih.gov/health_and_medical/disorders/amyotrophiclateralsclerosis_doc.htm#What_is_Amyotrophic_Lateral_Sclerosis
29. Aventis Rilutek Fact Sheet, at: http://www.alsinfo.com/patients/RilutekPatientFactSheet.jsp; Drugs Approved by the FDA, "Drug Name: Rilutek (riluzole), at: http://www.centerwatch.com/patient/drugs/dru3.html

30. http://ww2.med.jhu.edu/deptmed/
31. See L. R. Boone and A. Brown, "Variants of the Hr Strain of Sindbis Virus Lethal for Mice," *Journal of General Virology* 31:2 (May 1976): 261–63.
32. Jonathan Bor, "Stem-Cell Research Offers Hope in Neural Disorders; Paralyzed Mice Able to Move Legs after Receiving Injections," *Baltimore Sun*, 6 November 2000, p. 1A.
33. Robert Langreth, "Refusing to Die," *Forbes* 167:9 (16 April 2001): 368; Neal Travis, "Tackling Gehrig Disease," *New York Post*, 15 May 1998, p. 9; http://www.mdausa.org/
34. As quoted in Bor, "Stem-cell Research Offers Hope in Neural Disorders."
35. As quoted in Press Release, "Stem Cells Graft in Spinal Cord, Restore Movement in Paralyzed Mice," 5 November 2000, at: http://www.hopkinsmedicine.org/press/2000/NOVEMBER/001105.htm
36. Liz Kay, "Stem Cells May Keep MS from Advancing," *Los Angeles Times*, 17 April 2002, p. 19.
37. As quoted in Peter Gorner, "Victims of MS Get Risky Solution," *Chicago Tribune*, 4 June 2002, p. 1.
38. Ibid.
39. "Italian Researchers Develop Stem Cell Technique to Repair Damaged Corneas," *Transplant News* 11:21 (12 November 2001); K. Venkateshwarlu, "India: Breakthrough in Stem Cell Transplantation of Eye," *The Hindu*, 31 October 2001, at: http://www.hinduonnet.com/2001/10/31/stories/0231000q.htm; Gwen Kinkead, "Stem Cell Transplants Offer New Hope in Some Cases of Blindness," *New York Times*, 15 April 2003, p. F7.
40. Edward J. Holland, A. R. Djalilian and G. S. Schwartz, "Management of Aniridic Keratopathy with Keratolimbal Allograft: A Limbal Stem Cell Transplantation Technique," *Ophthalmology* 110:1 (January 2003): 125–30.
41. Kinkead, "Stem Cell Transplants Offer New Hope in Some Cases of Blindness."
42. His personal website is: http://www.eyehealthvision.com/profiles/kenyon_k.htm
43. As quoted in Kinkead, "Stem Cell Transplants Offer New Hope in Some Cases of Blindness."
44. http://www.corneajrnl.com/article.asp?ISSN=0277-3740&VOL=20&ISS=1&PAGE=50

45. As quoted in Kinkead, "Stem Cell Transplants Offer New Hope in Some Cases of Blindness."
46. Ibid.
47. Ibid.
48. Ibid.
49. Ra Jui-Fang Tsai, Lien-Min Li and Jan-Kan Chen, "Reconstruction of Damaged Corneas by Transplantation of Autologous Limbal Epithelial Cells," *New England Journal of Medicine* 343:2 (13 July 2000): 86–93; Joseph P. Shovlin, "Update on Stem Cell Research," at: http://www.revoptom.com/archive/DEPT/RO1100clqa.htm
50. http://sbrc.stanford.edu/faculty/sbrc_fac_list/blau.html; Timothy R. Brazelton et al., "From Marrow to Brain: Expression of Neuronal Phenotypes in Adult Mice," *Nature Medicine* 290:5497 (1 December 2000): 1775–79.
51. Paul Recer, "Bone Marrow Cells Can Transform into Brain Neurons, Researchers Find," *Associated Press,* 1 December 2000.
52. Éva Mezey et al., "Turning Blood into Brain: Cells Bearing Neuronal Antigens Generated in Vivo from Bone Marrow," *Nature Medicine* 290:5497 (1 December 2000): 1779–82.
53. Éva Mezey et al., "Transplanted Bone Marrow Generates New Neurons in Human Brains," *Proceedings of the National Academy of Sciences USA 2003* 100:3 (4 February 2003): 1364–69; James M. Weimann et al., "Contribution of Transplanted Bone Marrow Cells to Purkinje Neurons in Human Adult Brains," *Proceedings of the National Academy of Sciences USA 2003* 100:4 (18 February 2003): 2088–93.
54. As quoted in Andy Coghlan, "Stem Cells Migrate from Bone to Brain," *New Scientist,* 20 January 2003, at: http://www.newscientist.com/news/print.jsp?id=ns99993286
55. "Of Purkinje Cells, Glutamate, a Puff and a Sound," UniSci, at: http://unisci.com/stories/20014/1114016.htm
56. Personal telephone communication of 13 February 2003.
57. http://www.ucihs.uci.edu/anatomy/
58. James Fallon et al., "*In Vivo* Induction of Massive Proliferation, Directed Migration, and Differentiation of Neural Cells in the Adult Mammalian Brain," *Proceedings of the National Academy of Sciences USA* 97:26 (19 December 2000): 14,686–91.
59. Press Release, "Stem Cells Stimulated by Natural Growth Factor Reverse Damage, Restore Some Function in Adult Brain," University of California, Irvine, 18 December 2000, at: http://www.communications.uci.edu/00releases/171ap00.html

60. http://www.csmc.edu/; the company is a subsidiary of Theratechnologies, also of Montreal, at: http://www.theratech.com/
61. Rick Weiss, "Stem Cell Transplant Works in California Case: Parkinson's Traits Largely Disappear," *Washington Post*, 9 April 2002, p. 8; Sylvia Pagán Westphal, "Re-implanted Stem Cells Tackle Parkinson's," 8 April 2002, *New Scientist*, at: http://www.newscientist.com/news/news.jsp?id=ns99992139
62. As quoted in Weiss, "Stem Cell Transplant Works in California Case."
63. As quoted in Ronald Kotulak, "Scientists Get Cells to Shift Identities, Raise New Hopes; Findings May Open Doors for Alzheimer's and Spinal Cord Treatments," *Chicago Tribune*, 15 August 2000, p. 1.
64. Darwin J. Prockop, "Marrow Stromal Cells as Stem Cells for Nonhematopoietic Tissues," *Science* 276:5309 (4 April 1997): 71–74.
65. http://rwjms.umdnj.edu/
66. As quoted in Gina Kolata, "Researchers Turn Bone Marrow Cells into Nerve Cells," *New York Times*, 15 August 2000, p. A10.
67. Dale Woodbury et al., "Adult Rat and Human Bone Marrow Stromal Cells Differentiate into Neurons," *Journal of Neuroscience Research* 61:4 (31 July 2000): 364–70.
68. Ibid.
69. As quoted in Thomas H. Maugh II, "Scientists Form Brain Cells from Bone Marrow; Health: Patient's Own Body Could Provide Treatment for Brain Disorders, Findings Suggest," *Los Angeles Times*, 15 August 2000, p. A1.
70. As quoted in ibid.
71. Rossella Galli et al., "Skeletal Myogenic Potential of Human and Mouse Neural Stem Cells," *Nature Neuroscience* 3:10 (October 2000): 986–91.
72. Amy Norton, "Scientists Turn Adult Brain Cells to Muscle," *Reuters*, 19 September 2000.
73. http://www.dept-med.pitt.edu/
74. As quoted in CBS *60 Minutes II*, "Holy Grail: Stem-Cell Transplant Saves Teen Life," 28 November 2001, at: http://www.cbsnews.com/now/story/0%2C1597%2C319351-412%2C00.shtml
75. As quoted in ibid.
76. Maryclaire Dale, "French Researchers Say Stem Cell Transplants Successful in Sickle Cell Patients," *Associated Press*, 8 December 2002.
77. http://www.med.nyu.edu/; http://www.yale.edu/medical/; http://www.mskcc.com

78. Neil Theise et al., "Liver from Bone Marrow in Humans," *Hepatology* 32:1 (July 2000): 11–16.
79. As quoted in Press Release, "Humans Can Regrow Liver from Bone Marrow," *MedNews,* 19 June 2000, at: http://www.newswise.com/articles/2000/6/THEISE.KMC.html
80. Neil Theise et al., "Liver from Bone Marrow in Humans," *Hepatology* 32:1 (July 2000): 11–16.
81. "Cirrhosis of the Liver," National Digestive Diseases Information Clearinghouse, at: http://www.niddk.nih.gov/health/digest/pubs/cirrhosi/cirrhosi.htm
82. From the Organ Procurement and Transplantation Network, at: http://www.optn.org/latestData/rptData.asp
83. Medline Plus, "Bone Marrow Transplant," at: http://www.nlm.nih.gov/medlineplus/ency/article/003009.htm
84. "Bone Marrow Donors Worldwide," at: http://www.bmdw.org/; "About the National Marrow Donor Program," National Bone Marrow Donor Program, at: http://www.marrow.org/NMDP/about_nmdp_idx.html
85. http://www.artecel.com/
86. http://www.genzymebiosurgery.com/opage.asp?ogroup=2&olevel=3&opage=109
87. Robert Davis, "Human Fat Cells Can Be Transformed into Cartilage," *USA Today,* 28 January 2001, p. 6D, at: http://www.usatoday.com/news/health/2001-02-28-fat-cartilage.htm
88. http://www.mc.duke.edu/
89. As quoted in Maggie Fox, "Limitless Source of Repair Cells Comes from Fat," *Reuters Health,* 27 February 2001, available at: http://www.wired.com/news/print/0,1294,42052,00.html
90. Geoffrey R. Erickson et al., "Chondrogenic Potential of Adipose Tissue-Derived Stromal Cells *in Vitro* and *in Vivo,*" *Biochemical and Biophysical Research Communications* 290:2 (18 January 2002): 763–69; Press Release, "Cartilage Made from Stem Cells Tested in Animals," The Whitaker Foundation, 11 April 2002, at: http://www.whitaker.org/news/guilak.html
91. Press Release, "Artecel Sciences Announces *in Vitro* Proof-of-Principle Studies Supporting the Utility of Adipose Tissue as a Source of Differentiated Human Cells: Discovery Could Provide Abundant Supply of Cells for Four Potential Clinical Applications," at: http://www.artecel.com/press/020701.html

92. "Researchers Turn Fat Cells into Cartilage," *ScienceDaily Magazine,* at: http://www.sciencedaily.com/releases/2001/02/010228075959.htm
93. Nina Flanagan, "Advances in Stem-Cell Therapy," *Genetic Engineering News* 21:9 (1 May 2001): 1; Rick Weiss, "Human Fat May Provide Stem Cells," *Washington Post,* 10 April 2001, p. A1.
94. American Academy of Cosmetic Surgery, "Cosmetic Surgery—A Comparison Study of Its Growth in the 1990s," at: http://www.cosmeticsurgery.org/consumer/stats/1990s_Growth/1990s_growth.html
95. Patricia A. Zuk et al., "Multilineage Cells from Human Adipose Tissue: Implications for Cell-Based Therapies," *Tissue Engineering* 7:2 (April 2001): 211–26; Patricia A. Zuk et al., "Human Adipose Tissue Is a Source of Multipotent Stem Cells," *Molecular Biology of the Cell* 13:12 (December 2002): 4279–95.
96. Personal telephone communication of 12 July 2001.
97. Personal telephone communication of 12 July 2001.
98. See Keith J. Allred, "Fetal Tissue Transplants: A Primer with a Look Forward," *Journal of Health Law* 28:4 (July/August 1995): 1995.
99. National Institutes of Health, *Stem Cells: Scientific Progress and Future Research Directions* (Washington, D.C.: Department of Health and Human Services, July 2001), ch. 2, at: http://www.nih.gov/news/stemcell/chapter2.pdf
100. For example, Suheir Assady et al., "Insulin Production by Human Embryonic Stem Cells," *Diabetes* 50:8 (August 2002): 1691–97; Jayaraj Rajagopal et al, "Insulin Staining of ES Cell Progeny from Insulin Uptake," *Science* 299:5605 (17 January 2003): 363.
101. As quoted in Gretchen Vogel, "Same Results, Different Interpretations," *Science* 299:5605 (17 January 2003): 324.
102. For a more detailed description in lay terms, see National Institutes of Health, "Stem Cells: A Primer," at: http://www.nih.gov/news/stemcell/primer.htm
103. His personal website is: http://www.som.tulane.edu/neurograd/prockhm.htm; http://www.som.tulane.edu/gene_therapy/overview.shtml; Darwin J. Prockop, "Stem Cell Research Has Only Just Begun," letter, *Science* 293:5528 (13 July 2001): 211–12.
104. Personal telephone communication of 14 February 2003.
105. Her personal website is: http://biosci.cbs.umn.edu/mcdbg/faculty/Verfaillie.html; http://www1.umn.edu/stemcell/sci/page/pg/patch2gar2v2-8_6.htm

106. His personal website is: http://www.biomed.uga.edu/mem_hess_david.htm
107. Personal telephone communication of 14 January 2003.
108. Yuehua Jiang et al., "Pluripotency of Mesenchymal Stem Cells Derived from Adult Marrow," *Nature* 418:6893 (4 July 2002): 41–49; Justin Gillis, "Study Finds Potential in Adult Cells," *Washington Post,* 21 June 2002, p. A1.
109. Athersys Inc. of Cleveland (http://www.athersys.com/); Andrew Pollack, " 'Politically Correct' Stem Cell Is Licensed to Biotech Concern," *New York Times,* 11 December 2002, p. C8.
110. Gene C. Kopen, Darwin J. Prockop and Donald G. Phinney, "Marrow Stromal Cells Migrate throughout Forebrain and Cerebellum, and They Differentiate into Astrocytes after Injection into Neonatal Mouse Brains," *Proceedings of the National Academy of Sciences USA 1999* 96:19 (14 September 1999): 10,711–16; D. Woodbury et al., "Adult Rat and Human Bone Marrow Stromal Cells Differentiate into Neurons," *Journal of Neuroscience Research* 61:4 (15 August 2000): 364–70; J. Sanchez-Ramos et al., "Adult Bone Marrow Stromal Cells Differentiate into Neural Cells *in Vitro,*" *Experimental Neurology* 164:2 (August 2000): 247–56; Éva Mezey et al., "Turning Blood into Brain: Cells Bearing Neuronal Antigens Generated *in Vivo* from Bone Marrow," *Science* 290:5497 (1 December 2000): 1779–82; T. R. Brazelton et al., "From Marrow to Brain: Expression of Neuronal Phenotypes in Adult Mice," *Science* 290:5497 (1 December 2000): 1775–79; B. E. Petersen et al. "Bone Marrow as a Potential Source of Hepatic Oval Cells," *Science* 284:5417 (14 May 1999): 1168–70; Neil D. Theise et al., "Derivation of Hepatocytes from Bone Marrow Cells in Mice after Radiation-Induced Myeloablation," *Hepatology* 31:1 (January 2000): 235–40; Neil D. Theise et al., "Liver from Bone Marrow in Humans," *Hepatology* 32:1 (July 2000): 11–16; Diane S. Krause et al., "Multi-organ, Multi-lineage Engraftment by a Single Bone Marrow–Derived Stem Cell," *Cell* 105:3 (4 May 2001): 369–77.
111. Catherine M. Verfaillie, Martin F. Pera and Peter M. Landsdorp, "Stem Cells: Hype and Reality," *Hematology,* 2002: 369–91, at: http://www.asheducationbook.org/cgi/content/full/2002/1/369
112. Dale Woodbury, Kathleen Reynolds and Ira Black, "Adult Bone Marrow Stromal Stem Cells Express Germline, Ectodermal, Endodermal and Mesodermal Genes Prior to Neurogenesis," *Journal of Neuroscience Research* 69:6 (15 September 2002): 908–17.

113. Personal telephone communication of 26 February 2003.
114. Yong Zhao, David Glesne and Eliezer Huberman, "A Human Peripheral Blood Monocyte-Derived Subset Acts as Pluripotent Stem Cells," *Proceedings of the National Academy of Sciences USA 2003* 100:5 (4 March 2003): 2426–31; "Do-It-Yourself Stem Cell Extraction? Monocyte Cells, Easily Accessible to Draw from Blood, Can Double for Dicey Embryonic Stem Cells," *BioWorld Today* 14:37 (25 February 2003). The website for Argonne National Laboratory is: http://www.anl.gov/
115. Personal telephone communication of 28 February 2003.
116. Masako Miura et al., "Stem Cells from Human Exfoliated Deciduous Teeth," *Proceedings of the National Academy of Sciences USA 2003* 100:10 (13 May 2002): 5807–12; Press Release, "Scientists Discover Unique Source of Postnatal Stem Cells," National Institute of Dental and Craniofacial Research, 21 April 2003, at: http://www.nidcr.nih.gov/news/04212003.asp
117. Martin Körbling et al., "Hepatocytes and Epithelial Cells of Donor Origin in Recipients of Peripheral-Blood Stem Cells," *New England Journal of Medicine* 346:10 (7 March 2002): 738–46; Rick Weiss, "Stem Cells in Human Blood Are Reported Potential Help in Tissue Repair; Regeneration Cited," *Washington Post,* 7 March 2002, p. A8; National Institutes of Health, *Stem Cells: Scientific Progress and Future Research Directions* (Washington, D.C.: Department of Health and Human Services, July 2001, ch. 4, at: http://www.nih.gov/news/stemcell/chapter4.pdf; Patricia A. Zuk et al., "Multilineage Cells from Human Adipose Tissue: Implications for Cell-Based Therapies," *Tissue Engineering* 7:2 (April 2001): 211–26; Press Release, "Artecel Sciences Announces *in Vitro* Proof-of-Principle Studies Supporting the Utility of Adipose Tissue as a Source of Differentiated Human Cells: Discovery Could Provide Abundant Supply of Cells for Four Potential Clinical Applications," at: http://www.artecel.com/press/020701.html; Andy Coghlan, "Hair Today, Skin Tomorrow," *New Scientist* 170:2296 (23 June 2001): 19; Press Release, "Modex Therapeutics Reports Second Quarter Results," 6 August 2001, at: http://cws.huginonline.com/M/132034/PR/200108/829476_5.html
118. Nicholas Wade, "A New Source for Stem Cells Is Reported," *New York Times,* 12 April 2001, p. A25; Pablo Rubinstein et al., "Outcomes among 562 Recipients of Placental-Blood Transplants from Unrelated Donors," *New England Journal of Medicine* 339:22 (26

November 1998): 1628–29; "What Is Cord Blood?" http://www.viacord.com/cordblood/what_is_cord_blood.html; Peggy Peck, "Amniotic Fluid Could Be Source of Fetal Cells," *Reuters Health*, 11 October 2001.

119. Kathy E. Mitchell et al., "Matrix Cells from Wharton's Jelly Form Neurons and Glia," *Stem Cells* 21:1 (January 2003): 50–60; "Study Finds Possible New Stem Cell Source," *United Press International*, 18 January 2003, at: http://www.applesforhealth/HealthyFeatures/stufinstec4.html; "Umbilical Cord Matrix, a Rich New Stem Cell Source, Study Shows," *ScienceDaily*, 16 January 2003, at: http://www.sciencedaily.com/releases/2003/01/ 030122072949.htm

120. Kohshi Ohishi, Barbara Varnum-Finney and Irwin D. Bernstein, "Delta-1 Enhances Marrow and Thymus Repopulating Ability of Human CD34+CD38–Cord Blood Cells," *Journal of Clinical Investigation* 110:8 (22 October 2002): 1165–74; Katrina Woznicki, "Stem Cells from Cord Blood Show Promise," *United Press International*, 22 October 2002.

121. Micha Drukker et al., "Characterization of the Expression of MHC Proteins in Human Embryonic Stem Cells," *Proceedings of the National Academy of Sciences USA 2002* 99:15 (23 July 2002): 9864–69.

122. "The Why Files Guide to Stem Cells," University of Wisconsin-Madison, at: http://whyfiles.org/127stem_cell/3.html

123. Sylvia Pagán Westphal, "One Cell to Heal Them All," *New Scientist* 172:2321 (15 December 2001): 4.

124. Michael J. Shamblott et al., "Human Embryonic Germ Cell Derivatives Express a Broad Range of Developmentally Distinct Markers and Proliferate Extensively *in Vitro*," *Proceedings of the National Academy of Sciences USA 2001* 98:1 (2 January 2001): 113–18. Also see for example Wakitani et al., "Embryonic Stem Cells Injected into the Mouse Knee Joint Form Teratomas and Subsequently Destroy the Joint," *Rheumatology* 42:1 (January 2003): 162–65.

125. As quoted in "New Type of Human Stem Cells Engineered at Hopkins," *UniSci*, 7 January 2001, at: http://unisci.com/stories/20011/0108011.htm

126. Robert Tsai and Ronald McKay, "A Nucleolar Mechanism Controlling Cell Proliferation in Stem Cells and Cancer Cells," *Genes and Development* 16:23 (1 December 2002): 2991–3003; Kendall Powell, "Stem and Cancer Cells Have Something in Common," *Nature ScienceUpdate*, 30 December 2002, at: http://www.nature.com/nsu/021230/021230-2.html

127. David Prentice, "Current Clinical Applications of Adult Stem Cells," 25 June 2001, at: http://www.stemcellresearch.org/info/currentaps.pdf
128. David Prentice, "Potential Applications of Adult Stem Cells—Research News," 25 June 2001, at: http://www.stemcellresearch.org/info/potentialaps.pdf
129. Michelle Meadows, "Bone Marrow Transplants Come of Age," *FDA Consumer* 34:4 (July/August 2000), at: http://www.fda.gov/fdac/features/2000/400_bone.html
130. University of Minnesota, "Umbilical Cord Blood Transplantation," at: http://www.cancer.umn.edu/page/research/trsplant/cord3.html; for a complete list see: http://www.marrow.org/MEDICAL/diseases_treatable_by_stem_cell_transplants.html
131. Pablo Rubinstein et al., "Outcomes among 562 Recipients of Placental-Blood Transplants from Unrelated Donors," *New England Journal of Medicine* 339:22 (26 November 1998): 1565–77; Vanderson Rocha et al., "Graft-versus-Host Disease in Children Who Have Received a Cord-Blood or Bone Marrow Transplant from an HLA-Identical Sibling," *New England Journal of Medicine* 342:25 (22 June 2000): 1846–54; see also, Robertson Parkman, "The Future of Placental-Blood Transplantation," *New England Journal of Medicine* 339:22 (26 November 1998): 1628.
132. Personal telephone communication of 13 August 2001.
133. Nashiro Terada et al., "Bone Marrow Cells Adopt the Phenotype of Other Cells by Spontaneous Cell Fusion," letter, *Nature* 416:6880 (4 April 2002): 542–45; Qi-Long Ying et al., "Changing Potency by Spontaneous Fusion," *Nature* 416:6880 (4 April 2002): 545–48.
134. " 'Breakthrough' in Adult Stem Cells Is Hype, Studies Warn," *Agence France-Presse,* 13 March 2002.
135. "New Research Tips Debate on Stem Cells," *Australian Associated Press,* 14 March 2002; Justin Gillis, "Questions Raised on Stem Cells; Adult Cells Found Less Useful than Embryonic Ones," *Washington Post,* 14 March 2002, p. A3.
136. For a longer discussion, see Michael Fumento, "Stem-Cell Political Science: Nature's Agenda," *National Review Online,* 28 March 2002, at: http://www.fumento.com/marchstem.html
137. Helen M. Blau, "A Twist of Fate," *Nature* 419:6906 (3 October 2002): 437.
138. Masahiro Masuya et al., "Hematopoietic Origin of Glomerular Mesangial Cells," *Blood* 101:6 (15 March 2003): 2215–18.
139. E-mail communication from David Hess of 17 February 2003.

140. Personal telephone communication of 14 January 2003.
141. Simon D. Tran et al., "Differentiation of Human Bone Marrow-Derived Cells into Buccal Epithelial Cells *in Vivo:* A Molecular Analytical Study," *Lancet* 361:9363 (29 March 2003): 1084–88.
142. Raymond F. Castro et al., "Failure of Bone Marrow Cells to Transdifferentiate into Neural Cells *in Vivo,*" letter, *Science* 297:5585 (23 August 2002): 1299; Press Release, "Stem Cells from Bone Marrow Cannot Transform into Brain Cells," at: http://public.bcm.tmc.edu/pa/stemcells-braincells.htm
143. "Promise of Adult Stem Cells Put in Doubt," *United Press International,* 5 September 2002.
144. Amy J. Wagers et al., "Little Evidence for Developmental Plasticity of Adult Hematopoietic Stem Cells," *Science* 297:5590 (27 September 2002): 2256–59.
145. As quoted in "Promise of Adult Stem Cells Put in Doubt."
146. Eric Lagasse et al., "Purified Hematopoietic Stem Cells Can Differentiate into Hepatocytes *in Vivo,*" *Nature Medicine* 6:11 (November 2000): 1212–13.
147. Helen M. Blau et al., "Something in the Eye of the Beholder," letter, *Science* 298:5592 (11 October 2002): 362–63. See also, W. F Hickey and H. Kimura, "Perivascular Microglial Cells of the CNS Are Bone Marrow-Derived and Present Antigen *in Vivo,*" *Science* 239:4837 (15 January 1988): 290–92; D. W. Kennedy and J. L. Abkowitz, "Kinetics of Central Nervous System Microglial and Macrophage Engraftment: Analysis Using a Transgenic Bone Marrow Transplantation Model," *Blood* 90:3 (1 August 1997): 986–93; W. J. Krall et al., "Cells Expressing Human Glucocerebrosidase from a Retroviral Vector Repopulate Macrophages and Central Nervous System Microglia after Murine Bone Barrow Transplantation," *Blood* 83:9 (1 May 1994): 2737–48; William Krivit et al., "Microglia: The Effector Cell for Reconstitution of the Central Nervous System Following Bone Marrow Transplantation for Lysosomal and Peroxisomal Storage Diseases," *Cell Transplantation* 4:4 (July/August 1995): 385–92; Payam Rezaie and David Male, "Colonisation of the Developing Human Brain and Spinal Cord by Microglia: A Review," *Microscopy Research and Technique* 45:6 (15 June 1999): 359–82; K. Imamoto and C. P. Leblond, "Presence of Labeled Monocytes, Macrophages and Microglia in a Stab Wound of the Brain Following an Injection of Bone Marrow Cells Labeled with 3H-uridine into Rats," *Journal of Comparative Neurology* 174:2 (15 July 1977): 255–79.

148. Personal telephone communication of 13 February 2003.
149. For example, David C. Hess et al., "Bone Marrow as a Source of Endothelial Cells and Neun-Expressing Cells after Stroke," *Stroke* 33:5 (May 2002): 1362–68.
150. Prepared Testimony of Irving L. Weissman, Chair, Panel on Scientific and Medical Aspects of Human Reproductive Cloning, before the Committee on the Judiciary U.S. Senate, 5 February 2002.
151. William Safire, "Stem Cell Hard Sell," *New York Times*, 5 July 2001, p. A17.
152. Press Release, "Stanford Biology Professor Named California Scientist of the Year," 9 May 2002, at: http://www.casciencectr.org/GenInfo/MediaRoom/PressReleases/ScientistOfTheYear2002/ScientistOfTheYear2002.php
153. http://www.stemcellsinc.com/index2.html; Neil Munro, "Doctor Who?" *Washington Monthly* 34:11 (November 2002), at: http://www.washingtonmonthly.com/features/2001/0211.munro.html
154. Paul Elias, "Stanford University Announces Human Embryonic Stem Cell Project," *Associated Press*, 11 December 2002; Press Release, "Stanford Launches Unique Cancer and Stem Cell Biology Institute," 12 December 2002, at: http://mednews.stanford.edu/news_releases_html/2002/decreleases/stem-cell-institute.html
155. Elias, "Stanford University Announces Human Embryonic Stem Cell Project."
156. His personal website is: http://www.dartmouth.edu/%7Eethics/faculty.html
157. Both as quoted in Elias, "Stanford University Announces Human Embryonic Stem Cell Project."
158. Paul Elias, "Bush Bioethics Advisers Criticize Stem Cell Research Plan as Cloning," *Associated Press*, 20 December 2002.
159. Neil Munro, "Mixing Business with Stem Cells," *National Journal* 33:29 (21 July 2002).
160. Dr. Ronald McKay, as quoted in Munro, "Mixing Business with Stem Cells."
161. Paul Recer, "Study Finds Adult Blood Stem Cells Will Not Transform into Other Tissue Cells," *Associated Press*, 6 September 2002.
162. "New Research Tips Debate on Stem Cells," *AAP Newsfeed*, 14 March 2002.
163. The company is Stem Cell Sciences; see Deborah Smith, "Local Scientists to Clone Embryos," *The Age* (Melbourne, Australia), 8

March 2002, at: http://www.theage.com.au/articles/2002/03/07/1015365731357.html

164. Recer, "Study Finds Adult Blood Stem Cells Will Not Transform into Other Tissue Cells."
165. Daniel S. Kaufman et al., "Hematopoietic Colony-Forming Cells Derived from Human Embryonic Stem Cells," *Proceedings of the National Academy of Sciences USA 2001* 98:19 (11 September 2001): 10,716–21; his personal website is: http://www1.umn.edu/stemcell/sci/page/fac-mbr/frst-facmbr.htm
166. As quoted in "Blood Cells Made from Stem Cells," *CNN.com,* 17 July 2002, at: http://www.cnn.com/2001/HEALTH/09/03/stem.cells/
167. For example, C. D. Dunn, "The Differentiation of Haemopoietic Stem Cells," *Series Haematologica* 4:4 (1971): 1–71.
168. Malcolm R. Alison et al., "Plastic Adult Stem Cells: Will They Graduate from the School of Hard Knocks?" *Journal of Cell Science* 116: Pt. 4 (15 February 2003): 599–603.
169. Personal telephone communication of 14 January 2003.
170. Ibid.
171. Rick Weiss, "NIH Braces for Slower Funding Growth," *Washington Post,* 2 February 2003, p. A14.
172. Personal telephone communication of 14 January 2003.
173. Scott Gottlieb, "Adult Cells Do It Better," *American Spectator* 34:5 (June 2001): 16, at http://www.gilderbiotech.com/ArticlesByScott/Op%20Ed/AdultCells.htm
174. For example, Paul Elias, "Stem Cell Companies Struggle to Survive," *Associated Press,* 14 November 2002.
175. http://www.dean-med.pitt.edu
176. As quoted in Rick Weiss, "Human Fat May Provide Stem Cells," *Washington Post,* 10 April 2001, p. A1.
177. Thomas H. Maugh II, "Scientists Form Brain Cells from Bone Marrow," *Los Angeles Times,* 15 August 2000, p. A1; http://www.ohsu.edu/
178. As quoted in Susan Okie, "Bone Marrow Cells Offer Hope for Liver Therapies," *Washington Post,* 27 June 2000, p. A1.
179. See for example, prepared testimony of Samuel M. Cohen, University of Nebraska Medical Center, before the Health & Environment Subcommittee on Fetal Tissue Research, 9 March 2000.
180. Personal telephone communication of 14 January 2003.

181. He is a co-founder of "Do No Harm," the most prominent group opposing ES cell research. Their website is: http://www.stemcellresearch.org/
182. For example, Anna Quindlen, "A New Look, An Old Battle," *Newsweek,* 9 April 2001, p. 72.

Xenotransplantation

1. Frank Sesno and Steve Harrigan, "5-Year-Old Russian Boy Allegedly Sold for His Organs by His Own Grandmother," *CNN News,* 28 November 2000, transcript no. 00112802V11.
2. Organ Procurement and Transplantation Network, "Organ Donation and Transplantation Trends in the United States, 2001," at: http://www.optn.org/data/ar2002/ar02_chapter_one.htm
3. Organ Procurement and Transplantation Network, "Reported Deaths and Annual Death Rates per 1,000 Patient Years at Risk Waiting List, 1992 to 2001," at: http://www.optn.org/data/ar2002/ar02_table_107_dh.htm
4. "Donate Life," *Washington Times,* 8 June 2001, p. A18; "Editorial Deputy at Times Dies at 44; Smith's Style Had 'Robust Elegance,'" *Washington Times,* 4 July 2001, p. A3.
5. See, for example, Jason Trahan, "Wait for Transplants Longer in Dallas than FW—Experts Also Say Blacks Take Longer to Get Organs," *Dallas Morning News,* 3 May 2000, p. 20A; see Peter Alig, "Children for Sale," *American Outlook* 3:2 (Spring 2000): 11; see also, Dmitry Chubashenk, "Selling of Kidneys Becomes Moldovan Cottage Industry," *Washington Post,* 7 December 2000, p. A49; "Report: Indian Farmer Selling Kidneys in Tough Times," *United Press International,* 19 May 2000.
6. Andy Coghlan, "Heartening," *New Scientist* 163:201 (28 August 1999): 20.
7. Cook Biotech Inc., "SIS Technology, Products and Applications," at: http://www.cooksis.com/products/index.html; Carol Potera, "Cook Biotech Inc. Culls Biomaterials from Pigs," *Genetic Engineering News* 21:3 (1 February 2001): 32.
8. See Gina Kolata, "Company Says It Cloned Pig in Effort to Aid Transplants," *New York Times,* 14 March 2000, p. A21.
9. http://www.baxter.com/investors/r_d/nextran/index.html
10. As quoted in Kolata, "Company Says It Cloned Pig In Effort to Aid Transplants."
11. Richard Woodman, "Company Announces Birth of Transgenic Cloned Pigs," *Reuters Health,* 11 April 2001.

12. Yifan Dai et al., "Target Disruption of the Alpha-galactosyltransferase Gene in Cloned Pigs," *Nature Biotechnology* 20:3 (March 2002): 2GI; Liangxue Lai et al., "Production of {alpha}-1,3-Galactosyltransferase Knockout Pigs by Nuclear Transfer Cloning," *Science* 295:5557 (8 February 2002): 1089–92; Rick Weiss, "Gene Alteration Boosts Pig-Human Transplant Feasibility," *Washington Post,* 4 January 2002, p. A11.
13. Naomi Aoki, "Breakthrough in Cloning Claimed PPL Therapeutics Says Piglets Lack Both Genes Tied to Tissue Rejection," *Boston Globe,* 23 August 2002, p. D1.
14. Natasha McDowell, "Mini-Pig Clone Raises Transplant Hope," 3 January 2003, *New Scientist,* at: http://www.newscientist.com/news/news.jsp?id=ns99993257. (The company is Immerge BioTherapeutics, a joint venture of Novartis Pharma AG and BioTransplant Incorporated.)
15. Weiss, "Gene Alteration Boosts Pig-Human Transplant Feasibility."
16. Marialuisa Lavitrano et al., "Efficient Production by Sperm-Mediated Gene Transfer of Human Decay Accelerating Factor (hDAF) Transgenic Pigs for Xenotransplantation," *Proceedings of the National Academy of Sciences USA 2002* 99:22 (29 October 2002): 14,230–35.
17. As quoted in Paul Recer, "Pigs Created That Carry Human Genes," *Associated Press,* 22 October 2002.
18. See Kennedy F. Shortridge, "Pandemic Influenza: A Zoonosis?" *Seminars in Respiratory Infections* 7:1 (March 1992): 11–25.
19. See Stefan Lovgren, "China Is Perfect Breeding Ground for Viruses Like SARS, Expert Says," *National Geographic News,* 6 May 2003, at: http://news.nationalgeographic.com/news/2003/05/0506_030506_sarschina.html
20. L. J. van der Laan et al., "Infection by Porcine Endogenous Retrovirus after Islet Xenotransplantation in SCID Mice," *Nature* 407:6800 (7 September 2000): 90–94; see also Emma Dorey, "PERV Data Renew Xeno Debate," *Nature Biotechnology* 18:10 (October 2000): 1032.
21. As quoted in Alan Dove, "Milking the Genome for Profit," *Nature Biotechnology* 18:10 (October 2000): 1045.
22. http://www.pharmiweb.com/novartis/
23. As quoted in Ricki Lewis, "Porcine Possibilities: Can Transgenic Technology Reduce Risks of Xenotransplants?" *Scientist* 14:20 (16 October 2000): 1.
24. http://www.biotransplant.com; Dove, "Milking the Genome for Profit"; B. A. Oldmixon et al., "Porcine Endogenous Retrovirus

Transmission Characteristics of an Inbred Herd of Miniature Swine," *Journal of Virology* 76:6 (March 2002): 3045–48.

25. As quoted in Christine Russell, "Scientists Say Longevity Promoters Are Dead Wrong; Numbers Suggest We May Reach an Expectancy of 100—By the 26th Century," *Washington Post*, 6 March 2001, p. T6.
26. See MIT's "Inventor of the Week Archives, The World Wide Web," at: http://web.mit.edu/invent/www/inventorsA-H/berners-lee.html
27. Leonard Hayflick, "The Future of Aging," *Nature* 408:6809 (November 2000): 267–69, specifically box 2, "How Long Will We Live?"
28. M. Tatar et. al., "A Mutant *Drosophila* Insulin Receptor Homolog That Extends Life-Span and Impairs Neuroendocrine Function," *Science* 292:5514 (6 April 2001): 106–10.
29. As quoted in Merritt McKinney, "Scientists Boost Fruit Fly Lifespan: Humans Next?" *Reuters Health*, 5 April 2001.
30. http://www.ellisonfoundation.org/; personal e-mail communication with Richard L. Sprott, Executive Director, The Ellison Medical Foundation, 17 July 2001; Matt Marshall, "Oracle CEO Larry Ellison Pumps a Fortune into Research on Aging," *San Jose Mercury News*, 1 May 2001.
31. Personal telephone communication of 9 October 2000.
32. As quoted in Andrew Hon et al., "Interview with Bruce Ames," *Berkeley Scientific Journal* 4:1 (Spring 2000): 13.
33. As quoted in Kathryn Brown, "How Long Have You Got?" *Scientific American* 11:2 (Summer 2000): 9.
34. As quoted in Timothy Egan, "Bill Gates Views What He's Sown in Libraries," *New York Times*, 6 November 2002, p. A18.

Part Three: More (and Better) Food for a Growing Population
The Defeat of Hunger

1. U.N. Newservice, "2001 Will Be a 'Tough Year' for the World's Hungry, U.N. Food Agency Warns," 8 January 2000, at: http://www.UN.org/News/dh/latest/page2.html#4; Christopher S. Wren, "U.N. Report Maps Hunger 'Hot Spots,'" *New York Times*, 9 January 2001, p. A8.
2. http://www.rockfound.org/
3. Xudong Ye et al., "Engineering the Provitamin A (Beta-carotene) Biosynthetic Pathway into (Carotenoid-Free) Rice Endosperm," *Science* 287:5451 (14 January 2000): 303–5; Ingo Potrykus, "The

'Golden Rice' Tale," *AgBioView,* at: http://www.biotech-info.net/GR_tale.html

4. "Foods: Yellow Rice Research Project Aims to Prevent Vitamin A Deficiency," *European Report* (Europe Information Service), Sec. No. 2430, 1 September 1999.
5. World Health Organization, "Combating Vitamin A Deficiency," at: http://www.who.int/nut/vad.htm; C. J. Murray and A. D. Lopez, *The Global Burden of Disease* (Geneva, Switz.: World Health Organization: 1996), as cited in David Pimental et al., "Ecology of Increasing Disease: Population Growth and Environmental Degradation," *BioScience* 48:10 (October 1998): 817; "Foods: Yellow Rice Research Project Aims to Prevent Vitamin A Deficiency."
6. "International Vitamin A Consultative Group Policy Statement on Vitamin A, Diarrhea, and Measles," at: http://www.ilsi.org/measles.html; U.S.A.I.D., "Vitamin A Programs," at: http://gaia.info.usaid.gov/pop_health/cs/csvita.htm
7. World Health Organization, "Combating Vitamin A Deficiency."
8. http://www.ipw.agrl.ethz.ch/; Ian Phillips, " 'Golden Rice to Be Donated to Poor," *Associated Press,* 16 May 2000.
9. Justin Gillis, "Monsanto Offers Patent Waiver," *Washington Post,* 4 August 2000, p. A1.
10. http://www.syngenta.com/
11. http://www.monsanto.com
12. Gillis, "Monsanto Offers Patent Waiver."
13. As quoted in ibid.
14. http://dbtindia.nic.in/; G. S. Mudur, "Demons in Disguise? India's Most Popular Cereal Is Headed for a Big Controversy," *Telegraph* (Calcutta), 13 November 2000, at: http://194.131.104.243/ news/archives/november2000/131100tele.html; Gillis, "Monsanto Offers Patent Waiver."
15. Christine K. Shewmaker et al., "Seed-Specific Overexpression of Phytoene Synthase: Increase in Carotenoids and Other Metabolic Effects," *Plant Journal* 20:4 (November 1999): 401–12; Edie Lau, "Calgene Seeks to Make a Better Canola Oil," *Sacramento Bee,* 29 March 2000, p. B1.
16. Shewmaker et al., "Seed-Specific Overexpression of Phytoene Synthase."
17. http://www.msu.edu/home/; http://www.teriin.org/; "Golden Mustard with High-Vitamin A Soon," *Times of India Online,* at: http://www.timesofindia.com/today/23hlth10.htm; Marianna Riley, "Michigan State U. Join in Project to Help Indian Children

with Deficiencies in Vitamin A; New Cooking Oil May Help Solve Problem," *St. Louis Post-Dispatch,* 8 December 2000, p. D13; "Researchers Seek 'Golden Mustard,'" *Food Ingredient News* 8:12 (December 2000).

18. For example, David Scott, "Monsanto Generosity Puts Genetically Modified Food Benefit on Display," *Associated Press,* 4 August 2000; "Monsanto Offers Rice Patents to Hungry," *United Press International,* 4 August 2000.
19. As quoted in Christopher Marquis, "Monsanto Plans to Offer Rights to Its Altered-Rice Technology," *New York Times,* 4 August 2000, p. A11.
20. Paul C. Manglesdorf, Richard S. MacNeish and Walton C. Galinat, "The Domestication of Corn," in *Prehistoric Agriculture,* ed. Stuart Struever (New York: Garden City, 1971), pp. 472–86.
21. See Barry Holstun, *Of Wolves and Men* (New York: Scribners, 1982).
22. For a simple explanation of why this happens, see Karen Freifeld, "Seed Money," *Forbes* 134 (2 December 1985): 219.
23. "Report of the Meeting of the Working Group on Vitamin A and Iron," (Geneva, Switz.: United Nations Sub-Committee on Nutritions 26th Session Reporting, 30 May 2000).
24. David Pimental et al., "Ecology of Increasing Disease: Population Growth and Environmental Degradation," *BioScience* 48:10 (October 1998): 817; C. J. Murray and A. D. Lopez, *The Global Burden of Disease* (Geneva, Switz.: World Health Organization, 1996), as cited in David Pimental et al., "Ecology of Increasing Disease: Population Growth and Environmental Degradation," *BioScience* 48:10 (October 1998): 817; "Foods: Yellow Rice Research Project Aims to Prevent Vitamin A Deficiency," *European Report* (Europe Information Service), Sec. No. 2430, 1 September 1999.
25. J. D. Cook et al., "The Influence of Different Cereal Grains on Iron Absorption from Infant Cereal Foods," *American Journal of Clinical Nutrition* 65:4 (April 1997): 964–69.
26. Trisha Guru, "New Genes Boost Rice Nutrients," *Science* 285:5430 (13 August 1999): 994.
27. Xueyong Li et al., "Control of Tillering in Rice," *Nature* 422:6932 (10 April 2003): 618-21; Rick Callahan, "Gene Find May Boost Rice Yields," *Independent Online* (IOL), 18 April 2003, at: http://www.iol.co.za/index.php?set_id=1&click_id=31&art_id=iol1050678651843R200; Wolfgang Spielmeyer, Mark. H. Ellis and Peter Chandler, "Nucleotide, Protein Semidwarf (sd-1), 'Green Revolution' Rice, Contains a Defective Gibberellin 20-oxidase

Gene," *Proceedings of the National Academy of Sciences USA* 99:13 (25 June 2002): 9043–48.

28. Sakae Agarie et al., "High-Level Expression of Maize Phosphoenolpyruvate Carboxylase in Transgenic Rice Plants," *Nature Biotechnology* 17:1 (January 1999): 76–80; Gerry Edwards, "Turning up Crop Photosynthesis," *Nature Biotechnology* 17:1 (January 1999): 22–23; Martin Abbugao, "American Scientist Unveils New Super Rice for Asia's Billions," *Agence France-Presse*, 29 March 2000.
29. Jeetha D'Silva, "Drought-Resistant Rice Can Cut Water Usage by 50%," *Economic Times* (New Delhi), 10 September 2002.
30. http://www.avesthagen.com/
31. D'Silva, "Drought-Resistant Rice Can Cut Water Usage By 50%."
32. Yvonne M. Pinto, Rosan A. Kok and David C. Baulcombe, "Resistance to Rice Yellow Mottle Virus (RYMV) in Cultivated African Rice Varieties Containing RYMV Transgenes," *Nature Biotechnology* 17:7 (July 1999): 702; Florence Wambugu, "Why Africa Needs Agricultural Biotech," *Nature* 400:6739 (1 July 1999): 15–16; http://www.ahbfi.org/index.htm
33. Kristin Harmel, "Lack of Antioxidants behind Third World Kwashiorkor," at: http://unisci.com/stories/20004/1122005.htm
34. J. C. Waterlow, "Childhood Malnutrition in Developing Nations: Looking Back and Looking Forward," *Annual Review of Nutrition* 14 (1994): 1–19.
35. http://peds.wustl.edu/div/emerg/
36. As quoted in Harmel, "Lack of Antioxidants behind Third World Kwashiorkor."
37. As quoted in ibid.; see also Leeuwenburgh's home page at: http://www.hhp.ufl.edu/ESS/FACULTY/cleeuwen/cleeuwen.htm
38. Oral statement of Suman Sahai, President, Gene Campaign, India, OECD (Organization for Economic Cooperation and Development) Edinburgh Conference on the Scientific and Health Aspects of Genetically Modified Foods, 28 February–1 March 2000; see also her slide presentation by clicking on her name at: http://www.oecd.org/subject/biotech/ed_prog_sum.htm
39. See, for example, Nancy Clark, "Eating for Vitamins," *Physician and Sportsmedicine* 25:7 (July 1997): 103.
40. J. C. Brand and I. Darnton-Hill, "Lactase Deficiency in Australian School Children," *Medical Journal of Australia* 145:7 (6 October 1986): 318–22.

41. Bruce Whitelaw, "Toward Designer Milk," *Nature Biotechnology* 17:2 (February 1999): 135–36.
42. http://www.inserm.fr; Bernard Jost et al, "Production of Low-Lactose Milk by Ectopic Expression of Intestinal Lactase in the Mouse Mammary Gland," *Nature Biotechnology* 17:2 (1999):160; Andy Coghlan, "Great Shakes," *New Scientist* 61:172 (6 February 1999):14, at: http://www.newscientist.com/ns/19990206/newsstory11.html
43. Thomas Stuttaford, "A Genetic Benefit for Thousands," *Times* (London), 29 July 1999.
44. Anne Simon Moffat, "Toting Up the Early Harvest of Transgenic Plants," *Science* 282:5397 (18 December 1998): 2176–78; Martha Groves, "The Cutting Edge," *Los Angeles Times,* 18 August 1997, p. D1. For an entire book about the Flavr Savr, see Belinda Martineau, *First Fruit* (New York: McGraw Hill, 2001).
45. Agricultural and Biotechnology Strategies (Canada) Inc., "Global Status of Approved Genetically Modified Plants," 15 March 2002, at: http://64.26.172.90/agbios/dbase.php?action=Synopsis
46. http://www.bio.org/food&ag/approvedag98.html
47. Steven G. Pueppke, "Agricultural Biotechnology and Plant Improvement," *American Behavioral Scientist* 44:8 (April 2001): 1233–45; reprinted in *Agricultural Biotechnology: The Public Policy Challenges* (Thousand Oaks, California: Sage Publications, 2001): 1233–45, see especially Table 1.
48. http://www.ag.iastate.edu/departments/plantpath/faculty/gmunkvold/gmunkvold.html
49. As quoted in "Health Risks Reduced by GM Corn," *BBC.com,* 22 October 1999, at: http://news2.thls.bbc.co.uk/hi/english/sci/tech/newsid_482000/482467.stm; see also, Gary P. Munkvold, Richard L. Hellmich and Larry G. Rice, "Comparison of Fumonisin Concentrations in Kernels of Transgenic *Bt* Maize Hybrids and Non-transgenic Hybrids," *Plant Disease* 83:2 (February 1999): 130–38.
50. For a full carcinogen list of International Agency for Research on Cancer and National Toxicology Program, see: http://www.ozemail.com.au/~paulr/cancer.html; "Carcinogenic Mycotoxins Frequently Present in Corn Are Greatly Reduced," *Food Chemical News* 42:12 (8 May 2000): 21.
51. See Editors of *AgWeb News,* "Americans Say World Hunger Is More Urgent than Pollution, Disease or Global Warming," 12 October 2000, at: http://www.agweb.com/news

52. "Summary of Findings: Public Sentiment about Genetically Modified Food," prepared by the Mellman Group and Public Opinion Strategies for the Pew Initiative on Food and Biotechnology, March 2001, at: http://pewagbiotech.org/research/survey3-01.pdf
53. Clive James, "Global Hectarage of GM Crops in 2001," International Service for the Acquisition of Agri-biotech Applications, Crop Biotech Brief 2:1 (2002), at: http://www.isaaa.org/kc/Services/Media/biotech_briefs/brief2-1.htm; Clive James, "Biotech Crops Continue Rapid Global Growth," Press Release, ISAAA, 15 January 2003, at: http://www.isaaa.org/Press_release/GA_Jan2003.htm
54. James, "Biotech Crops Continue Rapid Global Growth."
55. Ibid.
56. Clive James, *Preview, Global Review of Commercialized Transgenic Crops: 2000,* ISAAA Brief 21-2000 (Ithaca, New York: International Service for the Acquisition of Agri-biotech Applications, December 2000), at: http://www.isaaa.org/briefs/Brief21.htm
57. Peter Blackburn, "Brazil GM Soy Move Sparks Green Fury, Farmer Doubt," *Reuters,* 27 March 2003.
58. See for example, Reese Ewing, "Brazilian Farmers Ignore Ban on GM Crops," *Reuters,* 22 March 2002.
59. As quoted in Philip Brasher, "Farmers Plan to Cut Back on Biotech Plantings," *Associated Press,* 31 March 2000.
60. Brian Halweil, "After Four Seasons of High Growth, Transgenic Crops Are Now Wilting" *International Herald Tribune,* 22 February 2000, p. 7. (See also my response, Michael Fumento, "Biotech Progress," letter, *International Herald Tribune,* 1 March 2000, p. 11.
61. National Agricultural Statistics Service, "Prospective Plantings," 31 March 2003, pp. 202–21, at: http://usda.mannlib.cornell.edu/reports/nassr/field/pcp-bbp/psp10303.pdf; National Agricultural Statistics Service, "Prospective Plantings," 28 March 2002, at: http://usda.mannlib.cornell.edu/reports/nassr/field/pcp-bbp/pspl0302.pdf
62. James, "Global Hectarage of GM Crops in 2001."
63. Clive James, *Global Review of Commercialized Transgenic Crops: 1999* (Ithaca, New York: International Service for the Acquisition of Agri-biotechnology Applications, 1999), at: http://www.isaaa.org/Global%20Review%201999/briefs12cj.htm
64. See Bill Lambrecht, "India Gives Monsanto an Unstable Lab for Genetics in Farming," *St. Louis Post-Dispatch,* 22 November 1998, p. A1.

65. http://www.vshiva.net/
66. Ronald Bailey, "Dr. Strangelunch," *Reason* 32:8 (January 2001): 20, at: http:// reason.com/0101/fe.rb.dr.shtml
67. As quoted in "Environmentalist Warns Vitamin A Rice Studies Have Blind Spots," *BusinessWorld,* 13 March 2000, p. 28.
68. See L. F. Bauman and P. L. Crane, "Hybrid Corn—History, Development and Selection Considerations," *The National Corn Handbook* (Purdue University: West Lafayette, Indiana), at: http://www.growinglifestyle.com/article/s0/a88613.html
69. Dina Kraft, "Vandana Shiva: Indias Champion of Biodiversity through the Seed," *Associated Press,* 2 September 2002.
70. As quoted at: http://www.monsantoindia.com/
71. Prakash's website is at: http://www.agbioworld.org/prakash.html
72. T. Doetschman et al., "Targeted Correction of a Mutant HPRT Gene in Mouse Embryonic Stem Cells," *Nature* 330:6148 (10–16 December 1987): 576–78.
73. Jennifer Brown, "Amish Take Part in Gene Therapy," *Associated Press,* 16 July 1999.
74. For more, see "The Crigler-Najjar Syndrome Web Pages," at: http://www.crigler-najjar.com/index.html
75. B. T. Kren et al., "Correction of the UDP-glucuronosyltransferase Gene Defect in the Gunn Rat Model of Crigler-Najjar Syndrome Type I with a Chimeric Oligonucleotide," *Proceedings of the National Academy of Sciences USA* 96:18 (31 August 1999): 10,349–54; Brown, "Amish Take Part in Gene Therapy."
76. http://www.valigen.net/
77. See, for example, Andy Coghlan, "Look, No New Genes," *New Scientist* 163:2197 (31 July 1999): 4.
78. http://www.cambia.org.au/
79. As quoted in Ehsan Masood, "Seeds of Dissent," *New Scientist* 168:2261 (21 October 2000): 66, at: http://www.newscientist.com/opinion/opinion_226152.html
80. http://pom.ucdavis.edu/
81. For more see "*Arabidopsis thaliana,* Cultivation Notes," at: http://gardenbed.com/A/6874.cfm
82. Press Release, "Gene Silencing Produces Disease Resistance in Plants," University of California at Davis, at: http://www.pubcomm.ucdavis.edu/newsreleases/10.01/news_gene_silencing.html; his personal website is at: http://pom.ucdavis.edu/personnel/ faculty/dandekar.htm

83. Matthew A. Escobar et al., "RNAi-Mediated Oncogene Silencing Confers Resistance to Crown Gall Tumorigenesis," *Proceedings of the National Academy of Sciences* 98:23 (6 November 2001): 13,437–42.
84. J. Lai and J. Messing, "Increasing Maize Seed Methionine by mRNA Stability," *Plant Journal* 30:4 (May 2002): 395.
85. Joachim Messing, as quoted in "Rutgers Scientists Create High-Protein Corn with Third World Potential," *EurekaAlert*, 4 June 2002, at: http://www.thecampaign.org/News/june02b.htm#rutgers
86. Hank Becker, "Revolutionizing Hybrid Corn Production," *Agricultural Research* 46:12 (December 1998): 10–11, at: http://www.ars.usda.gov/is/AR/archive/dec98/corn1298.htm; Oliver Rautenberg, "Apomixis: A Useful Contribution to Plant Breeding?" *AgBioTech InfoNet*, November 1999, at: http://www.biotech-info.net/apomixis.html
87. Andy Coghlan, "The Next Revolution," *New Scientist* 168:2262 (28 October 2000): 5, at: http://www.newscientist.com/news/news.jsp?id=ns22624
88. Becker, "Revolutionizing Hybrid Corn Production"; Rautenberg, "Apomixis: A Useful Contribution to Plant Breeding?"
89. See: http://www.cambia.org/main/r_fg_6. htm#molecularapomixis
90. As quoted in Ehsan Masood, "Seeds of Dissent," *New Scientist* 168:2261 (21 October 2000): 66, at: http://www.newscientist.com/opinion/opinion_226152.html
91. As quoted in Coghlan, "The Next Revolution."
92. As quoted in ibid.
93. For a much more detailed and illustrated explanation, see Colorado State University, "How Do You Make a Transgenic Plant?" at: http://www.colostate.edu/programs/lifesciences/TransgenicCrops
94. H. J. Klee et al., "Mutational Analysis of the Virulence Region of an Agrobacterium tumefaciens Ti Plasmid," *Journal of Bacteriology* 153:2 (February 1983): 878–83.
95. T. M. Klein et al., "High-Velocity Microprojectiles for Delivering Nucleic Acids into Living Cells," *Nature* 327:7 (May 1987): 70–73.
96. For a longer description and an illustration of both methods, see: http://www.gbhap.com/magazines/scispectra/21-1-fig1.htm
97. L. Chen et al., "Expression and Inheritance of Multiple Transgenes in Rice Plants," *Nature Biotechnology* 16:11 (November 1998): 1060–64.

Safe Biopesticides

1. Cameron R. Currie, Ulrich G. Mueller and David Malloch, "The Agricultural Pathology of Ant Fungus Gardens," *Proceedings of the National Academy of Sciences USA* 96:4 (6 July 1999): 7998–8002.
2. Robert Cooke, "Ants' Lament: There's a Fungus among Us," *New York Newsday*, 27 July 1999, p. C7.
3. Leonard P. Gianessi et al., *Plant Biotechnology: Current and Potential Impact for Improving Pest Management in U.S. Agriculture: An Analysis of 40 Case Studies* (National Center for Food and Agricultural Policy: Washington, D.C.: June 2002), Executive Summary, p. 12, table 6, at: http://www.ncfap.org/40CaseStudies.htm
4. As quoted in Press Release, "Biotechnology Helps Protect U.S. Food Crops from Pests," National Center for Food and Agricultural Policy, June 2002, at: http://www.ncfap.org/40CaseStudies/FinalNCFAPpressrelease-221%20.pdf
5. Matin Qaim and David Zilberman, "Yield Effects of Genetically Modified Crops in Developing Countries," *Science* 299:5608 (7 February 2003): 900–2.
6. As quoted in Shaoni Bhattacharya, "KwaZulu Farmers Boosted by GM Cotton," *New Scientist*, 7 March 2003, at: http://www.newscientist.com/news/news.jsp?id=ns99993473
7. Alex Hetherington, "GM Pays the Bills," *Mail & Guardian* (Johannesburg), 27 March 2003, at: http://allafrica.com/stories/200303270678.html
8. As quoted in Shaoni Bhattacharya, "GM Crops Boost Yields More in Poor Countries," *New Scientist*, 6 February 2003, at: http://www.newscientist.com/news/news.jsp?id=ns99993364; Qaim's personal website is http://www.qaim.de/matin.html
9. http://www.uni-bonn.de/
10. E. Berliner, "Über die Schlafsucht der Mehlmottenraupe (Ephestia kuhniella.) und ihrem Erreger, *Bacillus thurengienesis* n. sp.," *Zeitschrift für angewandte Entomologie* 2 (1915): 29–56, as cited in Hari C. Sharma et al., "Prospects for Using Transgenic Resistance to Insects in Crop Improvement," *Electronic Journal of Biotechnology* 3:2 (15 August 2000), at: http://ejb.ucv.cl/content/vol3/issue2/full/3/index.html#14
11. J. S. Feitelson, "The Bacillus Thuringienesis Family Tree," in *Advanced Engineered Pesticides*, ed. L. Kim (New York: Marcel Dekker, 1993): 63–71.

12. Carrie Swadener, "Bacillus Thuringiensis (Bt)," *Journal of Pesticide Reform* 14:3 (Fall 1994): 13–20, at: http://eap.mcgill.ca/MagRack/JPR/JPR_22.htm
13. Cecil H. Yancy Jr., "Seeds of the Future," *Peanut Farmer* 36:2 (February 2000), at: http://www.peanutfarmer.com/backissues/February2000/story1.asp
14. This information is available from a database run by Information Systems for Biotechnology at Virginia Tech in Blacksburg, Virginia. To access the database, see: http://www.nbiap.vt.edu/cfdocs/fieldtests1.cfm
15. "CSIRO Entomology," at: http://www.ento.csiro.au/business/focus.html; CSIRO, "5,000 *Bt* Genes Join Stored Grain Pest Armory," *Ground Cover* 23 (Spring 1998), at: http://www.grdc.com.au/growers/gc/gc24/storage/5000bt.html
16. Shamley Productions, *Pyscho,* Alfred Hitchcock director, 1960; Janet E. Carpenter and Leonard P. Gianessi, *Agricultural Biotechnology: Updated Benefit Estimates* (Washington, D.C.: National Center for Food and Agricultural Policy, 2001), p. 6, figure 2, "European Corn Borer Densities in Illinois 1943–99," at: http://www.ncfap.org/pup/biotech/updatedbenefits.pdf
17. Carpenter and Gianessi, *Agricultural Biotechnology: Updated Benefit Estimates,* p. 9, at: http://www. ncfap.org/pup/biotech/updatedbenefits.pdf
18. Ibid., p. 15.
19. http://www.csiro.au/
20. As quoted in "Gene Modified Cotton Use Seen Growing Dramatically," *Reuters,* 10 November 2000.
21. Matin Qaim and David Zilberman, "Yield Effects of Genetically Modified Crops in Developing Countries," *Science* 299:5608 (7 February 2003): 900–2.
22. Clive James, "Biotech Crops Continue Rapid Global Growth," Press Release, International Service for the Acquisition of Agri-biotech Applications (ISAAA), 15 January 2003, at: http://www.isaaa.org/Press_release/GA_Jan2003.htm
23. http://www.klinegroup.com
24. Press Release, "Biotech-Based Crops Poised to Erode Insecticide Sales; Growers," *PR Newswire,* 24 October 2000.
25. Kristin Danley-Greiner, "Study Shows Runoff from Bt Cotton Has Limited Impact on Nearby Waterways," *AgWeb News,* 7 March 2001, at: http://www.agweb.com/news_show_news_article.asp?file=AgNewsArticle_2001371228_5313&newscat=GN

26. Bruce E. Paulsrud, "Quick Note: Methyl Bromide Phase-Out," *Illinois Pesticide Review* 1999:4 (July 1999), at: http://www.aces.uiuc.edu/~pse/newsletter/html/199904b.html
27. His personal website is: http://www.pawpaw.kysu.edu/entomology/sedlacek.htm; http://www.uky.edu/Ag/Entomology/enthp.htm; John D. Sedlacek et al., "Life History Attributes of Indian Meal Moth (*Lepidoptera: Pyralidae*) and Angoumois Grain Moth (*Lepidoptera: Gelechiidae*) Reared on Transgenic Corn Kernels," *Journal of Economic Entomology* 94:2 (April 2001): 586–92.
28. Linda McGraw, "Three New Crops for the Future," *Agricultural Research* 47:12 (1 December 1999): 18, at: http://www.ars.usda.gov/is/AR/archive/dec99/crop1299.htm
29. Daniel J. Moellenbeck et al., "Insecticidal Proteins from Bacillus Thuringiensis Protect Corn from Corn Rootworms," *Nature Biotechnology* 19:7 (July 2001): 668–72.
30. Leonard P. Gianessi et al., *Plant Biotechnology: Current and Potential Impact for Improving Pest Management in U.S. Agriculture: An Analysis of 40 Case Studies* (National Center for Food and Agricultural Policy: Washington, D.C.: June 2002), Executive Summary, p. 32, at: http://www.ncfap.org/40CaseStudies.htm
31. Ibid., table 1, p. 7.
32. See Richard W. Knapton and Pierre Mineau, "Effects of Granular Formulations of Terbufos and Fonofos Applied to Cornfields on Mortality and Reproductive Success of Songbirds," *Ecotoxicology* 4:2 (February 1995): 137–52; and Robert Steyer, "Tough Corn Pest Is Getting Tougher; Crop Rotation Is No Longer Foolproof; Rootworm Finds a Taste for Beans," *St. Louis Post-Dispatch*, 19 April 1998.
33. Stephen Johnson, the assistant EPA administrator in charge of pesticide regulation, as quoted in Justin Gillis, "In Key Test, U.S. Allows Sale of Genetically Engineered Corn," Washington, 26 February 2003, p. A1; Julianne Johnston, "YieldGard Rootworm Receives Clearance for 2003 Season," *AgWeb News*, 25 February 2003, at: http://www.agweb.com/news_show_news_article.asp?file=AgNewsArticle_200322591_1412&articleid=95597&newscat=GN; http://www.dowagro.com/; http://www.pioneer.com/; Press Releases, "In-Plant Protection Out-Performs Rootworm Insecticides," 9 August 2000, at: http://www.pioneer.com/pioneer_news/press_releases/in_plant_protection_99.htm
34. http://www.ific.org/; International Food Information Council, "U.S. Consumer Attitudes towards Food Biotechnology," August

2002, at: http://ific.org/proactive/newsroom/release.vtml?id=19981

35. As quoted in Craig S. Smith, "China Rushes to Adopt Genetically Modified Crops," *New York Times,* 9 October 2000, p. A3.
36. Jikun Huang et al., "Plant Biotechnology in China," *Science* 295:5555 (25 January 2002): 674–76; see also Jikun Huang (Center for Chinese Agricultural Policy, Chinese Academy of Sciences Institute of Geographical Sciences and Natural Resources Research), Qinfang Wang and James Keeley, "Agricultural Biotechnology Processes in China," August 2001, at: http://nt1.ids.ac.uk/ids/env/China.pdf
37. Smith, "China Rushes to Adopt Genetically Modified Crops."
38. Janet E. Carpenter and Leonard P. Gianessi, *Agricultural Biotechnology: Updated Benefit Estimates* (Washington, D.C.: National Center for Food and Agricultural Policy, 2001), p. 8, at: http://www.ncfap.org/pup/biotech/updatedbenefits.pdf
39. In the U.S., for example, cropland declined from 382 million acres to 343 million acres between 1980 and 2000, according to U.S. Census Bureau, *Statistical Abstract of the United States* (Washington, D.C.: U.S. Census Bureau, 2001), p. 536, table 825, at: http://www.census.gov/prod/2002pubs/01statab/agricult.pdf
40. "Sierra Club Executive Endorses High-Yield Agriculture, Biotech Crops," *U.S. Newswire,* 14 January 1999, quoting a letter of 28 December 1998, to *Philanthropy* magazine.
41. "Brazil Non-GM Soy Seen Threatening Rain Forests," *Agbioview & Reuters News Service,* 5 November 2002, at: http://www.lifesciencesnetwork.com/news-detail.asp?newsID=2751
42. For more on the Green Revolution, see Lennard Bickel, *Facing Starvation: Norman Borlaug and the Fight against Hunger* (New York: Readers Digest Press, 1974), or Gordon Conway, *The Doubly Green Revolution: Food for All in the Twenty-First Century* (New York: Comstock Publishing, 1999).
43. L. P. Yuan, "Hybrid Rice in China," in *Hybrid Rice Technology,* ed. M. I. Ahmed, B. C. Viraktamath and C. H. M. Vijaya (Directorate of Rice Research: Rajendranagar, Hyderabad, India: 1996), pp. 51–54.
44. S. S. Vermani, "Hybrid Rice," *Advanced Agronomy* 57 (1996): 328–462.
45. D. N. Duvick, "Heterosis: Feeding People and Protecting Natural Resources," in *The Genetics and Exploitation of Heterosis in Crops,* ed. G. James and Shivaji Pandey (American Society of Agronomy–Crop

Science Society of America–Soil Science of America: Madison, WI: 1999), pp. 19–29.

46. Jumin Tu et al., "Performance of Transgenic Elite Commercial Hybrid Rice Expressing *Bacillus Thurengienesis* Delta-Endotoxin," *Nature Biotechnology* 18:10 (October 2000): 1101–4.
47. Ibid.; see also "Genetically-Modified Rice Passes Key Chinese Test," *Agence France-Presse,* 29 September 2000.
48. See Ronald Wall, *Getting Smarter about Pests: Management of Arthropods Blazes the Trail for New Ways of Facing Some of Humanity's Oldest Enemies* (Philadelphia: The Academy of Natural Sciences, 2000), at: http://sapphire.acnatsci.org/erd/ea/ipm.ht.html#sect2
49. David Brown, "New Crops Attract Rare Birds, Say U.S. Farmers," *Daily Telegraph* (London), 1 June 1999, p. 11.
50. Uamdao Noikorn, "Study Finds Biotech Plant Kills Bollworms, Spares Other Pests: Yields Better Health, Economic Benefits," *Bangkok Post,* 12 November 2000, at: http://www.bangkokpost.com/121100/121100_News16.html
51. A. I. Aronson, W. Beckman and P. Dunn, "*Bacillus Thuringiensis* and Related Insect Pathogens," *Microbiology Reviews* 50:1 (March 1986): 1–24.
52. Andy Coghlan, "Out of Egypt," *New Scientist* 162:2191 (19 June 1999): 14.
53. As quoted in ibid.
54. Linda McGraw, "Avidin: An Egg-Citing Insecticidal Protein in Corn,"*Agricultural Research* 48:8 (August 2000), at: http://www.ars.usda.gov/is/AR/archive/aug00/egg0800.htm
55. Karl J. Kramer et al., "Transgenic Avidin Maize Is Resistant to Storage Insect Pests," *Nature Biotechnology* 18:6 (June 2000): 670–74.
56. As quoted in McGraw, "Avidin: An Egg-Citing Insecticidal Protein in Corn."
57. Kramer et al., "Transgenic Avidin Maize Is Resistant to Storage Insect Pests"; McGraw, "Avidin: An Egg-Citing Insecticidal Protein in Corn."
58. Tara Weaver-Missick, "Gene Vectors Agents of Transformation," *Agricultural Research* 47:4 (1 April 1999): 10.
59. Suzanne Cady et al., "Mediterranean Fruit Fly: What Floridians Need to Know," ENY-626, Entomology and Nematology Department, Florida Cooperative Extension Service, Institute of Food and Agricultural Sciences, University of Florida, 1997, at: http://hammock.ifas.ufl.edu/new/in08200.htm#ftlink1
60. Weaver-Missick, "Gene Vectors Agents of Transformation."

61. Scott Kilman, "Buzz Bomb: Bioengineered Bugs Stir Scientific Dreams, but Will They Fly?" *Wall Street Journal*, 26 January 2001, p. A1; Stephen Evans, "A Quiet Battle between Genetic Engineers Is Underway in the Cotton Fields of Arizona," *BBC News*, 20 June 2002, at: http://news.bbc.co.uk/go/em/-/hi/english/sci/tech/newsid_2053000/2053884.stm; John J. Peloquin et al., "Germ-Line Transformation of Pink Bollworm (Lepidoptera: Gelechiidae) Mediated by the piggyBac Transposable Element," *Insect Molecular Biology* 9:3 (August 2000): 323–33.
62. http://www.usd.edu/anth/epa/dust.html
63. See "Protecting Your Land from Erosion," Oregon Association of Conservation Districts, at: http://www.netcnct.net/community/oacd/fs13eros.htm
64. Nyle C. Brady, *The Nature and Properties of Soils* (New York: Macmillan, 1990).
65. Steven Greenhouse, "Quiet Revolution on the Farm," *New York Times*, 5 November 1984, p. D1.
66. J. F. Hebblethwaite, *The Contribution of No-Till to Sustainable and Environmentally Beneficial Crop Production: A Global Perspective* (West Lafayette, Indiana: Conservation Technology Information Center, 1995), as cited in Richard Fawcett and Dan Towery, *Conservation Tillage and Plant Biotechnology: How New Technologies Can Improve the Environment by Reducing the Need to Plow* (West Lafayette, Indiana: Conservation Technology Information Center, 1995), p. 4, at: http://www.ctic.purdue.edu/CTIC/BiotechPaper.pdf
67. Fawcett and Towery, *Conservation Tillage and Plant Biotechnology*, p. 11; http://www.greennature.com/article833.html; Janet Carpenter and Leonard Gianessi, "Herbicide Tolerant Soybeans: Why Growers Are Adopting Roundup Ready Varieties," *AgBioForum* 2:2 (Spring 1999), at: http://www.agbioforum.org/vol2no2/carpenter.html
68. U.S. Environmental Protection Agency, *The Quality of Our Nation's Water, 2000* (Washington, D.C.: Environmental Protection Agency, 2000), at: http://www.epa.gov/305b/98report/ 98brochure.pdf; Fawcett and Towery, *Conservation Tillage and Plant Biotechnology*, p. 8.
69. Rick Boydston, "A Herbicide 'Mode of Action' Primer," *Agrichemical and Environmental News* 178 (February 2001), at: http://www.tricity.wsu.edu/aenews/Feb01AENews/Feb01AENews.htm#anchor5338542; S. R. Padgette et al., "New Weed Control Opportunities: Development of Soybeans with a Roundup Ready Gene," in *Herbicide-Resistant Crops*, ed. S. O. Duke (Boca Raton: CRC Press, 1996), pp. 53–84.

70. Greenhouse, "Quiet Revolution on the Farm."
71. "Glyphosate," in *Guidelines for Drinking-Water Quality*, 2nd ed., Addendum to Vol. 2, *Health Criteria and Other Supporting Information* (Geneva: World Health Organization, 1998), pp. 219–27, at: http://www.oms.ch/water_sanitation_health/GDWQ/Chemicals/glyphofull.htm#Effects
72. Padgette et al, "New Weed Control Opportunities: Development of Soybeans with a Roundup Ready Gene."
73. "History of Herbicide-Tolerant Crops, Methods of Development and Current State of the Art—Emphasis on Glyphosate Tolerance," *Weed Technology* 6:3 (July/September 1992): 626–34.
74. http://www.monsanto.com/monsanto/layout/products/seeds_genomics/herbicide_resistance/default.asp; Scott Kilman, "Monsanto Delays Debut of Wheat Bioengineered to Resist Herbicide," *Wall Street Journal*, 27 February 2002, p. B2.
75. Alan Felsot, "Herbicide Tolerant Genes, Part 1: Squaring up Roundup Ready Crops," *Agrichemical and Environmental News*, Washington State University, September 2000, at: http://www.biotech-info.net/felsot1.html; R. S. Sidhu et al., "Glyphosate-Tolerant Corn: The Composition and Feeding Value of Grain from Glyphosate-Tolerant Corn Is Equivalent to That of Conventional Corn (*Zea mays L.*)," *Journal of Agricultural Food Chemistry* 48:6 (June 2000): 2305–12; N. B. Taylor et al., "Compositional Analysis of Glyphosate-Tolerant Soybeans Treated with Glyphosate," *Journal of Agriculture and Food Chemistry* 47:10 (October 1999): 4469–73; A. W. Burks and R. L. Fuchs, "Assessment of the Endogenous Allergens in Glyphosate-Tolerant and Commercial Soybean Varieties," *Journal of Allergy and Clinical Immunolology* 96:6 part 1 (December 1995): 1008–10; B. G. Hammond et al., "The Feeding Value of Soybeans Fed to Rats, Chickens, Catfish and Dairy Cattle Is Not Altered by Genetic Incorporation of Glyphosate Tolerance," *Journal of Nutrition* 126:3 (March 1996): 717–27; L. A. Harrison et al., "The Expressed Protein in Glyphosate-Tolerant Soybean, 5-enolpyruvulshikimate-3-phosphate Synthase from *Agrobacterium Sp.* Strain Cp4 Is Rapidly Digested in Vitro and Is Not Toxic to Acutely Gavaged Mice," *Journal of Nutrition* 126:3 (March 1996): 728–40; Dean D. Metcalfe et al., "Assessment of the Allergenic Potential of Foods Derived from Genetically Engineered Crop Plants," *Critical Reviews in Food Science and Nutrition* 36(S) (1996): S165–86; D. L. Nida et al., "Glyphosate-Tolerant Cotton: The Composition of the Cottonseed Is Equivalent to That of Conventional Cottonseed," *Journal of Agri-*

culture and Food Chemistry 44:7 (June 1996): 1967–74; S. R. Padgette et al., "The Composition of Glyphosate-Tolerant Soybean Seeds Is Equivalent to That of Conventional Soybeans," *Journal of Nutrition* 126:3 (March 1996): 702–16; Xavier Delannay et al., "Yield Evaluation of a Glyphosate-Tolerant Soybean Line after Treatment with Glyphosate," *Crop Science* 35:5 (September/October 1995): 1461–67; G. M. Kishore, S. R. Padgette and R. T. Fraley, "History of Herbicide-Tolerant Crops, Methods of Development and Current State of the Art—Emphasis on Glyphosate Tolerance," *Weed Technology* 6:3 (July/September 1992): 626–34.

76. Jim Quinn, "Transgenic Grass Mows down Lawn-Care Time," *Arizona Republic*, 18 November 2000, p. AH1; David Barboza, "Suburban Genetics: Scientists Searching for a Perfect Lawn," *New York Times*, 9 July 2000, p. 1; Scotts Company: http://www.scotts.com/homepage.cfm
77. National Agricultural Statistics Service, "Prospective Plantings," 31 March 2003, p. 21, at: http://www.usda.gov/nass/PUBS/TODAYRPT/pspl0303.pdf
78. Clive James, *Preview, Global Review of Commercialized Transgenic Crops: 2000*, ISAAA Brief 21-2000 (Ithaca, New York: International Service for the Acquisition of Agri-biotech Applications, December 2000), at: http://www.isaaa.org/briefs/Brief21.htm
79. Richard Fawcett and Dan Towery, *Conservation Tillage and Plant Biotechnology: How New Technologies Can Improve the Environment by Reducing the Need to Plow* (West Lafayette, Indiana: Conservation Technology Information Center, 1995), p. 15, at: http://www.ctic.purdue.edu/CTIC/BiotechPaper.pdf
80. Ibid., p. 16.
81. Nyle C. Brady, *The Nature and Properties of Soils* (New York: Macmillan, 1990).
82. Charles Benbrook, "Evidence of the Magnitude and Consequences of the Roundup Ready Soybean Yield Drag from University-Based Varietal Trials in 1998," Benbrook Consulting Services, *Ag BioTech InfoNet*, Technical Paper 113 (July 1999); see also Dan Ferber, "GM Crops in the Cross Hairs," *Science* 286:5445 (26 November 1999): 1662.
83. For example, J. Burke, *Results from 1998 Experimental Field Trials of Genetically Engineered Glycophosate Tolerant Sugarbeet Grown in Ireland* (Carlow, Ireland: Oak Park Research Center, 1998); A. S. Culpepper and A. C. York, "Weed Management in Glycophosate-Tolerant Cotton," *Journal of Cotton Science* 4:2 (1998): 174–85.

84. Janet E. Carpenter and Leonard P. Gianessi, *Agricultural Biotechnology: Updated Benefit Estimates* (Washington, D.C.: National Center for Food and Agricultural Policy, 2001), p. 39, at: http://www.ncfap.org/pup/biotech/updatedbenefits.pdf; Janet Carpenter and Leonard Gianessi, "Herbicide Use on Roundup Ready Crops," letter, *Science* 287:5454 (4 February 2000): 803–4; personal telephone communication with Janet Carpenter, 27 December 2000.
85. Janet Carpenter and Leonard Gianessi, "Herbicide Tolerant Soybeans: Why Growers Are Adopting Roundup Ready Varieties," *AgBioForum* 2:2 (Spring 1999), at: http://www.agbioforum.org/vol2no2/carpenter.html
86. Jorgé Fernandez-Cornejo and William D. McBride, *Genetically-Engineered Crops for Pest Management in U.S. Agriculture: Farm-Level Effects* (Washington, D.C.: USDA Economic Research Service, April 2000), p. 6 and p. 7, table 5. (Data from the 1998 crop year.) (Available at: http://www.ers.usda.gov/epubs/pdf/aer786/index.htm)
87. See "A Look at the Future," *Boston Herald,* 2 January 2000, citing data from the Census Bureau; National Center for Health Statistics; Bureau of Labor Statistics; and Department of Agriculture Economic Research Service.
88. http://www.dupont.com/; "Pioneer Brand Hybrids with the LibertyLinkI Gene," at: http://www.pioneer.com/canada/crop/fsllink.htm
89. http://www.1888aventis.com/Products/Buctril/Buctril.htm; http://www.cottonexperts.com/m03/m0305.html
90. "Expert," *Chemical Business Newsbase,* 12 December 2000.
91. Mike Wilkinson, "Gene Flow from Transgenic Plants," pp. 413–33, in *Biotechnology and Safety Assessment,* ed. John A. Thomas and Roy L. Fuchs, 3rd ed. (San Diego; Academic Press, 2002).
92. Dick Ahlstrom, "Superweed Fears 'Confirmed,'" *Irish Times,* 5 September 1998, p. 4; "Fears Grow as 'Superweed' Resists Killers," *Buffalo News,* 3 September 1998, p. 12A.
93. Jeremy Rifkin, "Apocalypse When?" *New Scientist* 160:2158 (31 October 1998): 34, at: http://www.newscientist.com/nsplus/insight/gmworld/gmfood/rifkin.html
94. http://www.massey.ac.nz/~imbs/
95. As quoted in John Saunders, "Genetic Engineering: Sifting the Facts from the Fiction," *Evening Standard* (London), p. 9.
96. Henry Daniell, "Molecular Strategies for Gene Containment in Transgenic Crops," *Nature Biotechnology* 20:6 (June 2002): 581–86;

Nick de Vetten et al., "A Transformation Method for Obtaining Marker-Free Plants of a Cross-Pollinating and Vegetatively Propagated Crop," *Nature Biotechnology* 21:4 (April 2003): 439–42.

97. Julian M. Hibberd et al., "A Galinstan Expansion Femtosyringe for Microinjection of Eukaryotic Organelles and Prokaryote," *Nature Biotechnology* 17:9 (September 1999): 906–9; Susan E. Scott and Mike J. Wilkinson, "Low Probability of Chloroplast Movement from Oilseed Rape (*Brassica Napus*) into Wild *Brassica Rapa*," *Nature Biotechnology* 17:4 (April 1999): 390–91; Susan E. Scott and Mike J. Wilkinson, "Transgene Risk Is Low," letter, *Nature* 393:320 (28 May 1998): 320; Dean Chamberlain and C. Neal Stewart Jr., "Transgene Escape and Transplastomics," *Nature Biotechnology* 17:4 (April 1999): 330–31; Alex Kirby, "Modified Genes That Stay Put," *BBC News* (London), 1 April 1999; Stephanie Ruf et al., "Stable Genetic Transformation of Tomato Plastids and Expression of a Foreign Protein in Fruit," *Nature Biotechnology* 19:9 (September 2001): 870–75; Pal Maliga, "Plastid Engineering Bears Fruit," *Nature Biotechnology* 19:9 (September 2001): 826–27; Mark Henderson, " 'Safe' GM Tomato," *Times* (London), 31 August 2001.
98. http://www.sembiosys.ca/
99. Personal telephone communication with Andrew Bauma, CEO of SemBioSys Genetics Inc., 21 November 2000.
100. Molly Ivins, "Let's Try Making Green Bluebonnets Instead," *Fort Worth Star Telegram,* 5 January 1999, p. 11; Christian Aid, "Selling Suicide," at: http://www.christian-aid.org.uk/indepth/9905suic/suicide1.htm
101. See L. F. Bauman and P. L. Crane, "Hybrid Corn—History, Development and Selection Considerations," *The National Corn Handbook* (Purdue University: West Lafayette, Indiana), at: http://www.growinglifestyle.com/article/s0/a88613.html
102. Kurt Kleiner, "Blowing in the Wind," *New Scientist,* 3 April 2001, at: http://www.newscientist.com/news/news.jsp?id=ns9999586
103. As quoted in Alice Giordano, "New Technology Sows Seeds of Discontent," *Boston Globe,* 16 May 1999, p. D11; Delta & Pine Land: http://www.deltapineseed.com
104. Barnaby J. Feder, "Monsanto to Bar a Class of Seeds," *New York Times,* 5 October 1999, p. A1; see also Michael Fumento, "Biotech Companies Say 'Hasta la vista, baby!' to So-Called 'Terminators,' " *Global Food Quarterly* 28 (Summer 1999): 2, at: http://www.cgfi.com/pdf/SUMMER99.pdf
105. http://www.ucr.edu/

106. As quoted in Carol Kaesuk Yoon, "Some Biotech Upstarts Fizzle against Native Plants," *New York Times*, 20 February 2001, p. F2.
107. Mick J. Crawley et al., "Biotechnology: Transgenic Crops in Natural Habitats," *Nature* 409:6821 (8 February 2001): 682–83.
108. Carol Kaesuk Yoon, "Stalked by Deadly Virus, Papaya Lives to Breed Again," *New York Times*, 20 July 1999, p. F3.
109. "Papaya ringspot *potyvirus*," *Plant Viruses Online*, at: http://biology.anu.edu.au/research-groups/MES/vide/descr549.htm
110. Rod Thompson, "Papaya Virus Eating Away at Industry," *Honolulu Star-Bulletin*, 2 January 1995, p. A1.
111. University of Hawaii: http://www.hawaii.edu/
112. Leonard P. Gianessi et al., *Plant Biotechnology: Current and Potential Impact for Improving Pest Management in U.S. Agriculture: An Analysis of 40 Case Studies* (National Center for Food and Agricultural Policy: Washington, D.C.: June 2002), p. 17, at: www.ncfap.org/40CaseStudies.htm
113. For a great photo of the bacterial beast, go to: http://onsona.lbi.ic.unicamp.br/xf/
114. See Andrew J. G. Simpson et al., "The Genome Sequence of the Plant Pathogen Xylella fastidiosa," *Nature* 406:6792 (13 July 2000): 151–57.
115. http://www.lbi.ic.unicamp.br/
116. As quoted in Lidia Wasowicz, "Crop-Destroying Bacterium Sequenced," *United Press International*, 13 July 2000; see also: http://www.fundecitrus.com.br/escvcus.html
117. Simpson et al., "The Genome Sequence of the Plant Pathogen Xylella fastidiosa."
118. Wasowicz, "Crop-Destroying Bacterium Sequenced."
119. http://www.fundecitrus.com.br/indiceus.html; Reese Ewing, "Brazil Makes Breakthrough against Citrus Disease," *Reuters*, 2 October 2000.
120. As quoted in ibid.
121. For more on the canker, see http://www.citruscanker.com/science.htm; Dean W. Gabriel, "Citrus Canker Disease," at: http://www.biotech.ufl.edu/~pcfcl/canker.htm; Steven Ford, "A Tenacious Enemy; Through Genetic Manipulation, a UF Professor Is Working to Create Canker-Resistant Citrus Trees," *Orlando Sentinel*, 10 March 2002, p. G6.
122. http://www.ipgenetics.com/; Ford, "A Tenacious Enemy; Through Genetic Manipulation, a UF Professor Is Working to Create Canker-resistant Citrus Trees."

123. Jeff Barnard, "For the First Time, Scientists Map the Genes of a Plant Disease," *Associated Press,* 12 July 2000.
124. Peter Feng, "A Summary of Background Information and Food-borne Illness Associated with the Consumption of Sprouts," U.S. FDA, Center for Food Safety and Applied Nutrition, 6 August 1997.
125. As quoted in Marcia Wood, "Safer Sprouts," *Agricultural Research* 48:8 (August 2000), at: http://www.ars.usda.gov/is/AR/archive/aug00/sprout0800.htm
126. U.S. Food and Drug Administration, "Interim Advisory on Alfalfa Sprouts," FDA Talk Paper, 31 August 1998.
127. For more on using jellyfish genes to track bacteria, see Marcia Wood, "Jellyfish Gene Lights Up *E. coli,*" 48:3 (March 2000): *Agricultural Research,* at: http://www.ars.usda.gov/is/AR/archive/mar00/gene0300.htm
128. As quoted in Wood, "Safer Sprouts."
129. As quoted in ibid.

Survival of the Transgenic

1. http://www.cgiar.org/
2. Press Release, "Global Study Reveals New Warning Signals: Degraded Agricultural Lands Threaten World's Food Production Capacity," 21 May 2000, at: http://www.worldbank.org/html/cgiar/press/dres0005.htm
3. *World Population Prospects: The 2000 Revision* (New York: United Nations, February 2001), at: http://www.un.org/esa/population/publications/wpp2000/highlights.pdf (This is the middle variant projection.)
4. Phil Padey, as quoted in Anita Manning, "Report Finds Much of World's Soil Is 'Seriously Degraded'; Food Production Could Be at Risk across the Globe," *USA Today,* 22 May 2000, p. 5A.
5. "Crop Innovations," *Futurist* 3293 (April 1998): 13; Dawn Worrall and Maggie Smallwood, "A Carrot Leucine-Rich-Repeat Protein That Inhibits Ice Recrystallization," *Science* 282:5386 (2 October 1998): 115.
6. Mie Kasuga et al., "Improving Plant Drought, Salt, and Freezing Tolerance by Gene Transfer of a Single Stress-Inducible Transcription Factor," *Nature Biotechnology* 17:3 (March 1999): 287–91.
7. Martha Groves, "Researchers Getting Warmer in Quest for Cold-Resistant Plants," *Los Angeles Times,* 18 January 1999, p. 1; "WeatherGard Genes," at: http://www.mendelbio.com/products/

8. http://www.queensu.ca/
9. Laurie A. Graham et al., "Hyperactive Antifreeze Protein from Beetles," *Nature* 388:6844 (21 August 1997): 727–28.
10. As quoted in Stephen Strauss, "Hint of Bug Blood May Make Food Taste Better," *Globe & Mail* (Toronto), 21 August 1997, at: http://www.afprotein.com/globe.htm
11. As quoted in Linda McGraw, "The Degreening of Canola," *Agricultural Research* 48:1 (January 2000), at: http://www.ars.usda.gov/is/AR/archive/jan00/canola0100.htm
12. Ibid.
13. http://www.biot.cam.ac.uk/
14. James A. H. Murray et al., "Cyclin D Control of Growth Rate in Plants," letter, *Nature* 405:6786 (1 June 2000): 575–79.
15. Idit Raviv, "Israeli Scientists Hope to Speed Tree Growth," *Reuters,* 21 June 1999; Judy Siegel, "Hebrew University Scientists Discover Gene to Accelerate Plant Growth," *Jerusalem Post,* 9 June 1999, p. 5.
16. As quoted in Beile Grunbaum, "The Secret of the Beanstalk?" *Jerusalem Report,* 2 August 1999, p. 18.
17. Ibid.
18. Leandro Peña et al., "Constitutive Expression of Arabidopsis LEAFY or APETALA1 Genes in Citrus Reduces Their Generation Time," *Nature Biotechnology* 19:3 (March 2001): 263–67; Mark Henderson, "GM Orange Tree Bears Early Fruit," *Times* (London), 1 March 2001; Spanish National Institute of Agricultural and Food Science and Technology: http://www.inta.gov.ar/
19. Yuuki Murakami et al., "Trienoic Fatty Acids and Plant Tolerance of High Temperature," *Science* 287:5452 (21 January 2000): 476–79; Alan Dove and Keely Savoie, "Hot Plants," *Nature Biotechnology* 18:2 (March 2000): 249; Kim Honey, "This New Breed of Plant Thrives out of the Spotlight," *Globe & Mail* (Toronto), 18 May 1999.
20. From a 36% estimate in W. Robert Rangeley, "Irrigation and Drainage in the World," in *Water and Water Policy in World Food Supplies,* ed. Wayne R. Jordon (College Station, Tex.: Texas A&M University Press, 1987) and a 47% estimate solely for grain in Montague Yudelman, "The Future Role of Irrigation in Meeting the Worlds Food Supply," in *Soil Science of America, Soil and Water Science: Key to Understanding Our Global Environment* (Madison, Wis: 1994), both as cited in Sandra Postel, "Redesigning Irrigated Agriculture," in *State of the World 2000,* ed. Lester R. Brown, Christopher Flavin and Hilary French (New York: W. W. Norton & Co., 2000), p. 40;

Wolf B. Frommer, Uwe Ludewig and Doris Rentsch, "Taking Transgenic Plants with a Pinch of Salt," *Science* 285:5431 (20 August 1999): 1222–23.

21. University of Toronto, Department of Botany: http://www.botany.utoronto.ca/
22. Maris P. Apse et al., "Salt Tolerance Conferred by Overexpression of a Vacuolar Na+/H+ Antiport in *Arabidopsis*," *Science* 285:5431 (20 August 1999): 1256–58; Frommer, Ludewig and Rentsch, "Taking Transgenic Plants with a Pinch of Salt"; Alan Dove and Andrew Marshal, "Mustard Plant Passes the Salt," *Nature Biotechnology* 17:10 (October 1999): 942.
23. Hong-Xia Zhang and Eduardo Blumwald, "Transgenic Salt-Tolerant Tomato Plants Accumulate Salt in Foliage but Not in Fruit," *Nature Biotechnology* 19:8 (August 2001): 765–68; Marc Kaufman, "A New Strain of Tomatoes, and Don't Hold the Salt," *Washington Post*, 31 July 2001, p. A3.
24. The website for the genome is http://zdna.micro.umass.edu/haloweb/; see also "First Salt-Loving Bug Sequenced," *BBC News*, 2 October 2000, at: http://news.bbc.co.uk/hi/english/sci/tech/newsid_953000/953356.stm
25. See "First Salt-Loving Bug Sequenced."
26. As quoted in Laureen Fagan, "Israelis Use Tree Gene to Fight Desertification," *Reuters*, 5 September 2000.
27. As quoted in Haim Watzman, "Hope for the Badlands," *New Scientist* 167:2255 (9 September 2000): 9.
28. As quoted in Fagan, "Israelis Use Tree Gene to Fight Desertification."
29. Rufus L. Chaney et al., "Phytoremediation of Soil Metals," *Current Opinions in Biotechnology* 8:3 (June 1997): 279–84; Leon V. Kochian, "Cellular Mechanisms of Aluminum Toxicity and Resistance in Plants," *Annual Review of Plant Physiology and Plant Molecular Biology* 46 (1995): 237; C. D. Foy, R. L. Chaney and M. C. White, "The Physiology of Metal Toxicity in Plants," *Annual Review of Plant Physiology* 29 (1978): 511–66.
30. Marcia Barinaga, "Making Plants Aluminum Tolerant," *Science* 5318:276 (6 June 1997): 1497.
31. As quoted in Hank Becker, "Phytoremediation: Using Plants to Clean up Soils," *Agricultural Research* 48:6 (June 2000): 4, at: http://www.ars.usda.gov/is/AR/archive/jun00/soil0600.htm

32. The World Bank Group, "New & Noteworthy in Nutrition," 24 (13 October 1994), at: http://www.worldbank.org/html/extdr/hnp/nutrition/nnn/nnn24.htm
33. Juan Manuel de la Fuente et al., "Aluminum Tolerance in Transgenic Plants by Alteration of Citrate Synthesis," *Science* 5318:276 (6 June 1997): 1566.
34. As quoted in Bob Holmes, "Rooting for Acid," *New Scientist* 163:2199 (14 August 1999): 13.
35. Anne Simon Moffat, "Engineering Plants to Cope with Metals; Testing Crops That Can Thrive in Metal-Rich Soils," *Science* 285:5426 (16 July 1999): 369.
36. Ibid.
37. Mary Lou Guerinot, "Improving Rice Yields—Ironing out the Details," *Nature Biotechnology* 19:5 (May 2001): 417–18.
38. Michiko Takahashi et al., "Enhanced Tolerance of Rice to Low Iron Availability in Alkaline Soils Using Barley Nicotianamine Aminotrensfrerase Genes," *Nature Biotechnology* 19:5 (May 2001): 466–69.
39. J. J. Hudson, W. D. Taylor and D. W. Schindler, "Phosphate Concentrations in Lakes," *Nature* 406:6791 (6 July 2000): 54–56; "Hypoxia," City of New York Department of Environmental Protection," at: http://www.ci.nyc.ny.us/html/dep/html/hypoxia.html
40. See Michael Fumento, "Hypoxia Hysteria," *Forbes* 164:12 (15 November 1999): 96.
41. Press Release, "UF Biotech Breakthrough: Algae Gene Boosts Crop Yields," University of Florida, Gainesville, Institute of Food and Agricultural Sciences, 28 January 1999; "Algae Improves Crops," *Stuart News/Port St. Lucie News* (Stuart, Florida), 28 January 1999, p. A22; personal telephone communication with Robert Schmidt, 22 October 1999; Vaclav Smil, "Long-Range Perspectives on Inorganic Fertilizers in Global Agriculture," Travis P. Hignett Memorial Lecture, 1999, International Fertilizer Development Center, Muscle Shoals, Alabama.
42. Charles J. Vorosmarty et al., "Global Water Resources: Vulnerability from Climate Change and Population Growth," *Science* 289:5477 (14 July 2000): 284–88.
43. Peter Hadfield, "An Enzyme for Surviving in the Desert . . ." *New Scientist* 149:2013 (20 January 1996): 21.
44. Her personal website is at: http://sfns.u-shizuoka-ken.ac.jp/pctech/kyoko.html

45. As quoted in Hadfield, "An Enzyme for Surviving in the Desert. . . ."
46. Sapa-DPA, "GM Seeds Won't Wither without Water," *News24* (South Africa), 11 November 2000, at: http://www.news24.co.za/News24/Technology/Science_ Nature/0,1113,2-13-46_947366,00.html
47. Jay K. Garg et al., "Trehalose Accumulation in Rice Plants Confers High Tolerance Levels to Different Abiotic Stresses," *Proceedings of the National Academy of Sciences USA* 99:25 (10 December 2002): 15,898–903; "GM Rice Can Tough It Out," *BBC News,* 26 November 2002, at: http://news.bbc.co.uk/1/hi/sci/tech/2512195.stm; Press Release, "Stress Relief: Engineering Rice Plants with Sugar-Producing Gene Helps Them Tolerate Drought, Salt and Low Temperatures, Cornell Biologists Report," at: http://www.news.cornell.edu/releases/Nov02/trehalose_stress.hrs.html
48. As quoted in Paul Kendall, "GM Potatoes That Glow Green When They Need Watering," *Daily Mail* (London), 18 December 2000, p. 34.
49. http://www.cgiar.org/irri/
50. Andy Coghlan, ". . . and a Gene for Resisting Flooding," *New Scientist* 149:2013 (20 January 20 1996): 21.
51. http://www.bayercropscience.com/CS/CSCMS.NSF/ID/Home_EN
52. Alan M. Jones et al., "Auxin-Dependent Cell Expansion Mediated by Overexpressed Auxin-Binding Protein 1," *Science* 282:5391 (6 November 1998): 1114–17. (Note: Not long after the publication of this paper, a colleague of Jones was found guilty of fraud in fabricating data. While this did affect some of Jones' papers, it was irrelevant to this one. For more, see Michael Balter, "Data in Key Papers Cannot Be Reproduced; Scientific Fraud at Max Planck Institute for Plant Breeding Research," *Science* 283:5410 (26 March 1999): 1987.
53. Jim Shamp, "Scientists Find Genetic 'Switch' for Cell Size: Antibiotic Feeding May Allow Farmers to Customize Plants," *Durham Herald-Sun* (N.C.), 6 November 1998, p. C1.
54. As quoted in ibid.
55. As quoted in David Williamson, "By Revving up Key Gene, Researchers Discover They Can Manipulate the Size of Cells in Plants," *University of North Carolina at Chapel Hill News Services,* 6 November 1998.

56. Ibid.
57. http://www.biology.leeds.ac.uk/
58. As quoted in Paul Stokes, "Gene Could Stop Plants from Wilting," *Daily Telegraph* (London), 5 April 2002.
59. Elena Zubko et al., "Activation Tagging Identifies a Gene from Petunia Hybrida Responsible for the Production of Active Cytokinins in Plants," *Plant Journal* 29:6 (March 2002): 799–810; Stokes, "Gene Could Stop Plants from Wilting"; Zubko's personal website is: http://www.biology.leeds.ac.uk/centres/liba/pmgrp/elena.htm

Foods That Taste Too Good to Be Healthy

1. As quoted in Christopher Connell, "No Veto Override Here—Executive Order Dooms Veggies," *Associated Press,* 23 March 1990.
2. See Bryan Christie, "Scots Think Healthy Food Is 'Boring and Tasteless,'" *British Medical Journal* 323:7326 (15 December 2001): 1386.
3. For a good general overview see Maureen A. Mackey and Roy L. Fuchs, "Plant Biotechnology Products with Direct Consumer Benefits," pp. 117–42, in *Biotechnology and Safety Assessment,* ed. John A. Thomas and Roy L. Fuchs, 3rd ed. (San Diego; Academic Press, 2002).
4. "Making Sense of Taste and Smell," *Tufts University Diet & Nutrition Letter* 13:9 (November 1995): 3; "The Flavor of Food?" (interview with Valerie B. Duffy), *Journal of the American Dietetic Association* 96:7 (July 1996): 655.
5. "The Flavor of Food?"
6. "Germans Don't Give a Lick for Peanut Butter," *San Francisco Chronicle,* 20 January 1993, Food p. 9; "Hazelnut Council," *Prepared Foods* 3:169 (1 March 2000): 88; "Growers Launch Selling Campaign," *Associated Press,* 24 May 1988.
7. http://www.vegemite.com.au/
8. See Linda Milo Ohr, "Science Conjures the Essence of Flavors," *Prepared Foods* 170:3 (1 March 2001): 45; "Making Sense of Taste and Smell," *Tufts University Diet & Nutrition Letter* 13:9 (November 1995): 3; "The Flavor of Food?"
9. Paul Grayson, chairman and chief executive of Senomyx Inc., as quoted in Justin Gillis, "A New Science: Accounting for Taste Genetics Could Provide Tools to Engineer New Flavors, Fragrances," *Washington Post,* 29 May 2001, p. A1.

10. Gillis, "A New Science: Accounting for Taste Genetics Could Provide Tools to Engineer New Flavors, Fragrances."
11. Ibid.
12. http://www.senomyx.com; Sergey Zozulya, Fernando Echeverri and Trieu Nguyen, "The Human Olfactory Receptor Repertoire," *Genome Biology* 2:6 (June 2001): research0018.1-0018.12.
13. Eric Niiler, "Cheeseburger? No, Broccoli. Mmmm!" *Christian Science Monitor,* 25 January 2001, p. 17.
14. As quoted in ibid.
15. http://www.gmabrands.com/
16. As quoted in Niiler, "Cheeseburger? No, Broccoli. Mmmm!"
17. U.S. Department of the Census, *Statistical Abstract of the United States 2001,* p. 129, table 202, at: http://www.census.gov/prod/2002pubs/01statab/health.pdf
18. A. H. Mokdad et al., "Diabetes Trends in the U.S.: 1990–1998," *Diabetes Care* 23:9 (September 2000): 1278–83.
19. http://www.plant.wageningen-ur.nl/about/Biodiversity/Cgn
20. As quoted in Eric Onstad, "Healthy Sugar? Dutch Concoct Dieter's Dream," *Reuters,* 20 November 1999.
21. See "Fat," in HealthCentral's *General Health Encyclopedia,* at: http://www.healthcentral.com/mhc/top/002468.cfm
22. Elizabeth Parle, "GM Crops: More Food, or Thought?" *Chemical Market Reporter* 12:257 (20 March 2000): FR10.
23. Ibid.
24. Linda Milo Ohr, "Embattled Biotech Charges On," *Prepared Foods* 168:6 (1 June 1999): 109.
25. Edie Lau, "Calgene Seeks to Make a Better Canola Oil," *Sacramento Bee,* 29 March 2000, p. B1.
26. Marc T. Facciotti, Paul B. Bertain and Ling Yuan, "Improved Stearate Phenotype in Transgenic Canola Expressing a Modified Acyl-Acyl Carrier Protein Thioesterase," *Nature Biotechnology* 17:6 (June 1999): 593–97; Sanyin Siang, "Canola Oil Gusher," *Daily InScight,* at: http://www.academicpress.com/inscight/05281999/grapha.htm
27. "Flavonoids, the Next New Thing?" *Harvard Health Letter* 26:2 (1 December 2000): 1052–77; Kristi Steinmentz and John D. Potter, "Vegetables, Fruit, and Cancer Prevention: A Review," *Journal of the American Dietetic Association* 96:10 (October 1996): 1027.
28. "Three New Tomato Breeding Lines," *Food Chemical News* 40:41 (30 November 1998).

29. Susanne Romer et al., "Elevation of the Provitamin A Content of Transgenic Tomato Plants," *Nature Biotechnology* 18:6 (June 2000): 666–69; "GM Tomatoes 'Fight Cancer,'" *BBC News,* 30 May 2000, at: http://news.bbc.co.uk/hi/english/health/newsid_769000/769507.stm
30. "Collaboration Produces Successful Biotech Tomato Product in the U.K.," *Food & Drink Weekly,* 6 July 1998.
31. Shelagh R. Muir et al., "Overexpression of Petunia Chalcone Isomerase in Tomato Results in Fruit Containing Increased Levels of Flavonols," *Nature Biotechnology* 19:5 (May 2001): 470–74.
32. Melanie C. Sze and Patricia Van Arnum, "Getting Food Output through Genetically Engineered Crops," *Chemical Market Reporter* 253:25 (22 June 1998): FR3; Elizabeth Parle, "GM Crops: More Food, or Thought?" *Chemical Market Reporter* 12:257 (20 March 2000): FR10.
33. Roshni A. Mehta et al., "Engineered Polyamine Accumulation in Tomato Enhances Phytonutrient Content, Juice Quality, and Vine Life," *Nature Biotechnology* 20:6 (June 2002): 613–18; Gary Tucker and Graham Seymour, "Life on the Vine," *Nature Biotechnology* 20:6 (June 2002): 558–59.
34. Sze and Arnum, "Getting Food Output Through Genetically Engineered Crops"; Parle, "GM Crops: More Food, or Thought?"; "Australian Scientists Working to Make Pineapples Cheaper, Tastier," *AAP Newsfeed* (Australian Associated Press), 2 March 2000.
35. Linda Milo Ohr, "Embattled Biotech Charges On," *Prepared Foods* 168:6 (1 June 1999): 109.
36. Kathryn Barry Stelljes, "New Raspberries Use Virus against Itself," *Agricultural Research* 48:10 (October 2000), at: http://www.ars.usda.gov/is/AR/archive/oct00/virus1000.htm
37. See, for example, Natalie Kurinij, Mark A. Klebanoff and Barry A. Graubard, "Dietary Supplement and Food Intake in Women of Childbearing Age," *Journal of the American Dietetic Association* 86:11 (November 1986): 1536–40.
38. See Kristine Napier, "Facts and Fiction about Vitamin E," *Harvard Health Letter* 22:1 (November 1996): 1052–77.
39. As quoted in Sanjida O'Connell, "Vitamin E and the Case for GM," *Guardian* (Johannesburg), 9 March 2000, at: http://www.mg.co.za/mg/news/2000mar1/9mar-vitamin.html
40. David Shintani and Dean DellaPenna, "Elevating the Vitamin E Content of Plants through Metabolic Engineering," *Science* 282:5396 (11 December 1998): 2098–100.

41. The Arabidopsis Genome Initiative, "Analysis of the Genome Sequence of the Flowering Plant *Arabidopsis thaliana*," *Nature* 408:6814 (14 December 2000): 796–815; Rick Weiss, "Plant's Genetic Code Deciphered," *Washington Post*, 14 December 2000, p. A3; see Gregg Hillyer and Des Keller, "The Hunt for Genes," *Progressive Farmer*, 12 October 1998.
42. Shintani and DellaPenna, "Elevating the Vitamin E Content of Plants through Metabolic Engineering."
43. As quoted in "Engineered Plant Has More Vitamin E," *Reuters Health*, 11 December 1998.
44. Zhong Chen et al., "Increasing Vitamin C Content of Plants through Enhanced Ascorbate Recycling," *Proceedings of the National Academy of Sciences USA* 100:6 (18 March 2003): 3525–30; Robert Cooke, "Gene Tinkering Gives Plants a Vitamin Boost," *New York Newsday*, 20 May 2003, p. A35.
45. As quoted in Sanjida O'Connell, "Vitamin E and the Case for GM," *Guardian* (Johannesburg), 9 March 2000, at: http://www.mg.co.za/mg/news/2000mar1/9mar-vitamin.html
46. As quoted in Matthew Fordahl, "Researchers Identify Caffeine-Making Gene," *Associated Press*, 30 August 2000.
47. Nancy Clark, "Caffeine: A User's Guide," *Physician and Sportsmedicine* 25:11 (November 1997): 109; Diane Welland, "As Caffeine Controversy Rages On, What's a Coffee Lover to Do?" *Environmental Nutrition* 19:1 (January 1996): 1.
48. Clark, "Caffeine: A User's Guide"; Welland, "As Caffeine Controversy Rages On, What's a Coffee Lover to Do?"
49. Misako Kato et al., "Plant Biotechnology: Caffeine Synthase Gene from Tea Leaves," *Nature* 406:6799 (31 August 2000): 956–57; University of Glasgow: http://www.gla.ac.uk/
50. As quoted in Richard Starnes, "Geneticists Work on Coffee without the Kick," *Ottawa Citizen*, 31 August 2000, p. A9.
51. As quoted in Faye Flam, "Regular or Decaf? Scientists Find Way to Block Caffeine Gene in Java," *Philadelphia Inquirer*, 31 August 2000.
52. As quoted in Starnes, "Geneticists Work on Coffee without the Kick."
53. Ben DiPietro, "Hawaii Scientists Attempt to Grow Decaf Coffee Bean," *Associated Press*, 12 September 1999.
54. His personal website is at: http://www4.ncsu.edu:8030/%7Ehobantj/

55. As quoted in Terence Chea, "Going Whole Hog for Cloning; Biotech Firm Plans Its Use in Breeding Farm Animals," *Washington Post,* 5 August 2000, p. E1.
56. For a detailed description of the cloning of adult animals, see Robert G. McKinnell and Marie A. Di Bernardino, "The Biology of Cloning: History and Rationale," *BioScience* 49:11 (November 1999): 875–85, at: http://t2.technion.ac.il/~shkh/articles/article1.html
57. Robert P. Lanza, Betsy L. Dresser and Philip Damiani, "Cloning Noah's Ark," *Scientific American* 283:5 (October 2000): 84.
58. See Sandy Bauers, "Fish Farming on the Fast Track," *Philadelphia Inquirer,* 12 July 1999.
59. Choy L. Hew and Garth Fletcher, "Transgenic Fish for Aquaculture," *Chemistry and Industry* (United Kingdom) 8 (21 April 1997): 311; Scott Allen, "Some Aren't Hooked on Superfish 'Revolution,'" *Boston Globe,* 23 August 1999, p. A1.
60. Carol Kaesuk Yoon, "Redesigning Nature," *New York Times,* 1 May 2000, p. A1.
61. See Monte Burke, "Cannery Roe," *Forbes* 167:4 (19 February 2001): 106, at: http://www.forbes.com/asap/2001/0219/106.html
62. See Aqua Bounty Farms website, at: http://webhost.avint.net/afprotein/bounty.htm
63. Yoon, "Redesigning Nature."
64. For example, see Marc Kaufman, "'Frankenfish or Tomorrow's Dinner?" *Washington Post,* 17 October 2000, p. A1; "Critics Worry as Researchers Tinker with 'Frankenfish,'" *Associated Press,* 17 October 2000; Frederic Golden and Dick Thompson, "Make Way for Frankenfish!" *Time,* 6 March 2000, p. 62.
65. Allen, "Some Aren't Hooked on Superfish 'Revolution.'"
66. As quoted in Tom Spears, "Canada's GM Salmon Grow Five Times Faster than Normal," *Ottawa Citizen,* 13 April 2000, p. A1.
67. Ibid.; see also William M. Muir and Richard D. Howard, "Possible Ecological Risks of Transgenic Organism Release When Transgenes Affect Mating Success: Sexual Selection and the Trojan Gene Hypothesis," *Proceedings of the National Academy of Sciences USA* 96:24 (23 November 1999): 13,853–56.
68. Rex. A. Dunham et al., "Transfer, Expression, and Inheritance of Salmonid Growth Hormone Genes in Channel Catfish, *Ictalurus Punctatus,* and Effects on Performance Traits," *Molecular and Marine Biology and Biotechnology* 1:4–5 (August/October 1992): 380–89; Rex A. Dunham, "Predator Avoidance, Spawning and Foraging Ability

of Transgenic Catfish," http://www.nbiap.vt.edu/brarg/brasym94/dunham.htm

69. Jon Ward, "State Poisons Pond, Suffocating Snakeheads," *Washington Times,* 5 September 2002, p. B1.
70. Anita Huslin, "U.S. Moves to Ban Import of Snakeheads," *Washington Post,* 23 July 2002, p. B3.
71. Anita Huslin, "Freakish Fish Causes Fear in Maryland," *Washington Post,* 27 June 2002, p. B3.

Developing the Underdeveloped Nations

1. See Fred Pearce, "We Need More Babies," *Sunday Times* (London), 17 March 2002; Nicholas Eberstadt, "The Population Implosion," *Foreign Policy,* March/April 2001, at: http://www.foreignpolicy.com/issue_marapr_2001/eberstadt.html
2. Food and Agricultural Organization of the United Nations, "The State of Food Insecurity in the World 1999," at: http://www.fao.org/focus/e/sofi/home-e.htm
3. Paul R. Ehrlich, *The Population Bomb* (New York: Sierra Club–Ballantine Books, 1968).
4. U.N. Food and Agricultural Organization database, at: http://apps.fao.org/page/form?collection=FS.CropsAndProducts&Domain=FS&servlet=1&language=EN&hostname=apps.fao.org&version=default
5. Norman E. Borlaug, "Ending World Hunger: The Promise of Biotechnology and the Threat of Antiscience Zealotry," *Plant Physiology* 124:2 (October 2000): 487–90, at: http://www.plantphysiol.org/cgi/content/full/124/2/487?view=full&pmid=11027697
6. "Nobel Laureate Borlaug on Food Security, Biotechnology," speech at Kasetsart University in Thailand, 7 March 2000.
7. Adapted from Indur M. Goklany, "Applying the Precautionary Principle to Genetically Modified Crops," Policy Study No. 157 (St. Louis: Center for the Study of American Business, August 2000), p. 15.
8. As quoted in Fred Pearce, "Feeding Africa," *New Scientist* 166:2240 (27 May 2000): 40–43; see also Florence Wambugu, "Why Africa Needs Agricultural Biotech," *Nature* 400:6739 (1 July 1999): 15–16.
9. Judith Achieng, "African Scientists Endorse Genetically Modified Food," *Inter Press Service,* 20 July 1999.

10. http://www.cid.harvard.edu/; Laura Tangley, "Engineering the Harvest: Biotech Could Help Fight Hunger in the World's Poorest Nations—but Will It?" *U.S. News & World Report* 128:10 (13 March 2000): 46.
11. "World Population Nearing 6 Billion, Projected Close to 9 Billion by 2050," United Nations Population Division, Department of Economic and Social Affairs, tables 1 and 3, at: http://www.popin.org/pop1998/1.htm
12. Christian Aid, "Selling Suicide," at: http://www.christian-aid.org.uk/indepth/9905suic/suicide1.htm
13. T. J. Buthelezi as quoted in Clive James, "Biotech Crops Continue Rapid Global Growth," Press Release, International Service for the Acquisition of Agri-biotech Applications (ISAAA), 15 January 2003, at: http://www.isaaa.org/Press_release/GA_Jan2003.htm
14. Deepak Lal, "The New Cultural Imperialism: The Greens and Economic Development, lecture before the Liberty Institute, New Delhi, India, 9 December 2000, based upon a copy received from the author; see also, "Are Green Activists the New Imperialists?" *Times of India Online,* 7 January 2001, at: http://www.timesofindia.com/070101/07busi29.htm
15. From a statement at BIO 2001, San Diego, California, as quoted in Ronald Bailey, "Those Who Protest Biotechnology Do So with a Full Belly," *Reason Online,* 26 June 2001, at: http://www.reason.com/rb/rb-current.html#26
16. Ranjeni Munusamy, "Zambia Rejects Mbeki Maize Offer," *Sunday Times* (South Africa), 6 October 2002, p. 4.
17. Paul Martin and Nicole Itano, "Greens Accused of Helping Africans Starve," *Washington Times,* 30 August 2002, p. A1.
18. Press Release, "Zambian GM Food Aid Decision Highlights Global Problems," Friends of the Earth, 1 September 2002, at: http://www.foe.co.uk/pubsinfo/infoteam/pressrel/2002/20020901103634.html; Press Release, "Eat This or Die: The Poison Politics of Food Aid," *Greenpeace,* 30 September 2002, at: http:// www.greenpeace.org/news/details?news%5fid=40528
19. As quoted in Davan Maharaj and Anthony Mukwita, Zambia Rejects Gene-Altered U.S. Corn, *Los Angeles Times,* 28 August 2002, p. 1; Paul Martin and Nicole Itano, "Greens Accused of Helping Africans Starve," *Washington Times,* 30 August 2002, p. A1.
20. As quoted in "Food Ambassador to U.N. Defends Safety of U.S. Crops," *Dayton Daily News,* as quoted in Rick Weiss, "Starved for

Food, Zimbabwe Rejects U.S. Biotech Corn," *Washington Post,* 31 July 2002, p. A12.

21. As quoted in Martin and Itano, "Greens Accused of Helping Africans Starve."
22. For example, Miguel A. Altieri and Peter Rosset, "Ten Reasons Why Biotechnology Will Not Ensure Food Security, Protect the Environment and Reduce Poverty in the Developing World," *AgBio-Forum* 2:3–4 (Summer/Fall 1999), at: http://www.agbioforum.org/vol2no34/altieri.htm ("Most innovations in agricultural biotechnology have been profit-driven rather than need-driven. The real thrust of the genetic engineering industry is not to make Third World agriculture more productive, but rather to generate profits.")
23. Clive James, *Global Review of Commercialized Transgenic Crops: 1999* (Ithaca, New York: International Service for the Acquisition of Agri-biotechnology Applications, 1999), at: http://www.isaaa.org/Global%20Review%201999/briefs12cj.htm
24. Clive James, "Biotech Crops Continue Rapid Global Growth," Press Release, International Service for the Acquisition of Agri-biotech Applications (ISAAA), 15 January 2003, at: http://www.isaaa.org/Press_release/GA_Jan2003.htm
25. James, *Global Review of Commercialized Transgenic Crops: 1999.*
26. "Scientists Achieve Major Breakthrough in Rice," *Seed World* 138:6 (May 2000): 42.
27. Jun Yu et al., "A Draft Sequence of the Rice Genome (*Oryza sativa L. ssp indica*)," *Science* 296:5565 (5 April 2002): 79–92; Stephen A. Goff et al., "A Draft Sequence of the Rice Genome (*Oryza sativa L. ssp japonica*)," *Science* 296:5565 (5 April 2002): 92–100; University of Washington Genome Center: http://www.genome.washington.edu/UWGC/
28. Stephen Goff, as quoted in Philip Cohen, "Rice Genome Boosts All Crop Research," *New Scientist,* 2 April 2002, http://www.newscientist.com/news/news.jsp?id=ns99992132
29. As quoted in ibid.
30. John Whitfield, "Rice Genome Unveiled: Draft DNA Sequences of the World's Most Important Crop Announced," *Nature Science Update,* 5 April 2002, at: http://www.nature.com/nsu/020402/020402-6.html
31. Press Release, "DuPont Shares Significant Wheat Genome Information," at: http://www1.dupont.com/NASApp/dupontglobal/corp/index.jsp?page=/content/US/en_US/news/releases/2002/nr12_09_02.html

32. Personal communication with Maud Henche, Monsanto Company, 10 March 2000; Laura Tangley, "Engineering the Harvest," *U.S. News & World Report* 128:10 (13 March 2000): 46.
33. Fred Pearce, "Feeding Africa," *New Scientist* 163:2201 (27 May 2000): 40.
34. Zipporah Musau, "Modified Sweet Potato Launched in Kenya," *Africa News,* 19 August 2000.
35. Mark Turner, "African Trials Herald Biotech Food," *Financial Times* (London), 4 November 2000, p. 8.
36. http://www.kari.org/
37. Personal telephone communication of 18 September 2000.
38. Monsanto, "Technology Cooperation Backgrounder," at: http://www.monsantoafrica.com/monsantoafrica/partnership/tech_cooperation.html
39. Personal telephone communication of 18 September 2000.
40. Karen Pennar, "Gordon Conway, Green Revolutionary," *Business Week* 3604 (16 November 1998): 191.
41. "USDA Launches Biotech Research Project for Sub-Saharan Africa," Release No. 0307.99, at: http://www.wip.usda.gov/news/releases/1999/07/0307
42. Anne Simon Moffat, "Toting up the Early Harvest of Transgenic Plants," *Science* 282:5397 (18 December 1998): 2176–78; personal e-mail communication from C. S. Prakash, Tuskegee University.
43. As quoted in Michela Wrong, "Field of Dreams," *Financial Times,* 24 February 2000, p. 18.
44. As quoted in ibid.
45. Press Release, "Scientists Rebuke Critics of Golden Rice; Biotech Rice Can Benefit Developing World Says AgBioWorld Foundation," *PR Newswire,* 13 February 2001.
46. As quoted in Wrong, "Field of Dreams."
47. Michael Fumento, "Crop Busters," *Reason* 33:8 (January 2000): 44 , at: http://reason.com/0001/fe.mf.crop.shtml
48. Benedikt Haerline, "Modified Rice," letter, *Independent* (London), 17 February 2001, p. 2. (Haerline is the Greenpeace International Genetic Engineering Campaign Coordinator in Berlin.)
49. Union of Concerned Scientists mission statement, at: http://www.ucsusa.org/about/mission.html
50. John Carlisle, "No Room for Science at the Union of Concerned Scientists," National Policy Analysis No. 292, June 2000, at: http://www.nationalcenter.org/NPA292.html; John K. Carlisle, "Long on Activism, Short on Science," *Washington Times,* 10 June

2000, p. A13. (Carlisle is director of the environmental policy task force at the National Center for Public Policy Research in Dallas, Texas.)

51. As quoted in Wrong, "Field of Dreams."
52. "ICAR to Engineer Pest-Resistant Crops," February 2000, at http://www.agbiotechnet.com/topics/devco.asp. (For more on ICAR activities: http://www.nic.in/icar/)
53. S. Chakraborty, N. Chakraborty and A. Datta, "Increased Nutritive Value of Transgenic Potato by Expressing a Nonallergenic Seed Albumin Gene from Amaranthus Hypochondriacus," *Proceedings of the National Academy of Sciences USA* 97:7 (28 March 2000): 3724–29.
54. Andy Coghlan, "Genetically Modified 'Protato' to Feed India's Poor," *New Scientist.com*, 2 January 2003, at: http://www.newscientist.com/news/news.jsp?id=ns99993219; "JNU Develops Protein-Rich Potato," *Times of India Online*, at: http://www.timesofindia.com/today/02hlth22.htm
55. "Aussie Gene Put in Indian Wheat to Resist Weed Killer," *Times of India Online*, 21 November 2000, at: http://www.timesofindia.com/today/21hlth9.htm/
56. Hwang Jang-jin, "Genetic Engineering Creates Rice with 26% Higher Yield," *Korea Herald* (Seoul), 13 October 2000, at: http://koreaherald.co.kr/news/2000/10/__10/20001013_1035.htm
57. As quoted in ibid.
58. David Stipp, "China's Biotech Is Starting to Bloom," *Fortune* 146:4 (2 September 2002): 126.
59. Craig S. Smith, "China Rushes to Adopt Genetically Modified Crops," *New York Times*, 9 October 2000, p. A3.
60. Karby Leggett and Ian Johnson, "China Bets Farm on Promise (and Glory) of Genetic Engineering," *Wall Street Journal*, 29 March 2000, p. A17.
61. See "Bt Seed to Make China a Major Player in Cotton Market," *Business Recorder* (London), 1 March 2001.
62. International Service for the Acquisition of Agri-biotech Applications, "Global GM Crop Area Continues to Grow and Exceeds 50 Million Hectares for First Time in 2001," 10 January 2002, at: http://www.isaaa.org/press%20release/Global%20Area_Jan2002.htm
63. See "Bt Seed to Make China a Major Player in Cotton Market."
64. Jikun Huang et al., "Plant Biotechnology in China," *Science* 295:5555 (25 January 2002): 674–76; see also Jikun Huang (Cen-

ter for Chinese Agricultural Policy, Chinese Academy of Sciences Institute of Geographical Sciences and Natural Resources Research), Qinfang Wang,and James Keeley, "Agricultural Biotechnology Processes in China," August 2001, at: http://nt1.ids.ac.uk/ids/env/China.pdf

65. "China Breeds New Cotton from Rabbit Genes," *Reuters,* 10 June 1999.
66. As quoted in Smith, "China Rushes to Adopt Genetically Modified Crops."
67. As quoted in ibid.
68. Stipp, "China's Biotech Is Starting to Bloom."
69. Leggett and Johnson, "China Bets Farm on Promise (and Glory) of Genetic Engineering."
70. C. James and F. Krattiger, "Global Review of the Field Testing and Commercialization of Transgenic Plants, 1986 to 1995: The First Decade of Crop Biotechnology," Brief No. 1 (Ithaca, New York: ISAAA, 1996).
71. Clive James and Anatole F. Krattiger, "Global Review of the Field Testing and Commercialization of Transgenic Plants, 1986 to 1995: The First Decade of Crop Biotechnology," Brief No. 1, at: http://isaaa.org/frbrief1.htm
72. Judith Achieng, "Development: Africa and the Biotechnology Debate," *Inter Press Service,* 1 March 2000.
73. S. L. Murray et al., "Transformation of Potatoes (Cv Late Harvest) with the Potato Leafroll Virus Coat Protein Gene, and Molecular Analysis of Transgenic Lines," *South African Journal of Science* 94:6 (June 1998): 263–68; D. K. Berger et al., "Field Trial of Transgenic Potatoes Transformed with the Potato Leafroll Virus Coat Protein Gene," paper presented at the 36th Congress of the South African Society for Plant Pathology, Drakensberg, South Africa, 1998.
74. http://www.potatoes.co.za/; J. J. B Van Zijl and R. Botha, "Conservation of Plant Genes III," in *Conservation and Utilization of African Plants,* ed. R. P Adams and J. E. Adams (St. Louis, MO: Missouri Botanical Garden Press, 1998): 147–53.
75. J. B. J. Van Rensberg and M. J. Ferreira, "Resistance of Elite Corn Inbred Lines to Isolates of *Stenocarpella Maydis," South African Journal of Plant and Soil* 14:2 (1997): 1–4.
76. Florence Wambugu, "Why Africa Needs Agricultural Biotech," *Nature* 400:6739 (1 July 1999): 15–16.
77. Ed Rybicki, "Agricultural Molecular Biotechnology in South Africa: New Developments from an Old Industry," *AgBiotechNet* 1 (1

August 1999), at: http://www.agbiotechnet.com/reviews/aug99/html/Rybicki.htm; for more on the institute, see http://www.arc.agric.za/

78. This has been extremely heavily documented, but see for example, Ann Leslie, "The Last This Child Heard of His Father Was a Shot and a Scream," *Daily Mail* (London), 14 May 1998, p. 32; Jonathan Manthorpe, "Agency's Departure Shows Intolerable Situation in N. Korea: Doctors Without Borders Is Removing Its 13-Person Team Because They Were Unable to Effectively Provide Famine Relief," *Vancouver Sun,* 1 October 1998, p. A14.
79. "Food—Kenya: Monsanto Develops Virus-Resistant Sweet Potato," *Inter Press Service,* 21 August 2000.
80. As quoted in Sylvia Carter, "One Potato, New Potato: Farmers and Biotech Companies Are Battling for Control," *New York Newsday,* 28 March 1999, p. A51.
81. International Service for the Acquisition of Agri-biotech Applications, "Global GM Crop Area Continues to Grow and Exceeds 50 Million Hectares for First Time in 2001," 10 January 2002, at: http://www.isaaa.org/press%20release/Global%20Area_Jan2002.htm
82. Editors of *AgWeb News,* "Americans Say World Hunger Is More Urgent than Pollution, Disease or Global Warming," 12 October 2000, at: http://www.agweb.com/news
83. Ibid.
84. Jimmy Carter, "Who's Afraid of Genetic Engineering?" *New York Times,* 26 August 1998, p. A21.

Frankenfoods?

1. http://www.laskerfoundation.org/news/weis/rdna.html
2. Mary Murray, "Battling Illness with Body Proteins; Recombinant DNA Technology Has Enabled Medical Researchers to Manufacture Several Human Proteins That Can Bolster the Body's Natural Defenses against Disease," *Science News* 131:42 (17 January 1987): 42.
3. See generally Donald L. Uchtman and Gerald C. Nelson, "Regulatory Oversight of Agricultural and Food-Related Biotechnology," *American Behavioral Scientist* 14:3 (November 2000): 350–77, reprinted in *Agricultural Biotechnology: The Public Policy Challenges* (Thousand Oaks, California: Sage Publications, 2000), pp. 350–77.
4. Kelly Class, "Tropicana Offers Taste of 'Forbidden Fruit' Juice," *Adweek,* 3 July 1989.

5. "Monsanto vs. the Monarch," *St. Louis Post-Dispatch,* 23 May 1999, p. B2; Lee Bowman, "High-Tech Corn Killing Butterflies; Genetic Engineering Leads to Toxic Pollen," *Chicago Sun-Times,* 20 May 1999, p. 3; Laura Tangley, "Attack of the Killer Corn," *U.S. News & World Report* 126:21 (31 May 1999): 69.
6. His Royal Highness, The Prince of Wales, "My Ten Fears for GM Food," *Daily Mail* (London), 1 June 1999, pp. 10–11.
7. John Losey, Linda Rayor and Maureen Carter, "Transgenic Pollen Harms Monarch Larvae," letter, *Nature* 399:6733 (20 June 1999): 214; "Monarch Butterfly Researchers Urge Caution in Over-Interpreting Results," *PR Newswire,* 10 June 1999.
8. See Michael Fumento, "The World Is Still Safe for Butterflies," *Wall Street Journal,* 25 June 1999, p. A18, at: http://www.fumento.com/butterfly.html
9. Personal telephone communication of 2 June 1999.
10. Personal telephone communication of 4 June 1999.
11. Mark K. Sears et al., "Impact of *Bt* Corn Pollen on Monarch Butterfly Populations: A Risk Assessment," *Proceedings of the National Academy of Sciences USA* 98:21 (9 October 2001): 11,937–42; Richard L. Hellmich et al., "Monarch Larvae Sensitivity to *Bacillus Thuringiensis*–Purified Proteins and Pollen," *Proceedings of the National Academy of Sciences USA* 98:2 (9 October 2001): 11,925–30; John M. Pleasants et al., "Corn Pollen Deposition on Milkweeds in and near Cornfields," *Proceedings of the National Academy of Sciences USA* 98:21 (9 October 2001): 11,919–24; Diane E. Stanley-Horn et al., "Assessing the Impact of Cry1Ab-Expressing Corn Pollen on Monarch Butterfly Larvae in Field Studies," *Proceedings of the National Academy of Sciences USA* 98:21 (9 October 2001): 11,931–36; Karen S. Oberhauser et al., "Temporal and Spatial Overlap between Monarch Larvae and Corn Pollen," *Proceedings of the National Academy of Sciences USA* 98:21 (9 October 2001): 11,913–28.
12. Andrew Pollack, "Data on Genetically Modified Corn; Reports Say Threat to Monarch Butterflies Is 'Negligible,'" *New York Times,* 8 September 2001, p. C2; Kim Kaplan, "*Bt* Corn Poses 'No Significant Risk' to Monarchs," Agricultural Research Service News & Information, 6 February 2002, at: http://www.ars.usda.gov/is/pr/2002/020206.htm
13. Carol Kaesuk Yoon, "Storm in Mexico Devastates Monarch Butterfly Colonies," *New York Times,* 12 February 2002, p. A1.
14. George McGovern, "The Wonder of Fighting Famine with Biotechnology," *Minneapolis Star Tribune,* 6 November 2000, p. 13A,

at: http://webserv3.startribune.com/stOnLine/cgi-bin/article?thisSlug=MCGO06&date=06-Nov-2000&word=mcgovern&word=mcgoverns

15. George McGovern, *The Third Freedom: Ending Hunger in Our Time* (New York: Simon & Schuster, 2001).
16. See generally, Francis Fukuyama, *Our Posthuman Future: Consequences of the Biotechnology Revolution* (New York: Farrar, Straus & Giroux, 2002); for example, William Kristol and Jeremy Rifkin, "First Test of the Biotech Age: Human Cloning," *Los Angeles Times,* 6 March 2002, pt. 2, p. 11; see generally, Leon R. Kass, *Life, Liberty and the Defense of Dignity: The Challenge for Bioethics* (San Francisco: Encounter Books, 2002); for example, Dinesh D'Souza, "Staying Human: The Danger of Techno-Utopia," *National Review* 53:1 (22 January 2001): p. 36; and Dinesh D'Souza, *The Virtue of Prosperity: Finding Values in an Age of Techno-Affluence* (New York: Basic Books, 2000), p. 205.
17. Virginia Postrel, *The Future and Its Enemies* (New York: Free Press, 1998), pp. 160–69.
18. For example, Jonathan Leake and Guy Dennis, "Mutant Lobster Is Food of the Future," *Sunday Times* (London), 23 April 2000; Sean Poulter, "M&S Bows to Shoppers' Fears and Orders Ban on Frankenfoods," *Daily Mail* (London), 16 March 1999, p. 5.
19. George Gaskell, Nick Allum and Sally Stares, "Europeans and Biotechnology in 2002: Eurobarometer 58.0," 21 March 2003, p. 14, table 2, at: http://europa.eu.int/comm/public_opinion/archives/eb/ebs_177_en.pdf
20. As quoted in Jeff Otieno, "Biotechnology Controversy Rages," *Africa News Service,* 16 September 1999.
21. Aimé Guibert, as quoted in William Echikson et al., "Wine War," *Business Week* 3747 (3 September 2001): 54.
22. http://www.croplife.org/
23. Personal telephone communication of 4 April 2002.
24. Gaskell, Allum and Stares, "Europeans and Biotechnology in 2002: Eurobarometer 58.0," p. 21, table 6.
25. Thomas J. Hoban, "Social Controversy and Consumer Acceptance of Biotechnology," *Journal of Biolaw and Business* 3:3 (March 2000): 38–46.
26. See http://www.netspeed.com.au/ttguy/
27. Gaskell, Allum and Stares, "Europeans and Biotechnology in 2002: Eurobarometer 58.0," p. 39, table 17.
28. Personal telephone communication of 14 December 1999.

29. Prepared Testimony of Dan Glickman, U.S. Secretary of Agriculture, before the House Committee on Agriculture, 20 October 1999.
30. See Peter James Spielman, "EU Nations Agree to Restrict Rules on Sale of Genetically Modified Food," *Associated Press,* 25 June 1999.
31. Edmund L. Andrews, "Europe Refuses to Drop Ban on Hormone-Fed U.S. Beef," *New York Times,* 28 April 1999, p. C4.
32. For more information, see USDA, "Bovine Spongiform Encephalopathy (BSE)," at: http://www.aphis.usda.gov/oa/bse/
33. As quoted in "Greenpeace Blocks U.S. Shipment," *Associated Press,* 3 December 2000; see also Duane D. Freese, "Mad Cow Reality Confronts Phony Biotech Scare," *Tech Central Station,* 18 December 2000, at: http://www.techcentralstation.com/NewsDesk.asp?FormMode=PolicyTracksArticles&ID=85
34. For an interesting collection of articles on this phenomenon, all in one place, see Andy Goghlan, "How It All Went So Horribly Wrong," *New Scientist* 168:2263 (4 November 2000): 7, at: http://www.newscientist.com/nsplus/insight/bse/howitwent.html; Claire Ainsworth, "A Killer Is Born," *New Scientist* 168:2263 (4 November 2000): 7, at: http://www.newscientist.com/nsplus/insight/bse/akiller.html; Debora MacKenzie, "Mad Meat," *New Scientist* 168:2263 (4 November 2000): 8, at: http://www.newscientist.com/nsplus/insight/bse/madmeat.html; Debora MacKenzie, "The Human Tragedy May Just Be Beginning," *New Scientist* 168:2263 (4 November 2000): 9, at: http://www.newscientist.com/nsplus/insight/bse/humantragedy.html/
35. See Jonathan Leake and John Elliott, "BSE Inquiry Pins Blame on the Tories," *Sunday Times* (London), 1 October 2000; "European Dioxin-Contaminated Food Crisis Grows and Grows," *Lancet,* 353:9169 (12 June 1999): 2049; Walter P. von Wartburg and Julian Liew, *Gene Technology and Social Acceptance* (Lanham, Maryland: University Press of America, 1999).
36. Personal telephone communication of 15 December 1999. For an explanation of agricultural biotech regulation in the EU, see Margaret R. Grossman and Bryan Endres, *American Behavioral Scientist* 14:3 (November 2000): 378–434; reprinted in *Agricultural Biotechnology: The Public Policy Challenges* (Thousand Oaks, California: Sage Publications, 2000: 378–434.
37. George Gaskell, Nick Allum and Sally Stares, "Europeans and Biotechnology in 2002: Eurobarometer 58.0," 21 March 2003,

p. 30, table 13, at: http://europa.eu.int/comm/public_opinion/archives/eb/ebs_177_en.pdf

38. Personal telephone interview of 4 April 2002.
39. For a broader discussion, see Michael Fumento, "Why Europe Fears Biotech Foods," *Wall Street Journal,* 14 January 2000, p. A14.
40. Anne Swardson and Justin Gillis, "Facing the Backlash of Pushing the Envelope," *Sunday Star-Times* (Auckland, New Zealand), 7 November 1999, p. 5.
41. As quoted in Martin Enserink, "Ag Biotech Moves to Mollify Its Critics," *Science* 286:5445 (26 November 1999): 1666.
42. "Snow Brand Drink Adds to Sick List," *Mainichi Daily News* (Japan), 9 July 2000, p. 1.
43. As quoted in Akiko Kashiwagi, "Scandals Cause Japanese to Lose Their Appetites: Once-Revered Local Food in Doubt," *Washington Post,* 25 March 2002, p. A15.
44. See generally Donald L. Uchtman and Gerald C. Nelson, "Regulatory Oversight of Agricultural and Food-Related Biotechnology," *American Behavioral Scientist* 14:3 (November 2000): 350–77, reprinted in *Agricultural Biotechnology: The Public Policy Challenges* (Thousand Oaks, California: Sage Publications, 2000): 350–77; Amy W. Ando and Madhu Khanna, "Environmental Costs and Benefits of Genetically Modified Crops," *American Behavioral Scientist* 14:3 (November 2000): 435–63, reprinted in *Agricultural Biotechnology: The Public Policy Challenges* (Thousand Oaks, California: Sage Publications, 2000): 435–63.
45. Department of Health and Human Services, U. S. Food and Drug Administration, "Statement of Policy: Foods Derived from New Plant Varieties," *Federal Register* 57 (29 May 1992): 22,984–23,005; Environmental Protection Agency, "Proposed Policy; Plant-Pesticides Subject to the Federal Insecticide, Fungicide, and Rodenticide Act and the Federal Food, Drug, and Cosmetic Act," *Federal Register* 59:225 (23 November 1994): 60,496–60,518; U.S. Department of Agriculture, "Coordinated Framework for Regulation of Biotechnology," *Federal Register* 49 (31 December 1984): 50,856–50,907.
46. John Carey, Ellen Licking and Amy Barrett, "Are Bio-Foods Safe?" *Business Week* 3660 (20 December 1999): 70, at: http://www.businessweek.com/premium/99_51/b3660136.htm
47. Jane Rissler, as quoted in Gerard Aziakou, "Farm Biotechnology: Panacea or Dangerous Science?" *Agence France-Presse,* 16 December 1998; calculated from Jorgé Fernandez-Cornejo and William D.

McBride, *Genetically-Engineered Crops for Pest Management in U.S. Agriculture: Farm-Level Effects* (Washington, D.C.: USDA Economic Research Service, April 2000), p. 13, table 7. (Data from the 1998 crop year.) (Available at: http://www.ers.usda.gov/epubs/pdf/aer786/index.htm)

48. Kevin O'Sullivan, "Monsanto Pushes Benefits of GM Foods to the Environment," *Irish Times,* 1 March 1999, p. 3.
49. Dick Hogan, "Activists Destroy Monsanto GMO Beet; Attack Is Fourth on Genetically-Modified Food Field Trials," *Irish Times,* 17 August 1999, p. 7.
50. As quoted in Alison Healy, "Firm Claims Weed-Control Benefits of GM Sugar Beet," *Irish Times,* 10 October 2000, at: http://www.ireland.com/newspaper/ireland/2000/1010/hom8.htm
51. Ai-Guo Gao et al., "Fungal Pathogen Protection in Potato by Expression of a Plant Defensin Peptide," *Nature Biotechnology* 18:12 (December 2000): 1307–10.
52. A classic work on this is Cecil Woodham-Smith, *The Great Hunger: Ireland, 1845–1849* (New York: Penguin, 1995).
53. Judy Jamison, "Crop Fungal Protection," *Nature Biotechnology* 18:12 (December 2000): 1233; Maggie Fox, "Engineered Potatoes Said to Fight off Fungus," *Reuters,* 30 November 2000.
54. http://www.proagro.com/english.htm
55. John Stackhouse, "Designer Seeds a Growing Concern in India," *Globe & Mail* (Toronto), 25 January 1999, p. A16; "Proagro-PGS Set to Begin GM Mustard Field Trials," *World Reporter,* 29 September 2000. (*World Reporter* can be found at http://www.worldreporter.net/3rdworld/Index.htm)
56. For example, see David Wighton and John Willman, "Government Rejects Fresh Demands for Moratorium on Genetically Modified Foods," *Financial Times* (London), 13 February 1999, p. 6.
57. Michael Fumento, "Crop Busters," *Reason* 31:8 (January 2000): 44.
58. Oral statement of Doug Parr, OECD (Organization for Economic Cooperation and Development) Edinburgh Conference on the Scientific and Health Aspects of Genetically Modified Foods, 28 February–1 March 2000.
59. James D. Watson, "All for the Good: Why Genetic Engineering Must Soldier On," *Time,* 11 January 1999, p. 91.
60. Sean Munro, "GM Food Debate," letter, *Lancet* 354:9191 (12 November 1999): 1727–28.
62. Both statements as quoted in Michael Settle, "MPs Call for Rational Debate on Issues," *Herald* (Glasgow), 19 May 1999, p. 10.

62. Harry A. Kuiper, Hub P. J. M. Noteborn and Ad A. C. M. Peijenburg, "Adequacy of Methods for Testing the Safety of Genetically Modified Foods," *Lancet* 354:9187 (16 October 1999): 1315.
63. Royal Society, "Review of Data on Possible Toxicity of GM Potatoes," 18 May 1999 at: http://www.ncbe.rdg.ac.uk/NCBE/GMFOOD/royalsoc3.html
64. David Concar and Debora MacKenzie, "Mashed Potatoes," *New Scientist,* 6 March 1999, p 13.
65. Krista Thomas, "Hot Potato Debate on Biotechnology Comes to Canada Modified Foods," *National Post* (Canada), at: http://www.nationalpost.com/stories/20010220/479847.html
66. As quoted in Julianne Johnston, "FDA Official: Biotech Foods Safer than Hybridization," *AgWeb News,* 13 October 2000, at: http://www.agweb.com/getpage.asp?file=/main/agweb/articles/AgNewsArticle_200010171749_1.html&cat=News&date=10%2F17%2F2000
67. Kevin Knobloch, Chief Operation Officer, Union of Concerned Scientists, "Union of Concerned Scientists Defends Itself," letter, *Washington Times,* 2 July 2000, p. B2.
68. Department of Health and Human Resources, U.S. Food and Drug Administration, "Premarket Notice Concerning Bioengineered Foods," *Federal Register* 66:12 (18 January 2001): Docket No. 00N-1396.
69. See comments of Monsanto spokesman Bryan Hurley in Rebecca Osvath and Mary Ellen Butler, "FDA Will Require Pre-market Consultations under Proposed Biotech Policy," *Food Chemical News* 42:12 (8 May 2000): 3.
70. As quoted in Rick Weiss, "U.S. to Add Oversight on Biotech Food," *Washington Post,* 3 May 2000, p. 1.
71. Ibid.
72. Michael Koontz, "Clean Air and Transportation: the Facts May Surprise You," *Public Roads* 62:1 (July/August 1998), at: http://www.tfhrc.gov/pubrds/julaug98/clean.htm
73. See, for example, Paul R. Billings, "Modified Foods Are Like Drugs," *Boston Globe,* 28 August 1999, p. A19.
74. Oral remarks of Phil Hutton, U.S. EPA, Office of Pesticide Programs, at American Crop Protection Association New Issues Conference, Arlington, Virginia, 12–13 April 2000; see also, George A. Burdock, "Status and Safety Assessment of Foods and Food Ingredients Produced by Genetically Modified Microorganisms," pp. 54–56; and Bruce M. Chassey, "Food Safety Assessment of

Current and Future Plant Biotechnology Products," p. 96, in *Biotechnology and Safety Assessment,* ed. John A. Thomas and Roy L. Fuchs, 3rd ed. (San Diego: Academic Press, 2002).

75. As quoted in Larry Thompson, "Are Bioengineered Foods Safe?" *FDA Consumer* 34:1 (1 January 2000): 18.
76. Peter D. Hare and Nam-Hai Chua, "Excision of Selectable Marker Genes from Transgenic Plants," *Nature Biotechnology* 20:6 (June 2002): 575–80; David Ow, "Joint FAO/WHO Expert Consultation on Foods Derived from Biotechnology," 29 May–2 June 2000, http://www.who.int/fsf/gmfood/consultation_may2000/biotech_00_14.pdf; David W. Ow, "The Right Chemistry for Marker Gene Removal?" *Nature Biotechnology* 19:2 (February 2001): 115.
77. Tim Kunkel et al., "Inducible Isopentenyl Transferase as a High-Efficiency Marker for Plant Transformation," *Nature Biotechnology* 17:9 (September 1999): 916–19; for other examples see Muhammad Sarwar Khan and Pal Maliga, "Fluorescent Antibiotic Resistance Marker for Tracking Plastid Transformation in Higher Plants," *Nature Biotechnology* 17:9 (31 August 1999): 910–15, and Andy Coghlan, "On Your Markers," *New Scientist* 164:2213 (20 November 1999): 10.
78. For example, Jianru Zuo et al., "Chemical-Regulated, Site-Specific DNA Excision in Transgenic Plants," *Nature Biotechnology* 19:2 (February 2001): 157.
79. Jeremy Rifkin, "Unknown Risks of Genetically Engineered Crops," *Boston Globe,* 7 June 1999, p. A13; Green-Watch.com, "Foundation on Economic Trends," at: http://www.green-watch.com/Washington,%20DC/foundation_on_economic_trends.htm
80. Bruce E. Tabashnik et al., "Field Development of Resistance to *Bacillus Thuringiensis* in Diamondback Moth (Lepidoptera: Plataleidae)," *Journal of Economic Entomology* 83:5 (1990): 1671–76; Anthony M. Shelton et al., "Resistance of Diamondback Moth to *Bacillus Thuringiensis* Subspecies in the Field," *Journal of Economic Entomology* 86:3 (1993): 697–705; Bruce E. Tabashnik et al., "Frequency of Resistance to *Bacillus Thuringiensis* in Field Populations of Pink Bollworm," *Proceedings of the National Academy of Sciences USA* 97:24 (1 November 2000): 12,980–84; Jim Kling, "Stories of Modern Science . . . from UPI," *United Press International,* 22 November 2000.
81. As quoted in Susan McGinley, "Pest Resistance to Genetically Modified Cotton Not Seen," *UniSci.com,* at: http://unisci.com/stories/20004/1121005.htm/

82. As quoted in Bill Lambrecht, "India Gives Monsanto an Unstable Lab for Genetics in Farming," *St. Louis Post-Dispatch,* 22 November 1998, p. A1.
83. Ibid.
84. Philip Brasher, "EPA Placing Restrictions on Biotech Corn Plantings," *Associated Press,* 16 January 2000.
85. For a field study of the efficacy of such refuges, see Anthony M. Shelton et al., "Field Tests on Managing Resistance to *Bt*-Engineered Plants," *Nature Biotechnology* 18:3 (March 2000): 339–42.
86. See Owen Taylor, "First Look at Bollgard II," *Soybean Digest,* January 2000.
87. D. R. Porter et al., "Efficacy of Pyramiding Greenbug (Homoptera: Aphididae) Resistance Genes in Wheat," *Journal of Economic Entomology* 93:4 (August 2000): 1315–18.
88. R. D. Perry and J. D. Fetherston, "*Yersinia pestis*—Etiologic Agent of Plague," *Clinical Microbiological Reviews* 10:1 (January 1997): 35–66.
89. For the classic work on this, see William H. McNeill, *Plagues and Peoples* (New York: Anchor Press/Doubleday, 1976), pp. 149–72.
90. "The Amazing Story of Kudzu," at: http://www2.geocities.com/Heartland/Estates/8176/kudzu.html
91. Georgia Wildlife Federation, "Norway Rat," at: http://gwf.org/library/wildlife/ani_nrat.htm
92. Scott Camazine and Roger Morse, "The Africanized Honeybee," *American Scientist* 76:5 (September/October 1988): 465–71.
93. William F. Lyon and James E. Tew, "Africanized Honey Bee," Ohio State University Extension Fact Sheet (HYG-2124-97), at: http://ohioline.ag.ohio-state.edu/hyg-fact/2000/2124.html
94. "New Flora and Fauna for Old," *Economist,* 23 December 2000, http://www.economist.com/displayStory.cfm?Story_ID=457311
95. Indur M. Goklany, "Saving Habitat and Conserving Biodiversity on a Crowded Planet," *BioScience* 48:11 (November 1998): 941.
96. Cary Fowler and Pat Mooney, *Shattering: Food, Politics and the Loss of Genetic Diversity* (Tucson: University of Arizona Press, 1990).
97. Tom Clarke, "Seed Money for Future Food," *Nature Science Update,* 13 June 2002, at: http://www.nature.com/nsu/020603/020603-20.html; http://www.ars-grin.gov
98. Press Release, "U.S. Signs International Plant Genetic Resources Treaty," Department of State, 6 November 2002, at: http://usinfo.state.gov/cgi-bin/washfile/display.pl?p=%2Fproducts%2Fwashfile%2Flatest&f=02110602.clt&t= %2Fproducts%2Fwashfile%2Fnewsitem.shtml

99. As quoted in "Billions Served: An Interview with Norman Borlaug," *Reason* 31:11 (April 2000): 30.
100. As quoted in Mikkel Pates, "North Dakota Harvests Huge Canola Crop," *Agweek*, 21 August 2000.
101. Agricultural and Biotechnology Strategies (Canada) Inc., "Global Status of Approved Genetically Modified Plants," 15 March 2002, at: http://64.26.172.90/agbios/dbase.php?action=Synopsis
102. Paul Raeburn, *The Last Harvest: The Genetic Gamble That Threatens to Destroy American Agriculture* (New York: Simon & Schuster, 1995), pp. 233–37.
103. As quoted in ibid., p. 237; see also Andre Beattie and Paul Ehrlich, *Wild Solutions: How Biodiversity Is Money in the Bank* (New Haven: Yale University Press, 2001).
104. Reed Karaim, "Variety, the Vanishing Crop," *Washington Post*, 11 April 1999, p. B1.
105. Jo Revill, "GM Food: Why You Can't Trust Anybody," *London Evening Standard*, 21 May 1999, p. 10; see also Andrew R. Watkinson, "Predictions of Biodiversity Response to Genetically Modified Herbicide-Tolerant Crops," *Science* 289:5484 (1 September 2000): 1554–57.
106. John Hodgson, "GMO Roundup," *Nature Biotechnology* 18:10 (October 2000): 1023.
107. Sam Kazman, "Deadly Overcaution: FDA's Drug Approval Process," *Journal of Regulation and Social Costs* 1:1 (September 1990): 35–54.
108. U.S. Food and Drug Administration, "Statement of Policy: Foods Derived from New Plant Varieties," *Federal Register* 57 (1992): 22,984.
109. Julie A. Nordlee et al., "Identification of a Brazil-Nut Allergen in Transgenic Soybeans," *New England Journal of Medicine* 334:11 (14 March 1996): 688–92.
110. Mara Bovsun, "Allergy Causing Proteins Jump from Nuts to Soybeans in Gene Transfer," *Biotechnology Newswatch*, 18 March 1996, p. 1.
111. For more detailed information, see Greg A. Bannon, "Using Plant Technology to Reduce Allergens in Food: Status and Future Potential," pp. 1–11, and Steve L. Taylor and Susan L. Hefle, "Food Allergy Assessment for Products Derived through Plant Biotechnology," pp. 325–45, in *Biotechnology and Safety Assessment*, ed. John A. Thomas and Roy L. Fuchs, 3rd ed. (San Diego: Academic Press, 2002).

112. Personal telephone communication of 4 December 2000.
113. "Allergy-Causing Taco Bell Taco Shells Found in Groceries," *United Press International,* 18 September 2000; Julian Borger "Banned GM Corn Reported in Taco Snack," *Guardian* (London), 19 September 2000, p. 14; James Chapman, "Alarm at 'Harmful' GM Corn in Snacks," *Daily Mail* (London), 26 September 2000, p. 27.
114. Melinda Fulmer, "EPA, FDA to Review Data before Deciding on Taco-Shell Recall," *Los Angeles Times,* 19 September 2000, p. C1.
115. "Kraft Foods Announces Voluntary Recall of All Taco Bell Taco Shell Products from Grocery Stores: Tests Indicate Presence of Unapproved Variety of Corn," 22 September 2000, at: http://www.kraftfoods.com/corporate/news/tln_09222000.html; U.S. FDA, "Product Recalls, Alerts, and Warnings," at: http://www.fda.gov/opacom/7alerts.html (Site may be temporary.)
116. On their website, there are two groups they encourage viewers to join, Greenpeace and ActForChange. Both are virulently antibiotech; Ben & Jerry's and The Food Allergy Network, "Tree Nut Alert," 27 October 2000. (The Food Allergy Network is at http://www.foodallergy.org)
117. Press Release, "Aventis CropScience Taking Immediate Action to Assure Confidence in StarLink Corn Distribution," 26 September 2000, at: http://www.us.cropscience.aventis.com/AventisUS/CropScience/stage/html/starlink1.htm
118. See their website at: http://www.foe.org. They don't explicitly call for banning biotech food, but rather seek to do what was done to the nuclear power industry: regulate it to death.
119. Genetically Engineered Food Alert, "Contaminant Found in Taco Bell Taco Shells, Food Safety Coalition Demands Recall by Taco Bell, Philip Morris," *PR Newswire,* 18 September 2000, at: http://www.foe.org/act/getacobellpr.html
120. Centers for Disease Control and Prevention, *Investigation of Human Illness Associated with Potential Exposure to Genetically Modified Corn* (Washington, D.C.: Department of Health and Human Services, 11 June 2001), at: http://www.cdc.gov/nceh/ehhe/Cry9cReport/cry9creport.pdf; Philip Brasher, "CDC: Biotech Corn Didn't Cause Attacks," *Associated Press,* 14 June 2001; see also Andrew Pollack, "No Altered Corn Found in Allergy Samples," 11 July 2001, p. C8.
121. U.S. Senate Committee on Agriculture, Nutrition and Forestry, "The Science of Biotechnology and Its Potential Applications to Agriculture," Hearing before the Committee on Foreign Relations,

106th Cong., 1st sess., 6 October 1999, at: www.senate.gov/~agriculture/Hearings/Hearings_1999/buc99106.htm

122. James D. Astwood and Roy L. Fuchs, "Preventing Food Allergy: Emerging Technologies," *Trends in Food Science & Technology* 7:7 (July 1996): 219; Hugh A. Sampson, "Food Hypersensitivity: Manifestations, Diagnosis, and Natural History," *Food Technology* 46 (1992), as cited James D. Astwood and Roy L. Fuchs, "Preventing Food Allergy: Emerging Technologies," *Trends in Food Science & Technology* 7:7 (July 1996): 219.
123. Janet Rodrigues, "Peanut Pariah," *Houston Chronicle,* 6 September 2001, p. A1.
124. Personal telephone communication of 4 December 2000; for examples, see Y. Tada et al., "Reduction of 14-16 kDa Allergenic Proteins in Transgenic Rice Plants by Antisense Gene" *FEBS Letters* 391:3 (12 August 1996): 341–45; Prem L. Bhalla, Ines Swoboda and Mohan B. Singh, "Antisense-Mediated Silencing of a Gene Encoding a Major Ryegrass Pollen Allergen," *Proceedings of the National Academy of Sciences USA* 96:20 (28 September 1999): 11,676–80.
125. Andrew Pollack, "Gene Jugglers Take to Fields for Food Allergy Vanishing Act," *New York Times,* 15 October 2002, p. D2.
126. Jan Suszkiw, "Researchers Develop First Hypoallergenic Soybeans," *Agricultural Research* 50:9 (September 2002): at: http://www.ars.usda.gov/is/AR/archive/sep02/soy0902.htm
127. Rick M. Helm, associate professor at the University of Arkansas for Medical Sciences in Little Rock, Arkansas, as quoted in Jan Suszkiw, "Researchers Develop First Hypoallergenic Soybeans," *Agricultural Research* 50:9 (September 2002): at: http://www.ars.usda.gov/is/AR/archive/sep02/soy0902.htm
128. Personal telephone communication of 4 December 2000; see also James Astwood et al., "Identification and Characterization of IgE Binding Epitopes of Patatin, a Major Food Allergen of Potato," *Journal of Allergy and Clinical Immunology* 105:1, pt. 2 (January 2000): 555; Murtaza Alibhai et al., "Re-engineering Patatin (Sol t 1) Protein to Eliminate IgE Binding," *Journal of Allergy and Clinical Immunology* 105:1, pt. 2 (January 2000): 239.
129. Personal telephone communication of 4 December 2000.
130. Gregorio del Val et al., "Thioredoxin Treatment Increases Digestibility and Lowers Allergenicity of Milk," *Journal of Allergy and Clinical Immunology* 103:4 (April 1999): 690–97; Bob B. Buchanan et al., "Thioredoxin-Linked Mitigation of Allergic Responses to

Wheat," *Proceedings of the National Academy of Sciences USA* 97:10 (13 May 1997): 5372–77; Prepared Statement of Bob B. Buchanan before the Senate Committee on Agriculture, Nutrition and Forestry, 6 October 1999; Kathleen Scalise, "New Solution for Food Allergies Effective with Milk, Wheat Products, Maybe Other Foods, UC Researchers Discover," Press Release, University of California at Berkeley, 19 October 1997, at: http://amber.berkeley.edu:5010/news/media/releases/97legacy/10_19_97a.html

131. Patricia Callahan and Scott Kilman, "Some Ingredients Are Genetically Modified, Despite Labels Claims," *Wall Street Journal*, 5 April 2001, p. A1.
132. His Royal Highness the Prince of Wales, "Seeds of Disaster," *Daily Telegraph*, 8 June 1998, p. 16.
133. Bishop Elio Sgreccia, Vice President of the Pontifical Academy for Life, referring to the academy's report on ethics and genetic technology, presented 12 October 1999, Vatican City, Rome, as reported in "Designer Babies, Anyone?" *National Catholic Reporter* 36:1 (22 October 1999): 21.
134. Church of England, "The Church of England Statement on Genetically Modified Organisms," April 1999, available at: http://www.agbioworld.org/articles/church_england.html; "The Report of the Ethical Investment Advisory Group on GMOs," March 2000, p. 5, available as download at: http://www.cofe.anglican.org/view/environ.html
135. Grape Escape owner Allan Addison-Saipe as quoted in "Nelson Restaurants Want Clean Genes," NZ PressRoom / Press Release from the Nelson Environment Centre, New Zealand.
136. H. D. Thurston, "Andean Potato Culture: 5,000 Years of Experience with Sustainable Agriculture," in *Advances in Potato Pest Biology and Management of American Phytopathology*, ed. G. W. Zehnder et al. (St. Paul, Minn.: Society Press, 1994), pp. 6–13.
137. Paul C. Manglesdorf, Richard S. MacNeish and Walton C. Galinat, "The Domestication of Corn," in *Prehistoric Agriculture*, ed. Stuart Struever (New York: Garden City: 1971), pp. 472–86.
138. Dick Ahlstrom, "Radiation Used to Develop Varieties of Plants—Geneticist," *Irish Times*, 21 July 1999, p. 2.
139. Jared Diamond, *Guns, Germs, and Steel* (New York: W. W. Norton & Co., 1999), p. 118.
140. Ibid., p. 119.
141. Ibid., p. 122.

142. Ibid., p. 120.
143. As quoted in Ron Bailey, "Billions Served: An Interview with Norman Borlaug," *Reason* 31:11 (April 2000): 30.
144. Nicole T. Perna et al., "Genome Sequence of Enterohaemorrhagic *escherichia Coli* O157:H7," *Nature* 409:6819 (25 January 2001): 529–33.
145. H. H. Kazazian and J. V. Moran, "The Impact of L1 Retrotransposons on the Human Genome," *Nature Genetics* 19:1 (May 1998): 19–24; Matt Ridley, *Genome: The Autobiography of a Species in 23 Chapters* (New York: HarperCollins, 1999), p. 125.
146. Steven G. Pueppke, "Agricultural Biotechnology and Plant Improvement," *American Behavioral Scientist* 44:8 (April 2001): 1233–45; reprinted in *Agricultural Biotechnology: The Public Policy Challenges* (Thousand Oaks, California: Sage Publications, 2001): 1233–45.
147. Press Release, "Statement by Dr. Keith Steele, CEO of AgResearch," 3 May 2001, at: http://www.agresearch.cri.nz/agr/media/press/16_press.htm; Rodney Joyce, "Human Gene in Cows' Milk Part of MS Treatment Test Wellington," *Reuters*, 24 August 1999; Kelly McAra, "Agresearch Pleads for Right to Genetic Work," *Waikato Times* (Hamilton, New Zealand), 25 August 1999, p. 1.

Part Four: Biotech Brooms—Letting Nature Clean up Man's Messes

Tackling Toxic Waste

1. For a nice little poem on this subject, see R. N. Barr, "Spermatorrhea," at http://www.cts.com/sd/poo/spermatorrheaZB.html
2. U.S. Environmental Protection Agency, "Superfund Law and Policy," at: http://www.epa.gov/superfund/action/law/
3. U.S. Environmental Protection Agency, "Final National Priorities List (NPL) Sites," 19 March 2001, at: http://www.epa.gov/superfund/sites/query/queryhtm/nplfin1.htm; EPA, "New Final NPL Sites," 13 September 2001, kept updated at: http://www.epa.gov/ superfund/sites/newfin.htm
4. U.S. Environmental Protection Agency, "Brownfields Glossary of Terms," at: http://www.epa.gov/swerosps/bf/glossary.htm
5. U.S. Environmental Protection Agency, "USTFields Initiative," at: http://www.epa.gov/swerust1/ustfield/index.htm; a "brownfield site" is defined by the EPA as "Abandoned, idled, or under-used industrial and commercial facilities where expansion or redevelop-

ment is complicated by real or perceived environmental contamination."

6. If you really feel you won't be able to get to sleep tonight until you've read some of them, see "Superfund and the Insurance Issues Surrounding Abandoned Hazardous Waste Sites," Insurance Services Offices Inc., December 1995, at: http://www.iso.com/docs/stud003.htm
7. David J. Glass, "U.S. and International Markets for Phytoremediation, 1999–2000," at: http://www.channel1.com/users/dglass/INFO/phy99exc.htm
8. See Michael Fumento, "Roiling the Water," *Forbes* 164:2 (26 July 1999): 66.
9. For a primer on bioremediation, see: http://www.bugsatwork.com/SSWM/PRIMER.HTM; for an expansive website on the subject see: the Department of Energy's "Natural and Accelerated Bioremediation Research," at: http://www.lbl.gov/NABIR/
10. See Janet Raloff, "An Alaskan Feast for Oil-Eating Microbes; Hydrocarbon-Eating Aquatic Microbes Used to Clean up Beaches Contaminated by Exxon Valdez Oil Spill," *Science News* 143:16 (17 April 1993): 253.
11. *Brownfields Technology Primer: Selecting and Using Phytoremediation for Site Cleanup* (U.S. EPA, No. EPA 542-R-01-006 (Washington, D.C.: July, 2001), at: http://www.brownfieldstsc.org/
12. As quoted in Hank Becker, "Phytoremediation: Using Plants to Clean up Soils," *Agricultural Research* 48:6 (June 2000): 4, at: http://www.ars.usda.gov/is/AR/archive/jun00/soil0600.htm
13. *Global Water Supply and Sanitation Assessment 2000 Report* (Geneva: World Health Organization, 2000), at: http://www.who.int/water_sanitation_health/Globassessment/Global1.htm#1.1
14. Douglas Farah, "Nigeria's Oil Exploitation Leaves Delta Poisoned, Poor," *Washington Post*, 18 March 2001, p. A22.
15. http://www.ensolve.com/ ; "Combination Busts Emulsified Bilge," *Waste Treatment Technology News* 15:8 (April 2000); "New Biotechnology Device from EnSolve Biosystems Cleans Marine Environment; U.S. Coast Guard Certifies First Bio-Based Oily Water Separator," *E-Wire*, 23 February 2000.
16. http://www.qinetiq.com/
17. Tom Clarke, "Train Toilet Recycles Water: Bacteria to Battle Waste on the Rails," *Nature ScienceUpdate*, 3 December 2002, at: http://www.nature.com/nsu/021202/021202-1.html; Press Release, "QinetiQ Launches Membrane Bioreactors for Wastewater Treat-

ment on Trains," 25 November 2002, at: http://www.qinetiq.com/applications/news_room/news_releases/show.asp?ShowID=530&category=0

18. Kellyn S. Betts, "Microbes and Fungi Help Clean the Air," *Environmental Science & Technology* 36:23 (1 December 2002): 454A–455A.
19. A. Bosmann et al., "Deep Desulfurization of Diesel Fuel by Extraction with Ionic Liquids," *Chemical Communications* 23 (2001): 2494–95; Philip Ball, "Fuels Clean Up: Two New Techniques Promise to Reduce Pollution from Fossil Fuels," *Nature Science Update*, 31 December 2001, at: http://www.nature.com/nsu/011227/011227-10.html

Phytoremediation and Mighty Microbes

1. Rufus Chaney, as quoted in Anne Simon Moffat, "Engineering Plants to Cope with Metals; Testing Crops That Can Thrive in Metal-Rich Soils," *Science* 285:5426 (16 July 1999): 369.
2. For an illustration of the process, see Phytokinetics Inc.'s website at: http://www.phytokinetics.com/defini.htm; for more detail, see U.S. EPA, Office of Solid Waste and Emergency Response, "A Citizen's Guide to Phytoremediation," (5102G) EPA 542-F-98-011, August 1998, at: http://www.epa.gov/swertio1/products/citguide/phyto2.htm; David J. Glass, "The 1998 United States Market for Phytoremediation: Executive Summary," at: http://www.channel1.com/dglassassoc/INFO/phytexec.htm
3. Ralinda R. Miller, "Phytoremediation," Groundwater Remediation Technologies Analysis Center (Pittsburgh, Penn., 1996), at: http://www.gwrtac.org/pdf/Phyto_o.pdf
4. "Phytovolatilization," Hawaii Bioremediation Website," http://www.svpa.hawaii.edu/abrp/Technologies/phyvola.html
5. Myrna E. Watanabe, "Can Bioremediation Bounce Back?" *Nature Biotechnology* 19:12 (2001): 1111.
6. For example, Greg Basky, "Grass Guzzlers," *Canadian Geographic* 119:6 (1 September 1999): 19; Alan Dove, "Pulling Green Biotechnology out of the Red," *Nature Biotechnology* 16:11 (October 1998): 1022–24; Joseph B. Verrengia, "Hybrid Poplar Trees Suck up Heavy Metals and Solvents," *Associated Press*, 28 September 1998.
7. Joe Bower, "Mother Nature May Hold the Keys to Cleaning up Toxic Messes," *National Wildlife*, June/July 2000, at: http://www.nwf.org/natlwild/2000/plantjj0.html
8. Fiona McWilliam, "Vegetables to Tap Pollutions Roots," *Financial Times* (London), 27 October 2000, p. 17.

9. "The Colorful History of Leadville Colorado," at: http://www.leadville.com/history.htm
10. See "Colorado Town Proud of Tailings, Fights Superfund Cleanup Remedy," *Hazardous Waste News* 13:22 (27 March 2000).
11. Steve Lipsher, "Tailings 'Lab' Pitched for Leadville; Underground System Would Clean Water," *Denver Post,* 29 September 1999, p. B4.
12. As quoted in McWilliam, "Vegetables to Tap Pollution's Roots."
13. As quoted in Steve Lipsher, "Plants Root out Mine Pollution; Underground Gardens Clean up Leadville Water," *Denver Post,* 8 September 1999, p. A1.
14. As quoted in McWilliam, "Vegetables to Tap Pollution's Roots."
15. Ibid.
16. Brian J. Taylor, "Toxic Carrots," *Reason* 4:32 (1 August 2000): 10, at: http://www.reason.com/0008/ci.html#bt.toxic
17. See "Western Yarrow: Kansas Wildflowers and Grasses," at: http://www.lib.ksu.edu/wildflower/yarrow.html
18. McWilliam, "Vegetables to Tap Pollution's Roots"; "Research Projects in the Pilon-Smits Lab," at: http://lamar.colostate.edu/~epsmits/projects.html
19. Sharon Lafferty Doty et al., "Enhanced Metabolism of Halogenated Hydrocarbons in Transgenic Plants Containing Mammalian Cytochrome P450 2e1," *Proceedings of the National Academy of Sciences USA* 97:12 (6 June 2000): 6287–91.
20. As quoted in Louis Porter, "University of Washington Researchers Clean up Contaminants with Chemical-Eating Plants," *The Daily,* University of Washington, 12 April 1999.
21. Julie A. Sutfin, "How Methane Injection Attacks Chlorinated Solvents," *International Ground Water* 2:4 (1996).
22. As quoted in Hank Becker, "Phytoremediation: Using Plants to Clean up Soils," *Agricultural Research* 48:6 (June 2000): 4, at: http://www.ars.usda.gov/is/AR/archive/jun00/soil0600.htm
23. S. Bidwell, "Heavy Metal Physiology: Physiological Ecology of Native Plants Growing on Metalliferous Soil," Plant Cell Biology Research Center, School of Botany, University of Melbourne, Australia, at: http://www.plantcell.unimelb.edu.au/metal.html
24. As quoted in Becker, "Phytoremediation: Using Plants to Clean up Soils."
25. N. S. Pence et al., "The Molecular Physiology of Heavy Metal Transport in the Zn/cd Hyperaccumulator Thlaspi caerulescens," *Proceedings of the National Academy of Sciences USA* 97:9 (25 April 2000): 4956–60.

26. As quoted in Becker, "Phytoremediation: Using Plants to Clean up Soils."
27. For more on rhizofiltration, see "Rhizofiltration," Hawaii Bioremediation Website, at: http://www.svpa.hawaii.edu/abrp/Technologies/rhizofi.html
28. As quoted in Becker, "Phytoremediation: Using Plants to Clean up Soils."
29. "Methylmercury (MeHg)," at http://www.epa.gov/iris/subst/0073.htm
30. Elizabeth A. H. Pilon-Smits and Marinus Pilon, "Breeding Mercury-Breathing Plants for Environmental Cleanup," *Trends in Plant Science* 5:6 (June 2000): 235–36; "Scientists Sitting on a Goldmine with Plants That Absorb Precious Metals," *The Scotsman* (Edinburgh), 8 October 1998, p. 4.
31. As quoted in Tom Mead, "Rooting out Contamination: As Natural Extractors, Plants Could Offer an Affordable Way to Clean up Soil," *Financial Times,* 14 August 1996, p. 8.
32. As quoted in ibid.
33. Scott P. Bizily et al., "Phytoremediation of Methylmercury Pollution: *Merb* Expression in *Arabidopsis Thaliana* Confers Resistance to Organomercurials," *Proceedings of the National Academy of Sciences USA* 99:12 (8 June 1999): 6808–13.
34. As quoted in Anne Simon Moffat, "Engineering Plants to Cope with Metals; Testing Crops That Can Thrive in Metal-Rich Soils," *Science* 285:5426 (16 July 1999): 369.
35. Allan H. Smith, Elena O. Lingas and Mahfuzar Rahman, "Contamination of Drinking-Water by Arsenic in Bangladesh: A Public Health Emergency," *Bulletin of the World Health Organization* 78:9 (1 September 2000): 1093.
36. Abdul Wakib Khan et al., "Arsenic Contamination in Groundwater and Its Effect on Human Health with Particular Reference to Bangladesh," *Journal of Preventive and Social Medicine* 16:1 (1997): 65–73; R. K Dhar et al., "Groundwater Arsenic Contamination and Sufferings of People in Bangladesh May Be the Biggest Arsenic Calamity in the World," paper presented at the International Conference on Arsenic Pollution of Groundwater in Bangladesh: Causes, Effects and Remedies, Dhaka, Bangladesh, 8–12 February 1998.
37. Andrew A. Meharg and M. D. Mazibur Rahman, "Arsenic Contamination of Bangladeshi Paddy Field Soils: Implications of Rice Contribution to Arsenic Consumption," *Environmental Science and*

Technology 37:2 (15 January 2003): 229–34; Charles F. Harvey et al., "Arsenic Mobility and Groundwater Extraction in Bangladesh," *Science* 298:5598 (22 November 2002): 1602–6; Tom Clarke, "Irrigation Taints Bangladeshi Rice with Arsenic," *Nature ScienceUpdate,* 22 November 2002, at: http://www.nature.com/nsu/021118/b1#b1

38. U.S. Environmental Protection Agency, "Arsenic, Inorganic," Integrated Risk Information System, at: http://www.epa.gov/iris/subst/0278.htm
39. "EPA Wants Dramatic Arsenic Reductions," *Water Technology News* 8:3 (June 2000).
40. C. Roussel, C. Neel and H. Bril, "Minerals Controlling Arsenic and Lead Solubility in an Abandoned Gold Mine Tailings," *Science of the Total Environment* 263:1–3 (18 December 2000): 209–19; Richard Cockle, "Headline: Mine Wastes Pose Threat to Waterways," *Portland Oregonian,* 26 January, p. C4.
41. U.S. Environmental Protection Agency, "National Primary Drinking Water Regulations; Arsenic and Clarifications to Compliance and New Source Contaminants Monitoring; Proposed Rules," *Federal Register* 66:194 (5 October 2001): 50,961–63, at: http://www.epa.gov/fedrgstr/EPA-WATER/2001/October/Day-05/w25047.htm; Edward Walsh, "Arsenic Drinking Water Standard Issued," *Washington Post,* 10 November 2001, p. A31.
42. H. Josef Hebert, "EPA Sharply Cuts Drinking Water Arsenic Levels," *Associated Press,* 17 January 2001.
43. Om Parkash Dhankher et al., "Engineering Tolerance and Hyperaccumulation of Arsenic in Plants by Combining Arsenate Reductase and-Glutamylcysteine Synthetase Expression," *Nature Biotechnology* 20:11 (November 2002): 114–45; "Genetically Modified Plant Cleans Arsenic," *United Press International,* 6 October 2002.
44. As quoted in "Researchers Create New Strategy for Removing Arsenic from Soil," *ScienceDaily Magazine,* at: http://www.sciencedaily.com/releases/2002/10/021007071810.htm
45. For more information and illustrations see: "Indiana Plants Poisonous to Livestock and Pets: Brake Ferns," at: http://www.vet.purdue.edu/depts/addl/toxic/plant23.htm
46. http://soils.ifas.ufl.edu/
47. Lena Q. Ma et al., "A Fern That Hyperaccumulates Arsenic," *Nature* 409:6820 (1 February 2001): 579.
48. Bryn Nelson, "Fern a Natural in Arsenic Cleanup," *New York Newsday,* 2 February 2001, p. A26.

49. Dan Johnson, "Flowers That Fight Pollution," *Futurist* 33:4 (April 1999): 6–7.
50. http://www.edenspace.com/
51. Tina Adler, "Botanical Cleanup Crews: Using Plants to Tackle Polluted Water and Soil; Phytoremediation," *Science News* 150:3 (20 July 1996): 42.
52. http://www.ornl.gov/
53. As quoted in Hank Becker, "Phytoremediation: Using Plants to Clean up Soils," *Agricultural Research* 48:6 (June 2000): 4, at: http://www.ars.usda.gov/is/AR/archive/jun00/soil0600.htm
54. Ibid.; see also M. M. Lasat et al., "Phytoremediation of a Radiocesium-Contaminated Soil: Evaluation of Cesium-137 Bioaccumulation in the Shoots of Three Plant Species," *Journal of Environmental Quality* 265:5 (January/February 1997): 165–69.
55. Becker, "Phytoremediation: Using Plants to Clean up Soils."
56. As quoted in Gabrile Marcotti, "A Harvest of Heavy Metal," *Financial Times* (London), 8 June 1998, p. 16.
57. Marcotti, "A Harvest of Heavy Metal."
58. As quoted in ibid.; see also Katherine Rizzo, "Genetically Altered Algae Extract Metals," *Associated Press*, 13 April 1999.
59. His personal website is: http://www.ag.usask.ca/departments/scsr/centre/people/faculty/germida.html
60. As quoted in Greg Basky, "Grass Guzzlers," *Canadian Geographic* 119:6 (1 September 1999): 19.
61. See, for example, "Aberdeen to Use Poplar Trees to Treat Ground Water Contamination," *Ground Water Monitor* 15:6 (1 June 1999).
62. Andrew C. Revkin, "New Pollution Tool: Toxic Avengers with Leaves," *New York Times*, 6 March 2001, p. D1, quoting Steven A. Rock, an environmental engineer in the National Risk Management Research Laboratory of the federal Environmental Protection Agency in Cincinnati.
63. http://www.phytokinetics.com/
64. "3,3-Dichlorobenzidine," at: http://www.epa.gov/ngispgm3/iris/subst/0504.htm#II
65. As quoted in Revkin, "New Pollution Tool: Toxic Avengers with Leaves."
66. Alan Dove, "Pulling Green Biotechnology out of the Red," *Nature Biotechnology* 16:11 (October 1998): 1022–24; Joseph B. Verrengia, "Hybrid Poplar Trees Suck up Heavy Metals and Solvents," *Associated Press*, 28 September 1998.

67. See S. D. Cunningham and D. W. Ow, "Promises and Prospects of Phytoremediation," *Plant Physiology* 110:3 (1996): 715–19.
68. John W. Radin, "Using Superplants to Clean up Our Environment," *Agricultural Research Magazine* 48:6 (June 2000), at: http://www.ars.usda.gov/is/AR/archive/jun00
69. http://yuma.colostate.edu/Depts/NatSci/
70. As quoted in "Remediating with Mustard," *Outlook Magazine,* Spring 1999, at: http://yuma.colostate.edu/Depts/NatSci/html/mustard.html
71. As quoted in ibid.; see also EPA, "Selenium and Compounds," at: http://www.epa.gov/iris/subst/0472.htm
72. Bernard T. Nolan and Melanie L. Clark, "Selenium in Irrigated Agricultural Areas of the Western United States," *Journal of Environmental Quality* 26:3 (May/June 1997): 849–57.
73. For example, Cass Peterson, "Toxic Time Bomb Ticks in San Joaquin Valley; Farm Evaporation Ponds Killing Waterfowl," *Washington Post,* 19 March 1989, p. A3.
74. P. D. Whanger, "China, a Country with Both Selenium Deficiency and Toxicity: Some Thoughts and Impressions," *Journal of Nutrition* 119:9 (September 1989): 1236–39.
75. Spectrum Laboratories, "Selenium," at: http://www.speclab.com/elements/selenium.htm; Donald G. Barceloux, "Selenium," *Journal of Toxicology and Clinical Toxicology* 37:2 (1 March 1999): 145–72.
76. As quoted in Tina Adler, "Botanical Cleanup Crews: Using Plants to Tackle Polluted Water and Soil; Phytoremediation," *Science News* 150:3 (20 July 1996): 42; Terry's personal website is: http://plantbio.berkeley.edu/~terry/html/norman_terry_.html
77. Lidia Wasowicz, "Plants Can Help Clean up Toxic Waste," *United Press International,* 24 April 1995, citing Galilee Society microbiologist Hassan Azaizeh.
78. As quoted in Christine Hanley, "Wetlands Could Offer Remedy against Water Contamination," *Associated Press,* 22 October 1998.
79. D. Hansen et al., "Selenium Removal by Constructed Wetlands: Role of Biological Volatilization," *Environmental Science and Technology* 32:5 (1 March 1998): 591–97; Hanley, "Wetlands Could Offer Remedy against Water Contamination"; Nancy Vogel, "Cattails Clean Wastewater of Selenium, Biologist Says," *Sacramento Bee,* 29 January 1998, p. A1; Norman Terry, "Phytoremediation and Other Aspects of Environmental Plant Biology," at: http://plantbio.berkeley.edu/faculty/faculty_pages/Terry.html

80. As quoted in Alex Barnum, "Study Sparks Debate over Wetland Use," *San Francisco Chronicle,* 30 January 1998, p. A17.
81. Andrew C. Revkin, "New Pollution Tool: Toxic Avengers with Leaves," *New York Times,* 6 March 2001, p. D1; Joe Bower, "Mother Nature May Hold the Keys to Cleaning up Toxic Messes," *National Wildlife,* June/July 2000, at: http://www.nwf.org/natlwild/2000/plantjj0.html
82. As quoted in "DaimlerChrysler Uses Flower Power on Lead," *Waste Treatment Technology News* 15:7 (March 2000).
83. Robert Steyer, "Funding Lags behind Advances in Using Plants to Clean Earth; Methods Must Be Made Profitable to Gain Corporate Acceptance, Researchers Say," *St. Louis Post-Dispatch,* 5 August 1999, p. A10.
84. Katherine Rizzo, "Genetically Altered Algae Extract Metals," *Associated Press,* 13 April 1999; Kevin Parks, "Algae Shows Potential as Cleansing Agent," *This Week,* at: http://www.thisweeknews.com/7.12–7.19/hed/hednews9.html; Kathie Canning, "Genetically Altered Algae Remove Heavy Metals from Water, Sediment," *Pollution Engineering Online,* July 1999, at: http://www.pollutioneng.com/archives/1999/pol0701.99/pol9907news.htm
85. As quoted in Parks, "Algae Shows Potential as Cleansing Agent."
86. As quoted in Rizzo, "Genetically Altered Algae Extract Metals."
87. For an illustration which, alas, laypeople won't find very useful, see "Bioremediation Using *Chlamydomonasm,*" at: www.biosci. ohio-state.edu/~rsayre/BIOREM.htm
88. As quoted in Parks, "Algae Shows Potential as Cleansing Agent."
89. Marc Valls et al., "Engineering a Mouse Metallothionein on the Cell Surface of the *Ralstonia Eutropha* CH34 for Immobilization of Heavy Metals in Soil," *Nature Biotechnology* 18:6 (June 2000): 661–65; Derek R. Lovley and Jon R. Lloyd, "Microbes with a Mettle for Bioremediation," *Nature Biotechnology* 18:6 (June 2000): 600.
90. Amanda Onion, "Letting Nature Do the Dirty Work May Be Best in Cleaning Man's Messes," *Foxnews.com,* 25 February 2000, at: http://www.foxnews.com/science/022500/water.sml
91. As quoted in ibid.
92. EnviroTools Fact Sheets, "Bioremediation," at: http://www.envirotools.org/factsheets/bioremediation.shtml
93. http://www.regenesis.com

94. Personal telephone communication with Stephen Koenigsberg, Vice President of Research and Development at Regenesis Bioremediation Products Inc., 16 April 2003.
95. Press Release, "HRC-X™ New Product from Regenesis," 3 January 2003, at: http://www.regenesis.com/new.htm
96. http://cba.bio.utk.edu/
97. For more on this bacterium, see: http://murex.micro.umass.edu/~geosulf/geoindex.html
98. Personal telephone communication of 11 April 2003.
99. R. G. Riley, J. M. Zachara and F. J. Wobber, *Chemical Contaminants on DOE Lands and Selection of Contaminant Mixtures for Subsurface Science Research* (Washington, D.C.: U.S. Department of Energy, Office of Energy Research, Subsurface Science Program, 1992).
100. Colin Macilwain, "Science Seeks Weapons Clean-up Role," *Nature* 383:6599 (3 October 1996): 375–79; Department of Energy, *The 1996 Baseline Environmental Management Report* (Washington, D.C.: U.S. Department of Energy, 1996), at: http://www.em.doe.gov/bemr96
101. http://deinococcus.allbio.org/
102. Sue Vorenberg, "Radiation-Resistant Bacteria Clean Mixed Waste Contaminants," *Nuclear Waste News* 18:44 (29 October 1998); Joseph B. Verrengia, "Superbug for Toxic Cleanup: Scientists Engineer Bacteria to Digest Nuclear Waste," *Associated Press*, 28 December 1999, at: http:www.abcnews.go.com/sections/science/DailyNews/superbug991228.html
103. As quoted in Vorenberg, "Radiation-Resistant Bacteria Clean Mixed Waste Contaminants."
104. Hassan Brim et al., "Engineering *Deinococcus Radiodurans* for Metal Remediation in Radioactive Mixed Waste Environments," *Nature Biotechnology* 18:1 (January 2000): 85–90.
105. *Global Water Supply and Sanitation Assessment 2000 Report* (Geneva: World Health Organization, 2000), at: http://www.who.int/water_sanitation_health/Globassessment/Global1.htm#1.1
106. Jonathan Swift, "A Description of a City Shower," (from 1710) at: http://www.netpoets.com/classic/poems/062001.htm
107. Theodore Dalrymple, *Mass* Listeria (Suffolk, U.K: André Deutsch, 1998), pp. 91–92.
108. Water Infrastructure Network, *Clean and Safe Water for the Twenty-first Century* (Washington, D.C.: WIN, June 2000), 3-1, at: http://www.amsa-cleanwater.org/advocacy/winreport/chap3.pdf; see also

Katherin Rizzo, "Aging Water, Sewer Systems Present a Looming $1 Trillion Problem," *Associated Press*, 13 June 2000.

109. http://milieu-nomics.com/inpipe.html
110. As quoted in Andy Coghlan, "Stench Bugs," *New Scientist* 169:2275 (27 January 2001): 15, at: http://www.newscientist.com/news/news.jsp?id=ns227538
111. Ibid.
112. Janet Ginsburg, "Bacteria That Make Sewage Smell Sweeter," *Business Week* 3715 (15 January 2001): 85.
113. Michael Richman, "Microbial System Controls Odors in Florida Collection System," *Water Environment & Technology* 9:11 (October 1997), at: http://milieu-nomics.com/casestudies.html
114. "Development and Application of Bioremediation Technologies for Contaminated Soils," *Industrial Health & Hazards Update* 97:8 (1 August 1997); Curt Suplee, "Bacteriums Taste for Toxics Offers Food for Thought on Cleanups," *Washington Post*, 23 June 1997, p. A3.
115. http://www.vet.orst.edu/
116. Norman Miller, "Where There's Muck; There Is a New Way of Dealing with Hazardous Toxic Waste—and It Has Great Green Credentials," *Independent* (London), 30 March 1997, p. 45.
117. Ibid.
118. As quoted in David N. Leff, "Rare Bacterium Gets to Root of Bioremediation for Chlorinated Hydrocarbon Toxic-Waste Sites," *BioWorld Today* 11:131 (10 July 2000).
119. See David N. Leff, "A Potential Bioremediator, New Bacterium Defangs Hazardous Chlorinated Hydrocarbons," *BioWorld Today* 8:110 (9 June 1997).
120. Leff, "Rare Bacterium Gets to Root of Bioremediation for Chlorinated Hydrocarbon Toxic-Waste Sites."
121. Personal e-mail communication with Thomas Wood, 11 August 2000.
122. Doohyun Ryoo et al., "Aerobic Degradation of Tetrachloroethylene by Toluene-o-xylene Monooxygenase of Pseudomonas," *Nature Biotechnology* 18:7 (July 2000): 775–78.
123. Personal e-mail communication with Thomas Wood, 11 August 2000.
124. As quoted in Leff, "Rare Bacterium Gets to Root of Bioremediation for Chlorinated Hydrocarbon Toxic-Waste Sites."

125. Craig S. Marxsen, "Costs of Remediating Underground Storage Tank Leaks Exceed Benefits," *Oil & Gas Journal,* 9 August 1999, p. 21.
126. Christopher E. French et al., "Biodegradation of Explosives by Transgenic Plants Expressing Pentaerythritol Tetranitrate Reductase," *Nature Biotechnology* 17:5 (May 1999): 491–94; EPA, "Integrated Risk Management System (IRIS)," at: http://www.epa.gov/iris/subst/0269.htm#I.A
127. As quoted in Jill Lee, "Microbes Clean up Toxic Waste," *Agricultural Research* 45:3 (March 1997): 18, at: http://www.ars.usda.gov/is/AR/archive/mar97; http://www.tulane.edu/~cellmol/faculty.htm
128. Nerissa Hannink et al., "Phytodetoxification of TNT by Transgenic Plants Expressing a Bacterial Nitroreductase," *Nature Biotechnology* 19:12 (December 2001): 1168; Richard B. Meagher, "Pink Water, Green Plants, and Pink Elephants," *Nature Biotechnology* 19:12 (December 2001): 1120.
129. http://www.remedios.uk.com/
130. Remedios, "Technology: How Our Biosensors Work," at: http://www.remedios.uk.com/biosensor.htm; "British Researchers Light up the Dark to Track Toxics in Soil," *Engineering News-Record* 245:13 (2 October 2000): 19.
131. As quoted in "British Researchers Light up the Dark to Track Toxics in Soil," *Engineering News-Record* 245:13 (2 October 2000): 19.
132. http://www.umbi.umd.edu/
133. As quoted in Amanda Onion, "Letting Nature Do the Dirty Work May Be Best in Cleaning Man's Messes," *Foxnews.com,* 25 February 2000, at: http://www.foxnews.com/science/022500/water.sml
134. *ScienceDaily Magazine,* "Scientists Decode Genes of Microbe That Thrives in Toxic Metals," 4 December 2000, at: http://www.sciencedaily.com/releases/2000/12/001205071815.htm
135. http://www.biology.bnl.gov/
136. *ScienceDaily Magazine,* "Scientists Decode Genes of Microbe That Thrives in Toxic Metals."
137. Ibid.
138. Craig S. Marxsen, "Costs of Remediating Underground Storage Tank Leaks Exceed Benefits," *Oil & Gas Journal,* 9 August 1999, p. 21.
139. For example, Joe Bower, "Mother Nature May Hold the Keys to Cleaning up Toxic Messes," *National Wildlife,* June/July 2000, at: http://www.nwf.org/natlwild/2000/plantjj0.html

140. As quoted in Tina Adler, "Botanical Cleanup Crews: Using Plants to Tackle Polluted Water and Soil; Phytoremediation," *Science News* 150:3 (20 July 1996): 42.
141. As quoted in Onion, "Letting Nature Do the Dirty Work May Be Best in Cleaning Man's Messes."
142. "Economic Barriers Block Remediation of Brownfield Sites, Companies Say," *Hazardous Waste News* 22:41 (16 October 2000).
143. http://www.iacr.bbsrc.ac.uk/iacr/tiacrhome.html
144. As quoted in Andrew C. Revkin, "New Pollution Tool: Toxic Avengers with Leaves," *New York Times,* 6 March 2001, p. D1.

Conclusion: A Biotech Future

1. For more on this, see Gregory Timp, ed., *Nanotechnology* (New York: Springer Verlag, 1999); Markus Krummenacker and James Lewis, eds., *Prospects in Nanotechnology: Toward Molecular Manufacturing* (New York: John Wiley & Sons, 1995).
2. Rachel A. Roemhildt, "Many Patents Pending with Biotech Plans; Scientists Find Ways to Vary Food, Medicine," *Washington Times,* 1 November 1998, p. C12.
3. Michael Fumento, "I Know Why the Caged Genome Sings," *American Outlook* 4:12 (January/February 2001): 13, at: http://www.fumento.com/genome.html
4. See Thomas N. Robinson, "Effects of Reducing Children's Television and Video Game Use on Aggressive Behavior," *Archives of Pediatrics & Adolescent Medicine* 155:1 (January 2000): 17–23; Aine McCarthy, "License to Kill," *Irish Times,* 12 July 2000, p. 13; Vivi Hoang, "Kids, Video Games, Violence: No End in Sight," *Scripps Howard News Service,* 20 November 2000. For an example of how fast improvements are coming, see Michel Marriott, "Game Market Awaits Microsoft's Black Box," *New York Times,* 11 January 2001, p. G7.
5. Ray Kurzweil, *The Age of Spiritual Machines: When Computers Exceed Human Intelligence* (New York: Viking, 1999).
6. Bill Joy, "Why the Future Doesn't Need Us," *Wired* 8:4 (April 2000), at: http://www.wired.com/wired/archive/8.04/
7. "Interview: Hawking über die Veränderungsmöglichkeit des Menschen," *Focus* (Germany) 36 (3 September 2001): 136; see also Aine Harrington, "Hawkings Genetic Plan to Out-Think Computers," *Herald* (Glasgow), 3 September 2001, p. 3.

8. Both as quoted in Andrew Smith, "Science 2001: Net Prophets: Alas, Mankind, We Knew Him …" *Observer* (London), 31 December 2000, p. 18.
9. See, for example, Dinesh D'Souza, "Staying Human: The Danger of Techno-Utopia," *National Review* 53:1 (22 January 2001): 36.
10. See Mark Stewart, "Restoring Civility: Coarsening of Society Has Parents, Others Looking for Ways to Instill Finer Virtues," *Washington Times,* 17 October 2000, p. E1.
11. As quoted in Bruce Weber, "Swift and Slashing, Computer Topples Kasparov" *New York Times,* 12 May 1997, p. A1.
12. For example, a *USA Today* poll in early 2001 found that 43% of those surveyed oppose abortion. The newspaper reported that was an increase from 33% five years earlier. See "Americans Closer on Abortion Issue," *USA Today,* 24 January 2001, p. A1.

Index